Approximation Methods for Electronic Filter Design

Approximation Methods for Electronic Filter Design

With Applications to Passive, Active, and Digital Networks

Richard W. Daniels, Ph.D.
Bell Telephone Laboratories, Inc.

McGRAW-HILL BOOK COMPANY
New York St. Louis San Francisco Düsseldorf Johannesburg
Kuala Lumpur London Mexico Montreal New Delhi
Panama Paris São Paulo Singapore
Sydney Tokyo Toronto

Library of Congress Cataloging in Publication Data

Daniels, Richard W, date.
   Approximation methods for electronic filter design.

   Includes bibliographies.
   1. Electric filters. 2. Approximation theory.
I. Title.
TK7872.F5D36   621.3815'32   74-8091
ISBN 0-07-015308-6

Copyright © 1974 by Bell Telephone Laboratories, Incorporated.
All rights reserved.   Printed in the United States of America.
No part of this publication may be reproduced, stored in a
retrieval system, or transmitted, in any form or by any means,
electronic, mechanical, photocopying, recording, or otherwise,
without the prior written permission of the publisher.

1 2 3 4 5 6 7 8 9 0   KPKP   7 9 8 7 6 5 4

*The editors for this book were Tyler G. Hicks and Stanley
E. Redka, the designer was Naomi Auerbach, and its production
was supervised by George E. Oechsner. It was set in Modern 8A
by Bi-Comp, Inc.*

It was printed and bound by The Kingsport Press.

To Cynthia and Richard, Jr.
May your quest for knowledge be a happy one

Contents

Preface xiii

Symbols xv

1. Introduction 1

    1.1 Approximation theory   1
    1.2 Filter jargon   2
    1.3 Realizations   4
    1.4 Types of approximations   5
    REFERENCES   6

2. The Butterworth Approximation 7

    2.1 Introduction   7
    2.2 The characteristic function   7
    2.3 The lowpass characteristic function   8
    2.4 Butterworth polynomials   9
    2.5 Butterworth lowpass filters   10
    2.6 Determination of degree of $B_n(\omega)$ 11
    2.7 $H(s)$ for the Butterworth approximation   12
    2.8 Quality of Butterworth roots   14
    2.9 Conclusions   18
    REFERENCES   18
    PROBLEMS   19

3. The Tschebycheff Approximation … 20

- 3.1 Introduction   20
- 3.2 Equiripple approximations   21
- 3.3 The Tschebycheff lowpass approximation   23
- 3.4 Introduction to Tschebycheff polynomials   25
- 3.5 The Tschebycheff polynomial   27
- 3.6 The normalized Tschebycheff lowpass   30
- 3.7 Determination of degree of $T_n(x)$ 33
- 3.8 A program for the loss of Tschebycheff lowpass filters   34
- 3.9 An optimum property of Tschebycheff filters   36
- 3.10 $H(s)$ for the Tschebycheff approximation   36
- 3.11 Concluding remarks   40
  - REFERENCES   41
  - PROBLEMS   41

4. The Inverse Tschebycheff Filter … 43

- 4.1 Introduction   43
- 4.2 Manipulation of Tschebycheff characteristics   43
- 4.3 Maximally flat property of the inverse Tschebycheff filter   45
- 4.4 Determination of inverse Tschebycheff loss   45
- 4.5 Determination of degree of the inverse Tschebycheff filter   46
- 4.6 A program for the loss of inverse Tschebycheff lowpass filters   47
- 4.7 Concluding remarks   48

5. Elliptic Filters … 51

- 5.1 Introduction   51
- 5.2 Introduction to Tschebycheff rational functions   52
- 5.3 A basic form for $R_n(x,L)$ 55
- 5.4 A differential equation for $R_n(x,L)$ 58
- 5.5 The elliptic integral of the first kind   59
- 5.6 Elliptic functions   62
- 5.7 An alternative form for the elliptic integral   66
- 5.8 Elliptic functions and $R_n(x,L)$ 68
- 5.9 The periodic rectangle for $R_n(x,L)$ 69
- 5.10 Determination of degree of elliptic filters   72
- 5.11 Determination of L 73
- 5.12 A rational expression for $R_n(x,L)$ 75
- 5.13 A program for elliptic lowpass filters   78
- 5.14 Concluding remarks   81
  - REFERENCES   83
  - PROBLEMS   83

6. Frequency Transformations … 86

- 6.1 Introduction   86
- 6.2 Normalized-lowpass-to-unnormalized-lowpass transformation   86
- 6.3 Lowpass-to-highpass transformation   87
- 6.4 Lowpass-to-bandpass transformation   88
- 6.5 Lowpass-to-bandstop transformation   97

6.6 Lowpass-to-multiple-bandpass transformation 98
6.7 Reactance transformations 99
6.8 Other frequency transformations 99
6.9 Conclusions 104
    REFERENCES 106
    PROBLEMS 106

7. The Transformed Variable 108

7.1 Introduction 108
7.2 The transformed variable 109
7.3 Functions in terms of the transformed variable 111
7.4 $F(Z)$ and $Q(Z)$ for lowpass filters 114
7.5 The inverse transformation 116
7.6 Conclusions 119
    REFERENCES 120
    PROBLEMS 120

8. Attenuation Poles for Equiripple Passband Filters 122

8.1 Introduction 122
8.2 Poles and zeros 122
8.3 The loss function $L(Z)$ 124
8.4 The template method 126
8.5 Intuition development 129
8.6 A simple computer program 130
8.7 A computer template method 133
8.8 Terminology used for the general problem 133
8.9 Outline of solution 135
8.10 Determination of arc minimum 136
8.11 Determination of the minimum difference $D_{\min_i}$ 140
8.12 Determination of new poles 140
8.13 Summary of pole-placer program 142
8.14 Examples for the pole-placer program 143
8.15 Modifications of the pole-placer program 146
8.16 Lowpass pole-placer program 147
8.17 Highpass pole-placer program 152
8.18 Conclusion 154
    REFERENCES 154
    PROBLEMS 154

9. The Characteristic Function for Equiripple Passband Filters 156

9.1 Introduction 156
9.2 The characteristic function and the transformed variable 156
9.3 Determination of $Q^2(Z)$ 158
9.4 Determination of $F^2(Z)$ 158
9.5 Determination of $Q^2(Z)$ and $F^2(Z)$ for lowpass filters 161
9.6 The loss function $L(Z)$ 164
9.7 Stopband performance 165
9.8 Passband performance 166
    PROBLEMS 170

10. Natural Modes for Equiripple Passband Filters 172

    10.1   Introduction   172
    10.2 $H(s)$ and the transformed variable   173
    10.3   Determination of $E(Z)E^*(Z)$ 174
    10.4   Determination of $e(s)$ 176
    10.5   Determination of natural modes of equiripple lowpass filters   179
    10.6   Conclusions   183
           REFERENCE   183
           PROBLEMS   183

11. Maximally Flat Passbands 185

    11.1   Introduction   185
    11.2   Determination of the characteristic function   186
    11.3   Determination of $Q^2(Z)$ and $F^2(Z)$ for lowpass filters   189
    11.4   Stopband and passband performance   192
    11.5   Determination of attenuation poles for maximally flat filters   193
    11.6   Natural modes for maximally flat passband filters   202
    11.7   Comparison of maximally flat and equiripple filters   202
    11.8   Conclusions   207
           REFERENCES   207
           PROBLEMS   208

12. Parametric Filters 210

    12.1   Introduction   210
    12.2   The basic trick   211
    12.3   Addition of a pole at infinity   211
    12.4   Addition of a pole at the origin   214
    12.5   Stopband and passband performance   215
    12.6   Determination of attenuation poles for parametric filters   218
    12.7   Results from the pole-placer program   221
    12.8   Parametric lowpass filters   223
    12.9   Natural modes for parametric filters   224
    12.10 Conclusions   227
           REFERENCES   227
           PROBLEMS   227

13. Optimization Techniques for Approximation Theory 229

    13.1   Introduction   229
    13.2   System response and error criteria   230
    13.3   Initial parameters   232
    13.4   Minimization techniques   232
    13.5   Practical optimization programs for the approximation problem   234
    13.6   Arbitrary passband, equiminimum stopband   234
           REFERENCES   237

14. Delay and Related Subjects 238

    14.1   Introduction   238
    14.2   Definition of delay   238

14.3 Calculation of delay 240
14.4 A program for the calculation of delay 241
14.5 Relations between magnitude and delay 244
14.6 The Hilbert transformations 248
14.7 The Bessel approximation 249
14.8 The Gaussian magnitude approximation 256
14.9 Transitional Butterworth-Thomson filters 259
14.10 Tschebycheff approximation of constant delay 260
14.11 Delay considerations for bandpass filters 261
14.12 Addition of attenuation poles 264
14.13 Amplitude equalizers 265
14.14 Allpass networks 268
14.15 Allpass networks derived from lowpass delay approximations 272
14.16 Conclusions 275
REFERENCES 275
PROBLEMS 277

15. Time-Domain Response ... 282

15.1 Introduction 282
15.2 Definition of terms 283
15.3 Transient response and the Laplace transformation 284
15.4 Partial fractions and the inverse Laplace transformation 285
15.5 Taylor series and the inverse Laplace transformation 286
15.6 Comparison of transient response 289
REFERENCES 290
PROBLEMS 290

16. Approximation Methods and Passive Network Synthesis ... 292

16.1 Introduction 292
16.2 Insertion loss 293
16.3 The transmission function $H(s)$ 294
16.4 The reflection function $T_1(s)$ 298
16.5 The lossless coupling network 300
16.6 Synthesis of the lossless coupling network 302
16.7 The zero-shifting technique 303
16.8 Typical network configurations 307
16.9 Synthesis in terms of the transformed variable 311
16.10 Conclusions 314
REFERENCES 315
PROBLEMS 315

17. Approximation Methods and Active Filter Synthesis ... 319

17.1 Introduction 319
17.2 Negative-impedance-converter active filters 320
17.3 Gyrator active filters 322
17.4 Second-order transfer functions 324
17.5 Decomposition into second-order transfer functions 328
17.6 Some practical active circuits 331

xii   Contents

    17.7   Coupled active filters   340
    17.8   Conclusions   342
           REFERENCES   344

18. Approximation Methods and Digital Filter Synthesis 346

    18.1   Introduction   346
    18.2   The Z transformation   348
    18.3   Design of digital filters from continuous filters   349
    18.4   Nonrecursive digital filters   350
    18.5   Impulse-invariant method   350
    18.6   Matched Z-transform method   355
    18.7   Bilinear digital filters   357
    18.8   Realizations for digital filters   363
    18.9   Conclusion   366
           REFERENCES   366

Appendix A Telcomp   368

Appendix B Filter Design by the "Cookbook" Approach   373

Answers to Selected Problems   377

Index   381

Preface

This book was written to help fill a void that exists in the literature of filter theory. There are many electrical-engineering books that can be used to help synthesize passive, active, or digital filters; however, there is no satisfactory book that treats the approximation problem from the viewpoint of an electrical engineer—and the approximation problem must be solved before one can synthesize a filter. This book is concerned with finding approximations for attenuation and delay functions; the approximations are ratios of polynomials, and the polynomials are expressed in terms of the frequency variable s.

There are many approximation books written by mathematicians, but these are not directly applicable to filter design. Although some filter-synthesis books devote a chapter to the approximation problem, this is not nearly enough for a comprehensive treatment. Thus, electrical engineers have had to read many articles to become familiar with the approximation problem. This book presents various solutions to the problem in an organized and logical manner.

The first few chapters are concerned with what might be termed the classical approximation theory: they discuss Butterworth, Tschebycheff, inverse Tschebycheff, and elliptic filters. These chapters apply to low-pass filters; Chapter 6 presents transformations that can be used for other types of filters.

Chapter 7 discusses a special transformation. It introduces filter design in terms of the transformed variable and serves as a bridge between classical approximation theory and modern approximation theory. Modern approximation theory makes use of the transformed variable because it provides computational accuracy and simplifies many derivations. The transformed variable is used in Chapters 8 through 12 to investigate filters that have arbitrary stopbands and either equiripple or maximally flat passbands.

After treating approximations for magnitude response, the book discusses delay and transient response. Chapters 14 and 15 contain formulas that can be used to calculate the delay and transient response for any of the filters mentioned in previous chapters.

The remaining chapters discuss methods for synthesizing the transfer functions obtained in the first part of the book. Chapter 16 treats passive network synthesis, Chapter 17 active filter synthesis, and Chapter 18 digital filter synthesis. This material emphasizes the fact that the transfer functions obtained in the book are not just mathematical abstractions; networks can be constructed to perform these filtering functions.

Modern approximation theorists utilize computers to help solve their problems; therefore it seemed essential to include a substantial number of programs in the book. The programs were written in Telcomp II because that language contains statements that are easy to understand. It is not expected that the reader will ever program in Telcomp, but he should be able to write similar programs in other languages. An appendix discusses Telcomp II in enough detail so that the programs in this book can be comprehended.

Problems have been included at the ends of most chapters for two reasons: First, they can help test the reader's comprehension of the material and thus serve as a learning aid. Second, many of the problems are used to establish results that appear in the text. This approach keeps lengthy proofs from cluttering up the presentations. Answers to selected problems are given at the end of the text.

The working environment at Bell Telephone Laboratories helped make this book possible. There were numerous stimulating discussions with colleagues, and most of the material in the book was taught in a course to fellow employees. Special thanks are due them for their valuable comments and suggestions, as well as to the many people who helped prepare the manuscript.

Finally, to my wife and children, who found out firsthand how much time is required to write a book of this nature, my appreciation for your patience and understanding.

Richard W. Daniels

Symbols

A	attenuation, loss
A_i	attenuation to the right of FS_i
$A_{\max}$	maximum passband loss
$A_{\min}$	minimum stopband loss
$A_{\min_i}$	attenuation of ith arc at frequency $F_{\min_i}$
B_n	nth-order Butterworth polynomial
cn	elliptic cosine function
C_H	constant multiplier of H
dn	elliptic difference function
D	delay
$D_{\min_i}$	minimum difference (excess attenuation) of the ith arc
$D(Z)$	polynomial that determines the loss minimums of arcs
$D(\omega)$	delay
e	numerator of H, as in $H(s) = e(s)/q(s)$
e_e	even part of $e(s)$
e_o	odd part of $e(s)$
Ev	even part of
$E(Z)$	transformed version of $e(s)$
$E^*(Z)$	transformed version of $e(-s)$
f	frequency
f	numerator of K, as in $K(s) = f(s)/q(s)$
f_e	even part of $f(s)$

Symbols

f_i	location of ith attenuation pole
f_o	odd part of $f(s)$
F	frequency
FA	lower passband edge
FB	upper passband edge
FH	upper stopband edge
FL	lower stopband edge
$F_{\min_i}$	frequency at which ith arc is closest to the requirement
FS_i	ith attenuation step
$F(Z)$	transformed version of $f(s)$
H	transfer function (ratio of input to output)
$i(t)$	impulse response
IA	image attenuation
IL	insertion loss
IP	integer part of
j	$\sqrt{-1}$
k	modulus of the elliptic integral
k'	complementary modulus of the elliptic integral
K	characteristic function, complete elliptic integral
K'	complementary complete elliptic integral
$K_{\min}$	minimum stopband value of $K(s)$
L	minimum stopband value of $\lvert R_n(x,L) \rvert$
$L(Z)$	loss function
m	a parameter defined by $m = \mathrm{NZ} + \mathrm{NIN} + 2N$
n	indicates degree of a filter, polynomial etc.
N	number of attenuation poles (excluding those at zero and infinity)
NA	number of poles in lower stopband
NB	number of poles in upper stopband
NIN	number of attenuation poles at infinite frequency
NZ	number of attenuation poles at zero frequency
P	parametric multiplier
P_0	power dissipated in load when coupling network is replaced by a short circuit
P_1	power delivered to coupling network
P_2	power dissipated in load
P_m	maximum available power
P_r	reflected power
q	denominator of $H(s)$, as in $H(s) = e(s)/q(s)$
Q	quality of a root
Q_p	quality of a pole
Q_z	quality of a zero
$Q(Z)$	transformed version of $q(s)$
R_c	relative change in ripple
R_n	nth-order Tschebycheff rational function
s	complex frequency $s = \sigma + j\omega$
sn	elliptic sine function
S	normalized complex frequency

SA	number of attenuation steps in lower stopband
SB	number of attenuation steps in upper stopband
T	sampling time
T	transfer function (ratio of output to input)
T_1	reflection function
T_n	nth-order Tschebycheff polynomial
u	elliptic integral of the first kind
V_0	load voltage when coupling network is replaced by a short circuit
V_1	voltage at input of coupling network
V_2	load voltage
x_L	the first value of x at which $R_n(x,L) = L$
Y	admittance
z	variable of the Z transformation ($z = e^{sT}$)
Z	impedance
Z_1	input impedance of coupling network
Z_{fi}	transformed version of ω_{fi}
Z_i	transformed version of f_i
ZF_i	ith zero of $H(s)$
ZQ_i	quality of the ith zero of $H(s)$
α	attenuation function, parametric constant
β	phase function
β_i	angle of a typical term of $L(Z)$
$\delta(t)$	unit impulse function
ϵ	a constant uniquely determined by $A_{\max}$
σ	real part of complex frequency s
ϕ	amplitude of the elliptic integral
ω	imaginary part of complex frequency s
ω_A	lower passband edge
ω_B	upper passband edge
ω_H	upper stopband edge
ω_i	ith attenuation pole
ω_L	lower stopband edge
ω_0	frequency of maximum flatness, critical frequency
ω_s	sampling frequency
Ω	imaginary part of complex frequency S

Approximation Methods for Electronic Filter Design

CHAPTER ONE

Introduction

1.1 APPROXIMATION THEORY

The mathematical discipline known as *approximation theory* is very general and contains many useful theorems. The books *Theory of Approximation*[1],* and *Approximation Theory*[2] may be examined by those interested in the niceties of this area of mathematics.

This book is not an attempt to treat approximation theory in general. Instead, the approximation problem is investigated from the viewpoint of an electrical engineer interested in designing filters. Thus, while the treatment is not elementary, neither does it become bogged down with mathematical details. For example, in discussing how to measure approximation errors, we will not become involved in a lengthy discussion of norms. We will, instead, use our time and energy to develop the approximation theory that is the basis for modern network-synthesis computer programs.

Network synthesis is used to find networks that will perform a desired task. For example, in an AM communication system it is usually necessary to synthesize networks that attenuate certain unwanted frequencies.

---

* Superscript numbers indicate references listed at the end of the chapter.

The approximation problem of network synthesis refers to the determination of a system function that, when synthesized, will perform the desired task. There are usually many different approximating functions that could be used to solve a specific approximation problem; which one is "best" will depend on many factors, such as the complexity of the resulting network. Thus this book does not present a unique solution to the approximation problem; instead it offers tools that can be used by the engineer interested in filter design.

1.2 FILTER JARGON

This book is written from the viewpoint of a filter designer, and the material is described in terms commonly used by filter designers. Since some of the jargon might not be familiar to readers from other disciplines, this section provides a brief introduction to the terminology.

Most of the functions encountered in this book are expressed in terms of the complex frequency $s = \sigma + j\omega$. This is the variable commonly encountered in Laplace transformation theory; given a function $v(t)$ (for example, a voltage that is a function of time), its Laplace transformation is defined as

$$V(s) = \int_0^\infty v(t)e^{st}\,dt = \mathcal{L}[v(t)] \tag{1.1}$$

A lowercase letter is usually used to denote a time-domain function, and an uppercase letter is employed for its Laplace transformation. For example, for currents one writes

$$I(s) = \mathcal{L}[i(t)] \tag{1.2}$$

An impedance is defined to be a voltage-to-current ratio and is always expressed in terms of the complex variable s. It is usually denoted by $Z(s)$; that is,

$$Z(s) = \frac{V(s)}{I(s)} \tag{1.3}$$

The reciprocal of an impedance is defined to be an admittance:

$$Y(s) = \frac{I(s)}{V(s)} \tag{1.4}$$

All lumped network response functions—impedances, admittances, and dimensionless ratios—are rational functions of the complex variable s.[6] This is the major reason for introducing the complex frequency s; it allows us to consider only rational functions and to employ our previous knowledge about such functions. For example, a transfer function which is a ratio of voltages may be written as

$$\frac{V_{\text{out}}}{V_{\text{in}}} = \frac{a_0 + a_1 s + a_2 s^2 + \cdots + a_m s^m}{b_0 + b_1 s + b_2 s^2 + \cdots + b_n s^n} = T(s) \qquad (1.5)$$

For this transfer function to represent a stable system, the denominator of $T(s)$ must be Hurwitz; that is, the poles must be in the left half-plane.

If a transfer function is written in terms of the complex variable s, then it is very easy to investigate the steady-state response of the network. Consider a linear system that has an input which is a sinusoid of frequency ω:

$$v_{\text{in}}(t) = \sin \omega t \qquad (1.6)$$

The output will be a sinusoid of the same frequency:

$$v_{\text{out}}(t) = C \sin (\omega t + \phi) \qquad (1.7)$$

The amplitude C and phase ϕ arise because, in general, the output of a network will not have the same amplitude or phase as the input. The ratio of the output amplitude to the input amplitude can be found by simply evaluating the transfer function $T(s)$ at $s = j\omega$. Because $s = j\omega$ corresponds to a sinusoidal frequency, the letter ω is often said to represent *real* frequencies while the letter s is said to represent *complex* frequencies.

If the amplitude of a sinusoid at the output of a network is smaller than the amplitude at the input, then the signal is said to have been *attenuated* (i.e., it has encountered loss). The attenuation is usually expressed in terms of decibels (dB) as

$$A(\omega) \triangleq 20 \log \left| \frac{V_{\text{in}}(j\omega)}{V_{\text{out}}(j\omega)} \right| \qquad (1.8)$$

By our definition of the transfer function $T(s)$, this can also be written as

$$A(\omega) = -20 \log |T(j\omega)| \qquad (1.9)$$

In filter theory it is common practice to consider transfer functions that are ratios of input to output; i.e., we work with

$$H(s) = \frac{1}{T(s)} = \frac{b_0 + b_1 s + b_2 s^2 + \cdots + b_n s^n}{a_0 + a_1 s + a_2 s^2 + \cdots + a_m s^m} \qquad (1.10)$$

$$= \frac{e(s)}{q(s)} \qquad (1.11)$$

It follows that the attenuation can be expressed as

$$A(\omega) = 20 \log |H(j\omega)| \qquad (1.12)$$

The function $H(s)$ will be referred to simply as an *input/output* transfer function; it will not be given a special name such as "transducer function" because this term tends to be restricted to passive networks. As mentioned in the next section, the input/output transfer function $H(s)$ can be realized as either a passive, an active, or a digital filter.

One often needs a filter that will attenuate certain frequencies with respect to others; the requirements for such a filter are usually given in the frequency domain. Most of the approximations that we consider will be concerned with amplitude response, i.e., with the amplitude of $H(j\omega)$.

Filters are usually classified according to their amplitude response as lowpass, highpass, bandpass, or bandreject. A lowpass filter is one that "passes" low frequencies; i.e., below a certain frequency there is negligible attenuation. This cutoff frequency is commonly called the *passband edge*. Of course, negligible attenuation is a relative term—what is negligible will depend on the filter specifications. Similarly, a highpass filter passes high frequencies and attenuates low frequencies. In general, the regions of high attenuation are called *stopbands*, and the regions of low attenuation *passbands*. The nebulous region between a passband and a stopband is termed a *transition* region.

By analogy to the filters just mentioned, a bandpass filter passes a band of frequencies (it attenuates low frequencies with a lower stopband and high frequencies with an upper stopband). Similarly, a bandreject filter rejects a band of frequencies.

1.3 REALIZATIONS

As indicated, the transfer functions that are derived in this book are not just mathematical abstractions—they are indispensable in the design of practical filters. There are many different types of realizations; one way of classifying them depends on whether they are passive, active, or digital.

Passive filter realizations can be considered to be of the classical type. As the term is used in this book, a passive filter is one which does not require a power supply. Thus, although a gyrator is theoretically passive, it will not be considered as a passive circuit element because it requires the use of a power supply.

Recently, especially at low frequencies, active and digital filters have been offering competition to the passive filter. Both these types of filters require a power supply; the difference between them is that the active filter is used for analog signals, and the digital filter for digital signals.

There have been many books written about each of these different types of filters.[3-5] Thus, it would be presumptuous to claim to cover them thoroughly in this text. However, an introduction to passive, active, and digital filters is included. For those who have no synthesis background, it should help emphasize the practical aspects of approximation theory. For those who are familiar with one or more of the synthesis techniques, it should put that knowledge in broader perspective.

1.4 TYPES OF APPROXIMATIONS

In a specific problem we shall want to find a rational function $H(s)$ such that its amplitude or phase approximates a given shape. The $H(s)$ that is found will depend on what is considered to be an optimum approximation. For example, assume that we want to approximate the function $f(\omega)$ with $h(\omega)$ between ω_A and ω_B.*

A possible approximation is given in Fig. 1.1, which also shows the approximation error $e(\omega)$. Of course, the function $h(\omega)$ should be a good approximation of $f(\omega)$; one way to measure the "goodness" of the approximation would be to find the mean-square error

$$E \triangleq \int_{\omega_A}^{\omega_B} [f(\omega) - h(\omega)]^2 \, d\omega \qquad (1.13)$$

Thus, one way to optimize an approximation would be to minimize the mean-square error. If some frequencies are more important than others, we could minimize the weighted mean-square error

$$E \triangleq \int_{\omega_A}^{\omega_B} g(\omega)[f(\omega) - b(\omega)]^2 \, d\omega \qquad (1.14)$$

where $g(\omega)$ is some weighting function. Instead of the entire frequency range from ω_A to ω_B, we might be concerned primarily with a point ω_0.

* The variable $h(\omega)$ represents a general approximation function; it is not related to $H(s)$.

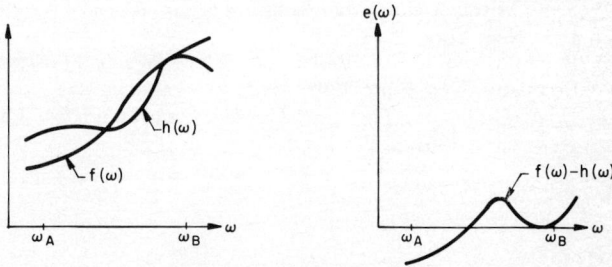

Fig. 1.1 Approximation of $f(\omega)$ by $h(\omega)$ and the resulting error.

If we want $h(\omega)$ to approximate $f(\omega)$ very closely at ω_0, we can expand $f(\omega)$ in a Taylor series about ω_0:

$$f(\omega) = f(\omega_0) + \frac{f'(\omega_0)}{1!}(\omega - \omega_0) + \frac{f''(\omega_0)(\omega - \omega_0)^2}{2!} + \cdots$$

Similarly,

$$h(\omega) = h(\omega_0) + \frac{h'(\omega_0)}{1!}(\omega - \omega_0) + \frac{h''(\omega_0)(\omega - \omega_0)^2}{2!} + \cdots$$

Thus the approximation will be very good at ω_0 if $f(\omega_0) = h(\omega_0)$ and as many derivatives as possible are equal. An approximating function found by satisfying a criterion such as this will be called a *Taylor approximation*. It might be noted here that the Butterworth approximation is a special case of a Taylor approximation. Taylor approximations are very good about some point ω_0, but they may not be very good over the frequency band of interest.

Instead of emphasizing the behavior at the point ω_0, we may simply require that the magnitude of the approximation error be as small as possible. Such approximations will introduce theory due to Tschebycheff.

The first few chapters present approximations that are useful in designing lowpass filters. Then some transformations are discussed that can be used to transform lowpass approximations into other types. It is shown that these transformations are often more restrictive than necessary, and that it may be better to approximate the desired shape (such as a bandpass shape) directly.

REFERENCES

1. N. I. Achieser, *Theory of Approximation*, Frederick Ungar Publishing Co., New York, 1956.
2. A. Talbot, *Approximation Theory*, Academic Press, Inc., New York, 1970.
3. E. A. Guillemin, *Synthesis of Passive Networks*, John Wiley & Sons, Inc., New York, 1957.
4. S. K. Mitra, *Analysis and Synthesis of Linear Active Networks*, John Wiley & Sons, Inc., New York, 1969.
5. B. Gold and C. M. Rader, *Digital Processing of Signals*, McGraw-Hill Book Company, New York, 1969.
6. E. A. Guillemin, *Theory of Linear Physical Systems*, John Wiley & Sons, Inc., New York, 1963, p. 249.

CHAPTER TWO

The Butterworth Approximation

2.1 INTRODUCTION

In the previous chapter we discussed the problem of approximating an arbitrary shape $f(\omega)$ in the frequency range ω_A to ω_B. In the passband of a filter, the arbitrary shape that we want to approximate is usually just a constant. For simplicity the constant value can be normalized to unity; that is, in the passband,

$$|H(j\omega)| \approx 1 \qquad (2.1)$$

In this chapter we shall discuss the simplest approximation that can be used for a lowpass filter of the type described by (2.1). Because it is simple it has some practical disadvantages, but by its simplicity it provides a good introduction to approximation theory.

2.2 THE CHARACTERISTIC FUNCTION

$H(s)$ is the input/output transfer function. We shall assume that in the passband the amplitude of $H(s)$ should approximate unity. It is convenient to eliminate this constant term and instead use the *characteristic function* $K(s)$, defined by

$$H(s)H(-s) = 1 + K(s)K(-s) \tag{2.2}$$

This equation indicates that $H(s)$ and $K(s)$ have the same denominator polynomial; thus if $H(s)$ is written as

$$H(s) = \frac{e(s)}{q(s)} \tag{2.3}$$

we can write $K(s)$ as

$$K(s) = \frac{f(s)}{q(s)} \tag{2.4}$$

The fact that $H(s)$ and $K(s)$ have the same denominator implies that they have the same attenuation poles. That is, the attenuation of the network is given by

$$\begin{aligned} A(\omega) &= 10 \log |H(j\omega)|^2 \\ &= 10 \log (1 + |K(j\omega)|^2) \end{aligned} \tag{2.5}$$

Thus, when either $H(j\omega)$ or $K(j\omega)$ is infinite, the attenuation is infinite. However, only when $K(j\omega)$ is zero is there an attenuation zero. The characteristic function is thus the more useful, because it eliminates the unity constant and focuses attention on the attenuation zeros.

2.3 THE LOWPASS CHARACTERISTIC FUNCTION

For an ideal lowpass filter, $|H(j\omega)|$ is unity in the passband and infinite in the stopband. Subtracting unity from $|H(j\omega)|^2$ produces $|K(j\omega)|^2$; thus the ideal lowpass characteristic function is as shown in Fig. 2.1a.*

* Bandpass filters, studied in later chapters, will have two passband edges, ω_A and ω_B. In general, the subscript A will imply that a variable is in the lower part of the passband, and B will imply that it is in the upper part. In Fig. 2.1, ω_H (H for high) is the stopband edge. A bandpass filter would also have a lower stopband edge ω_L.

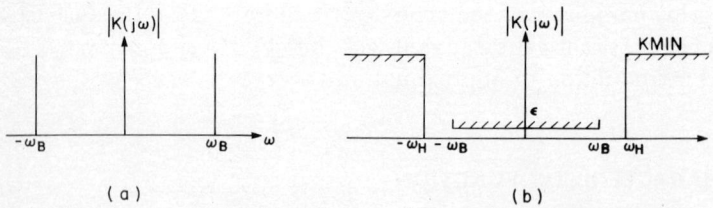

Fig. 2.1 (a) Characteristic function for an ideal lowpass filter; (b) symbolic representation of the characteristic function for a practical lowpass filter.

This characteristic function is impossible to realize because of the discontinuity in slope at the passband edge ω_B. Realizable characteristic functions must instead be as shown in Fig. 2.1b.

Figure 2.1b implies that in the passband $|K(j\omega)| < \epsilon$, while in the stopband $|K(j\omega)| > K_{\min}$. These values can be related to the loss of the filter by Eq. (2.5). Thus the maximum passband loss is

$$A_{\max} = 10 \log (1 + \epsilon^2) \tag{2.6}$$

whereas the minimum stopband loss is

$$A_{\min} = 10 \log (1 + K_{\min}^2) \tag{2.7}$$

2.4 BUTTERWORTH POLYNOMIALS[1]

Butterworth polynomials can be used to approximate the characteristic function shown in Fig. 2.1b. This section defines the general nth-order Butterworth polynomial, and the next section demonstrates how it can be used for lowpass filters.

The nth-order Butterworth polynomial $B_n(\omega)$ satisfies the following conditions:

1. $B_n(\omega)$ is an nth-order polynomial.
2. $B_n(0) = 0$.
3. $B_n(\omega)$ is maximally flat at the origin.
4. $B_n(1) = 1$.

Condition 1 implies that

$$B_n(\omega) = c_0 + c_1\omega + c_2\omega^2 + \cdots + c_n\omega^n \tag{1}$$

Condition 2 requires that $c_0 = 0$. "Maximally flat at the origin" implies that as many derivatives as possible are zero at the origin.* From (1),

$$\frac{dB_n}{d\omega} = c_1 + 2c_2\omega + \cdots + nc_n\omega^{n-1}$$

For this to be zero (at $\omega = 0$) we must have $c_1 = 0$. Similarly, higher-order derivatives can be made to vanish by making higher-order coefficients zero. Thus, condition 3 requires

$$B_n(\omega) = c_n\omega^n$$

Finally, condition 4 yields $c_n = 1$. Summarizing, we have:

---
* Since we are approximating a constant, making as many derivatives zero as possible implies that we are finding a Taylor approximation.

Theorem 2.1

The nth-order Butterworth polynomial $B_n(\omega) = \omega^n$ has the following properties:

1. $B_n(\omega)$ is a polynomial.
2. $B_n(0) = 0$.
3. $B_n(\omega)$ is maximally flat at the origin.
4. $B_n(1) = 1$.

2.5 BUTTERWORTH LOWPASS FILTERS

From (2.5), the loss in decibels is given by

$$A(\omega) = 10 \log (1 + |K(j\omega)|^2) \qquad (2.8)$$

For the Butterworth lowpass, the characteristic function is chosen such that

$$K(j\omega) = \epsilon B_n\left(\frac{\omega}{\omega_B}\right) \qquad (2.9)$$

where, from Theorem 2.1,

$$B_n\left(\frac{\omega}{\omega_B}\right) = \left(\frac{\omega}{\omega_B}\right)^n \qquad (2.10)$$

These three equations can be seen to yield a lowpass function, since combining them gives

$$A(\omega) = 10 \log \left[1 + \epsilon^2 \left(\frac{\omega}{\omega_B}\right)^{2n}\right] \qquad (2.11)$$

which has the following properties:

$$A(0) = 0 \qquad A(\omega_B) = 10 \log (1 + \epsilon^2) = A_{\max}$$

These results are summarized in the following theorem:

Theorem 2.2

The nth-order Butterworth lowpass filter described by

$$A(\omega) = 10 \log \left[1 + \epsilon^2 \left(\frac{\omega}{\omega_B}\right)^{2n}\right] \qquad (2.11)$$

has the following properties:

1. $H(s)$ is a polynomial.
2. $A(0) = 0$.

3. $A(\omega)$ is maximally flat at the origin.
4. $A(\omega_B) = 10 \log (1 + \epsilon^2) = A_{max}$.

The constant n (which determines the complexity of the filter) can be chosen so that the minimum stopband attenuation is A_{min}. However, before demonstrating this, it might be instructive to investigate a normalized form of (2.11):

$$A(\omega) = 10 \log (1 + \omega^{2n}) \qquad (2.12)$$

Figure 2.2 has values of $A(\omega)$ for various degrees n. The following observations can be made:

1. As n is increased, the passband is flat over a wider interval.
2. As n is increased, the stopband loss is increased.

2.6 DETERMINATION OF DEGREE OF $B_n(\omega)$

From Theorem 2.2, the loss of an nth-order Butterworth lowpass filter is given by

$$A(\omega) = 10 \log \left[1 + \epsilon^2 \left(\frac{\omega}{\omega_B} \right)^{2n} \right] \qquad (2.13)$$

The constants ϵ and n should be chosen such that the loss is as shown in Fig. 2.3.

Since the loss at ω_B is defined to be A_{max}, it follows that

$$\epsilon^2 = 10^{0.1 A_{max}} - 1 \qquad (2.14)$$

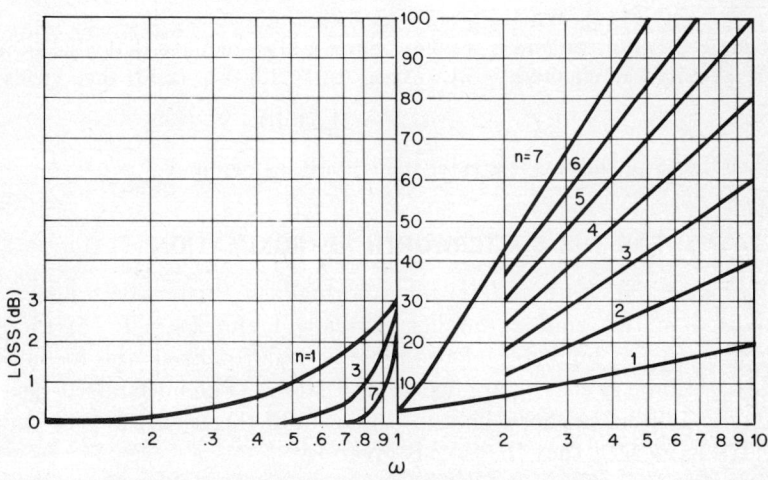

Fig. 2.2 Loss of a normalized Butterworth filter as a function of the degree n.

12 Approximation Methods for Electronic Filter Design

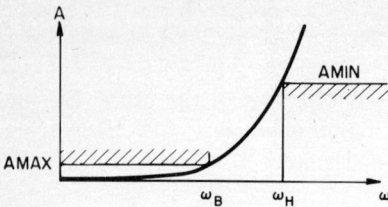

Fig. 2.3 Loss of a Butterworth filter.

Thus, the loss at ω_H is

$$A(\omega_H) = 10 \log \left[1 + (10^{0.1 A_{max}} - 1)\left(\frac{\omega_H}{\omega_B}\right)^{2n}\right] = A_{min}$$

Solving for the degree n yields

$$n = \frac{\log\left[(10^{0.1 A_{min}} - 1)/(10^{0.1 A_{max}} - 1)\right]}{2 \log(\omega_H/\omega_B)} \quad (2.15)$$

This formula can be used to calculate the order necessary for a Butterworth lowpass filter.

Example 2.1

In a Butterworth filter, what degree is necessary to meet the following lowpass requirements?

$$A_{max} = 0.1 \text{ dB} \qquad A_{min} = 30 \text{ dB} \qquad \frac{\omega_H}{\omega_B} = 1.3$$

Solution

Using Eq. (2.15), we find $n = 20.3$. Since a filter requires that n be an integer, we must instead choose $n = 21$. At $\omega_H/\omega_B = 1.3$, Eq. (2.13) then yields

$$A(\omega_H) = 31.53 \text{ dB}$$

Thus the stopband loss is greater than 30 dB, as required.

2.7 $H(s)$ FOR THE BUTTERWORTH APPROXIMATION

In this section we find $H(s)$ for a normalized Butterworth filter. Normalized in this context implies that $\epsilon = 1$ and $\omega_B = 1$. If these conditions do not hold, i.e., if the filter is unnormalized, the techniques of this section can still be applied to find $H(s)$. The normalized case is the only one discussed here, because the form of the results is very simple.

We have seen that $|H(j\omega)|^2$ is given by

$$|H(j\omega)|^2 = 1 + |K(j\omega)|^2$$

For the normalized Butterworth lowpass of degree n,

$$|K(j\omega)|^2 = \omega^{2n}$$

Thus,
$$|H(j\omega)|^2 = 1 + \omega^{2n}$$

By analytic continuation, we can replace ω^2 with $-s^2$ to obtain

$$H(s)H(-s) = 1 + (-s^2)^n \qquad (2.16)$$

The $2n$ zeros of this function are located on the unit circle. Their positions are given by

n even:
$$s_{1,2,\ldots,2n} = \exp\left[\frac{j\pi}{2n}(2k-1)\right] \qquad k = 1, 2, 3, \ldots \qquad (2.17)$$

n odd:
$$s_{1,2,\ldots,2n} = \exp\left(\frac{j\pi k}{n}\right) \qquad k = 1, 2, 3, \ldots$$

The cases $n = 2$ and $n = 3$ are shown in Fig. 2.4.

Since $H(s)$ is an input/output transfer function, its zeros must be located in the left half-plane for the network to be stable. Thus, the left half-plane (LHP) zeros of $H(s)H(-s)$ should be assigned to $H(s)$, and the RHP zeros to $H(-s)$.

Example 2.2
a. Find $H(s)$ for the normalized Butterworth filter of order 2.
b. Repeat part a for $n = 3$.

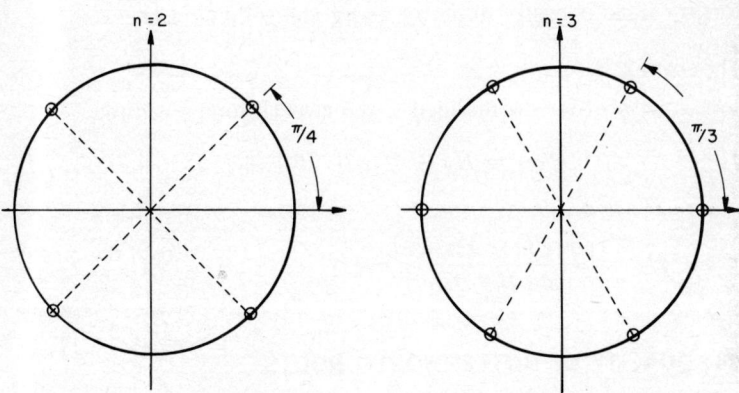

Fig. 2.4 Location of Butterworth roots for second- and third-degree cases.

Solution

a. From Fig. 2.4, the left half-plane zeros of $H(s)$ are located at

$$\frac{1}{\sqrt{2}}(-1 \pm j)$$

Thus,
$$H(s) = s^2 + \sqrt{2}s + 1.$$

b. Similarly, for $n = 3$,

$$H(s) = (s + 1)(s^2 + s + 1)$$

Generalizing the above results gives

n even:

$$H(s) = \prod_{k=1}^{n/2} [s^2 + 2(\cos \theta_k)s + 1]$$

where $\theta = (\pi/2n)(2k - 1)$ (2.18)

n odd:

$$H(s) = (s + 1) \prod_{k=1}^{(n-1)/2} [s^2 + 2(\cos \theta_k)s + 1]$$

where $\theta_k = \pi k/n$.

The form of $H(s)$ in (2.18) is especially useful for active filters, because active filters are usually designed by cascading second-order sections. That is, an active filter section can be used to produce

$$H_K(s) = s^2 + 2(\cos \theta_K)s + 1$$

Transfer functions such as these can be cascaded because of the non-interactive properties of active filter sections.

In designing a passive filter, it is useful to express $H(s)$ in polynomial form. This is easily done by using the following theorem:

Theorem 2.3[2]

For an nth-order normalized Butterworth lowpass filter,

$$H(s) = H_0 + H_1 s + H_2 s^2 + \cdots + H_n s^n$$

where $H_0 = 1$

$$H_k = \frac{\cos\,[(k - 1)\pi/2n]}{\sin\,(k\pi/2n)} H_{k-1}$$

2.8 QUALITY OF BUTTERWORTH ROOTS

This section defines the *quality* of a root. This is an important term because it indicates how much effect a particular root has on the response of a transfer function. For instance, consider the bandpass function which has the output/input transfer function*

$$T(s) = \frac{s}{s^2 + (\omega_p/Q_p)s + \omega_p^2} \quad (2.19)$$

The amplitude response is sketched for various values of Q_p, the quality of the pole, in Fig. 2.5. As indicated in the figure, $|T(j\omega_p)|$ has a maximum at ω_p. The maximum value is

$$|T(j\omega_p)| = \frac{Q_p}{\omega_p} \quad (2.20)$$

Thus, increasing Q_p increases the amount of peaking in the response.

For the bandpass function in (2.19), Q_p is related to the *half-power frequencies*, the locations where $|T(j\omega)|^2$ is one-half the peak value. The half-power frequencies ω_1, ω_2 are the solutions to

$$\left|\frac{T(j\omega)}{T(j\omega_p)}\right|^2 = \frac{1}{2} = \frac{\omega^2/[(\omega^2 - \omega_p^2)^2 + (\omega\omega_p/Q_p)^2]}{(Q_p/\omega_p)^2} \quad (2.21)$$

It can be shown that

$$\frac{\omega_p}{\omega_2 - \omega_1} = Q_p \quad (2.22)$$

That is, for a bandpass function, Q_p can be determined by dividing the peak frequency ω_p by the bandwidth $\omega_2 - \omega_1$.

The concept of quality is of interest even when the transfer function is more general than the bandpass function of (2.19). The following definition introduces the common notation.

Definition 2.1

If a transfer function is written as

$$T(s) = \frac{s^2 + (\omega_z/Q_z)s + \omega_z^2}{s^2 + (\omega_p/Q_p)s + \omega_p^2} \quad (2.23)$$

* The subscript p indicates "pole."

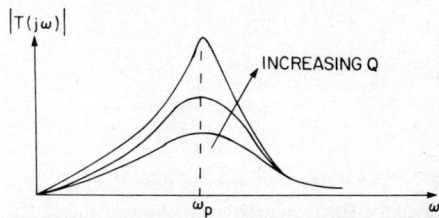

Fig. 2.5 Increasing the quality Q of a bandpass function increases the selectivity.

then ω_z (or ω_p) is called the *undamped natural frequency* of the zero (or pole) and Q_z (or Q_p) is the *quality* of that zero (or pole).

For the transfer function in (2.23), Q_p is no longer simply related to the half-power frequencies. This should be obvious, as the numerator polynomial must also influence the location of these half-power points.

The quality of a pole is very important in filter realizations, since it indicates how close the pole is to the imaginary axis. In passive circuits, increasing pole quality requires that the elements be of better quality (the inductors and capacitors should be less lossy). In active circuits, increasing pole quality requires more active elements in the realization, so that the network is not too sensitive.

To see that the quality of a pole (or zero) indicates how close the root is to the imaginary axis, consider the denominator of (2.23):

$$D(s) = s^2 + \frac{\omega_p}{Q_p} s + \omega_p^2 \tag{2.24}$$

The roots of this expression are

$$s_{1,2} = -\frac{\omega_p}{2Q_p} \pm j\omega_p \sqrt{1 - \left(\frac{1}{2Q_p}\right)^2} \tag{2.25}$$

Thus, if $Q_p > \frac{1}{2}$ the roots are complex. In fact, they are located on a circle, as indicated in Fig. 2.6. The radius of the circle is ω_p; it is not a function of the quality of the pole. However, the larger Q_p is, the closer the pole is to the imaginary axis. In fact, if Q_p is infinite, then the pole is on the imaginary axis.

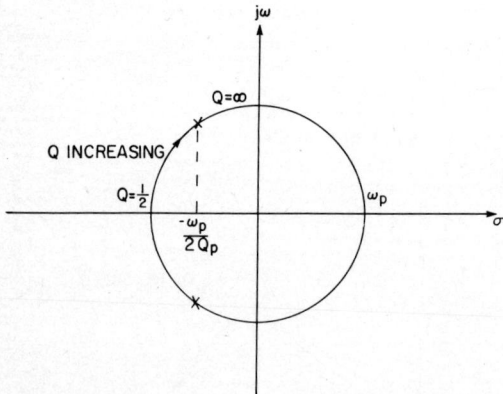

Fig. 2.6 Butterworth roots move on a circle as their quality is varied.

As a demonstration of the calculation of quality Q, we shall consider the normalized Butterworth transfer function (for n even) of (2.18):

$$H(s) = \prod_{k=1}^{n/2} [s^2 + 2(\cos \theta_k)s + 1] \tag{2.26}$$

where $\theta_k = (\pi/2n)(2k - 1)$.

Comparing this with (2.23) yields*

$$\omega_{z_k} = 1 \qquad Q_{z_k} = \frac{1}{2 \cos \theta_k} \tag{2.27}$$

This equation is also valid for odd-degree Butterworth filters, except that, in that case,

$$\theta_k = \frac{\pi k}{n}$$

* The subscript z is meant to imply that the roots are zeros of $H(s)$.

TABLE 2.1 Quality of Some Butterworth Roots*

Quality for n even							
2	4	6	8	10	12	14	16
0.71	0.54	0.52	0.51	0.51	0.50	0.50	0.50
	1.31	0.71	0.60	0.56	0.54	0.53	0.52
		1.93	0.90	0.71	0.63	0.59	0.57
			2.56	1.10	0.82	0.71	0.65
				3.20	1.31	0.94	0.79
					3.83	1.51	1.06
						4.47	1.72
							5.10

Quality for n odd							
3	5	7	9	11	13	15	
1.00	0.62	0.55	0.53	0.52	0.51	0.51	
	1.62	0.80	0.65	0.59	0.56	0.55	
		2.24	1.00	0.76	0.67	0.62	
			2.88	1.20	0.88	0.75	
				3.51	1.41	1.00	
					4.15	1.62	
						4.78	

* For n odd there is also a real root for which $Q = 0.5$.

It should be noted that, although the above formulas were stated for normalized Butterworth filters, they are also true for the unnormalized case, since frequency scaling does not change the quality of a root.

Using the above formulas, one can calculate the quality of the roots of a Butterworth filter of any degree. The results for some low-order Butterworth filters are given in Table 2.1. For example, consider the case $n = 4$. There are two pairs of complex roots ($Q > 0.5$). One pair has $Q = 0.54$, and the other pair has $Q = 1.31$. In general, it should be noted that as the degree n increases, the quality of the roots increases.

It is a general property of *any* type of filter that as the degree n is increased, the quality of the roots increases. Thus, if we want an abrupt transition between the passband and stopband, we must pay some penalties. First, increasing the degree n means that more elements will be required in any realization. Second, the higher quality of the roots implies that better elements will be needed in the realization. Also, the higher quality of the roots means that the resulting design will be more sensitive to element variations.

2.9 CONCLUSIONS

The Butterworth lowpass approximation yields an input/output transfer function $H(s)$ that is a polynomial. This polynomial exhibits maximally flat behavior at the origin. If, as implied by Fig. 2.3, one is only interested in the loss as a function of frequency, the only virtue of the Butterworth approximation is its mathematical simplicity. This simplicity is indicated by (2.16), which is repeated below:

$$H(s)H(-s) = 1 + (-s^2)^n$$

The price that is paid for this simplicity is a very slow transition between the passband and stopband. This was demonstrated in Example 2.1, which required a twenty-first-degree filter! This is a very high price to pay when other types of filters are readily available. The next chapter discusses another approximation that produces a much sharper transition region.

While not too practical itself, the Butterworth filter does provide a convenient foundation on which we can build to produce other more practical maximally flat filters.

REFERENCES

1. S. Butterworth, "On the Theory of Filter Amplifiers," *Wireless Eng.*, October 1936, pp. 536–541.

2. L. Weinberg, *Network Analysis and Synthesis*, McGraw-Hill Book Company, New York, 1962, pp. 494–497.

PROBLEMS

2.1 Prove that the quality of a root is not affected by frequency scaling.

2.2 Show that, for the bandpass function $T(s) = s/(s^2 + \omega_p s/Q_p + \omega_p^2)$, the quality Q_p is related to the peak frequency and half-power frequencies by

$$Q_p = \frac{\omega_p}{\omega_2 - \omega_1}$$

That is, prove Eq. (2.22).

2.3 Prove that the roots of $s^2 + \omega_p s/Q_p + \omega_p^2$ are located on a circle (this should be shown for ω_p fixed and Q_p a variable).

2.4 Find $H(s)$ for a fifth-order normalized Butterworth filter by applying (2.18). Check these results by using Theorem 2.3.

2.5 Find the quality of the roots of a fifth-order normalized Butterworth filter.

CHAPTER THREE

The Tschebycheff Approximation

3.1 INTRODUCTION

We want to approximate a given function $f(\omega)$ with another function $h(\omega)$. The approximation should be accurate in the frequency range $\omega_A \leq \omega \leq \omega_B$. If $h(\omega)$ approximates $f(\omega)$ in the Taylor sense, the adjustable parameters are chosen so that as many derivatives as possible are matched at some point ω_0. For example, the Butterworth lowpass filter has the first $n-1$ derivatives zero at $\omega_0 = 0$ and is thus maximally flat at the origin. Thus $h(\omega)$ approximates $f(\omega)$ very well at ω_0, but this is accomplished at the expense of the rest of the frequency band.

If we are interested in a band of frequencies, we will often use a Tschebycheff* type of approximation which considers the approximation error throughout the frequency interval of interest. We define a Tschebycheff approximation as follows:

Definition 3.1

A function $h(\omega)$ is a Tschebycheff approximation of $f(\omega)$ if the available parameters are adjusted so that the magnitude of the largest error is minimized.

---
* Tschebycheff is spelled in many different ways; another popular version is Chebyshev.

In the definition, the "available parameters" refer to the quantities that determine the function $h(\omega)$ (for example, they might be resistors in a specific network). Since the Tschebycheff approximation minimizes the maximum error, it is often called a *min-max* approximation.

3.2 EQUIRIPPLE APPROXIMATIONS[1]

A function $h(\omega)$ is said to be a Tschebycheff approximation to $f(\omega)$ if the available parameters are adjusted so that the magnitude of the largest error is minimized. In this section, we show that the error function

$$e(\omega) = f(\omega) - h(\omega)$$

has a special property: it is *equiripple*. That is, the error oscillates between maximums and minimums of equal amplitude. An example of an equiripple error function is given in Fig. 3.1. In the figure it is assumed that the approximation region is $-\omega_B \leq \omega \leq \omega_B$, so that the error oscillates about zero in this interval.

We want to show that a Tschebycheff approximation must have an equiripple error. This will be done by assuming that the error $e(\omega)$ has a maximum value at $\omega = \omega_1$. If this error can be reduced, then it is not a Tschebycheff approximation, because by definition a Tschebycheff approximation is such that the maximum error has been minimized. To determine whether or not the error at ω_1 can be reduced, consider the total differential

$$de(\omega_1) = \sum_{j=1}^{n} \frac{\partial e(\omega_1)}{\partial p_j} dp_j \qquad (3.1)$$

where the p_j are the adjustable parameters.

Equation 3.1 gives the incremental change in error de as a function of the incremental parameter changes dp_j. For any specific maximum $e(\omega_1)$, these parameter changes can be chosen so that the error is reduced. However, reducing the error at ω_1 might increase the error elsewhere. Thus, assume that the error at ω_1 is reduced until it becomes equal to

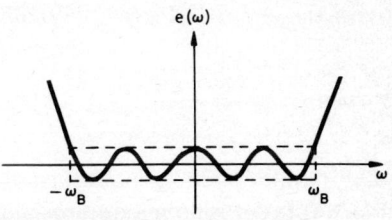

Fig. 3.1 Equiripple error function.

the error at some other frequency ω_2. Can we simultaneously reduce the errors at ω_1 and ω_2? To answer this question, consider the total differential

$$de(\omega_2) = \sum_{j=1}^{n} \frac{\partial e(\omega_2)}{\partial p_j} dp_j \qquad (3.2)$$

We want to choose the incremental parameter changes dp_j so that the errors at both ω_1 and ω_2 are reduced. From (3.1) and (3.2), we can choose a set of parameter changes to reduce the errors at ω_1 and ω_2 simultaneously if $n \geq 2$ (as then we have two equations and at least two adjustable parameters). Thus we can choose the parameter changes so as to reduce the maximum error at both ω_1 and ω_2. The error can be reduced until it becomes equal to the error at some third frequency ω_3.

The maximum errors can be simultaneously reduced until we have more equations [i.e., relations similar to Eqs. (3.1) and (3.2)] to satisfy than adjustable parameters. This leads to the following theorem.

Theorem 3.1

A Tschebycheff approximation is equiripple; that is, the approximation error $e(\omega)$ has maximums and minimums that are of equal magnitude. If there are n adjustable parameters, then the approximation error is zero at least n times, which implies that there are at least a total of $n + 1$ maximums and minimums.

As an example of the use of this theorem, consider again the approximation error in Fig. 3.1. This is equiripple and has eight zero crossings. Thus, if there are nine adjustable parameters, then this is not a Tschebycheff approximation.

The next logical question is: If there are eight adjustable parameters, is it a Tschebycheff approximation? To see that this question cannot be given a general answer, consider a specific case. Assume we are trying to approximate a third-degree polynomial

$$f(\omega) = 1 + 2\omega^2 + 3\omega^3$$

with the seventh-degree polynomial (i.e., eight adjustable parameters)

$$h(\omega) = \sum_{i=0}^{7} p_i \omega^i$$

The parameters can be adjusted so that the error function $f(\omega) - h(\omega)$ has a total of nine maximums and minimums. However, this would not be a Tschebycheff approximation, as the maximum error has not been minimized. The maximum error is minimized (made zero in this case) by choosing all p_i equal to zero except

$$p_0 = 1 \quad p_2 = 2 \quad p_3 = 3$$

The above example indicates why Theorem 3.1 says there must be *at least* $n + 1$ maximums and minimums if the approximation is to be of the Tschebycheff type. Examples can be concocted where there must be more than $n + 1$ maximums and minimums (the previous example had an infinite number, each maximum and minimum being of zero amplitude).

3.3 THE TSCHEBYCHEFF LOWPASS APPROXIMATION

The previous section defined a Tschebycheff approximation in general. That is, the function $f(\omega)$ that we wanted to approximate with $h(\omega)$ was an arbitrary function. In this section, we restrict ourselves to a specific function: we want to approximate zero loss (so that $|H(j\omega)| = 1$) in the passband $-\omega_B \leq \omega \leq \omega_B$.

We first define a normalized Tschebycheff lowpass approximation and then give an important property of this approximation. Finally we consider the "unnormalized" Tschebycheff lowpass approximation.

Definition 3.2

The input/output transfer function $H(s)$ is said to be an nth-order normalized Tschebycheff lowpass approximation if it has the following properties:

1. $H(s)$ is an nth-order polynomial.
2. $|H(j)|^2 = 2$.
3. $|H(j\omega)| \geq 1$.
4. The peak deviation of $|H(j\omega)|$ from unity is minimized over the interval $-\omega_B \leq \omega \leq \omega_B$ for ω_B between 0 and 1.

The nth-order approximation is said to be normalized because of condition 2. Since $A(\omega) = 20 \log |H(j\omega)|$, condition 2 is equivalent to stating that $\omega = 1$ is the half-power point; that is, $A(1) = 3$ dB. The half-power point can be shifted by frequency scaling if one does not want it located at $\omega = 1$.

Condition 3 is included so that the minimum passband loss is zero. Of course, if one is designing active filters and has a passband gain, one would change the level of the constant multiplier of $H(s)$.

The approximation is said to be of the Tschebycheff type because of condition 4. That is, we want to approximate 0 dB of loss in the passband interval $-\omega_B \leq \omega \leq \omega_B$. The approximation should be such that the maximum deviation from zero loss is minimized; it is thus a min-max approximation.

We know from Theorem 3.1 that a Tschebycheff filter is equiripple, as is stated in the following theorem:

Theorem 3.2

If the nth-order polynomial $H(s)$ is a normalized Tschebycheff lowpass approximation, then the passband is equiripple.

We shall now show that for any specific degree n there is only one nth-order normalized Tschebycheff lowpass approximation. That is, there cannot be two different loss curves $A(\omega)$ and $A_1(\omega)$. We shall show this for the case $n = 4$; the general case is an obvious extension. The proof will be by contradiction: We shall assume there is another curve $A_1(\omega)$ [different from $A(\omega)$] that satisfies conditions 1 to 4 of Definition 3.2, and then show that this assumption leads to a contradiction.

To show that $A(\omega)$ must be as shown in Fig. 3.2 for the case $n = 4$, first observe that $A(\omega)$ has three relative maximums for $-\omega_B < \omega < \omega_B$; any other curve $A_1(\omega)$ must intersect $A(\omega)$ at least six times in this region (twice for each maximum). $A_1(\omega)$ must also intersect $A(\omega)$ once near each passband edge and at $\omega = \pm 1$. Thus $A_1(\omega) - A(\omega)$ must have at least 10 zeros (for the general case there would be at least $2n + 2$ zeros). But $H_1(s)$ and $H(s)$ are fourth degree by assumption, so that $A_1(\omega) - A(\omega)$ can have only eight zeros. This establishes the contradiction.

We just saw that the fourth-order normalized Tschebycheff lowpass has the unique shape shown in Fig. 3.2. The fact that the curve has a unique shape implies that the maximum passband attenuation is uniquely determined. That is, if n and ω_B are specified, then $A_{\max}$ can be calculated. Alternatively, if $A_{\max}$ is specified, then ω_B can be found.

We have yet to give a formal definition of an unnormalized Tschebycheff lowpass filter. This could be done by paraphrasing Definition 3.2 for the unnormalized case, but the following definition is more convenient

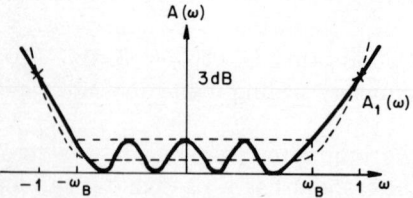

Fig. 3.2 Loss of fourth-order normalized Tschebycheff lowpass approximation.

(the preceding discussion has shown that a min-max criterion is equivalent to an equiripple criterion):

Definition 3.3

The transfer function $H(s)$ is said to be an nth-order Tschebycheff lowpass approximation if it and the corresponding loss function have the following properties:

1. $H(s)$ is an nth-degree polynomial.
2. $A(\omega)$ has $n - 1$ equal maximums in the interval $-\omega_B < \omega < \omega_B$. These maximums have the same value as the loss at $\pm \omega_B$, which is defined to be $A_{\max}$.
3. The loss $A(\omega)$ has a minimum value of zero n different times in the interval $-\omega_B \leq \omega \leq \omega_B$.

3.4 INTRODUCTION TO TSCHEBYCHEFF POLYNOMIALS

The nth-order Tschebycheff lowpass approximation was defined in the previous section. This section demonstrates how Tschebycheff polynomials can be used to produce Tschebycheff lowpass approximations. The nth-order Tschebycheff polynomial $T_n(x)$ is defined as follows:

Definition 3.4

The Tschebycheff polynomial $T_n(x)$ is the nth-degree polynomial that has the following properties:

1. T_n is even (odd) if n is even (odd).
2. T_n has all its zeros in the interval $-1 < x < 1$.
3. T_n oscillates between values of ± 1 in the interval $-1 \leq x \leq 1$.
4. $T_n(1) = +1$.

In the next section we shall develop a formula for $T_n(x)$ which can be used to find the shape of $T_n(x)$ for any value of n. The shapes of some Tschebycheff polynomials are shown in Fig. 3.3. As a step toward understanding how a Tschebycheff polynomial can be used for a Tschebycheff lowpass approximation, consider $[T_4(\omega/\omega_B)]^2$, which is shown in Fig. 3.4. This function is very similar to the loss shape that was drawn in Fig. 3.2; in fact, we can write

$$A = 10 \log \left\{ 1 + \left[\epsilon T_4 \left(\frac{\omega}{\omega_B} \right) \right]^2 \right\} \qquad (3.3)$$

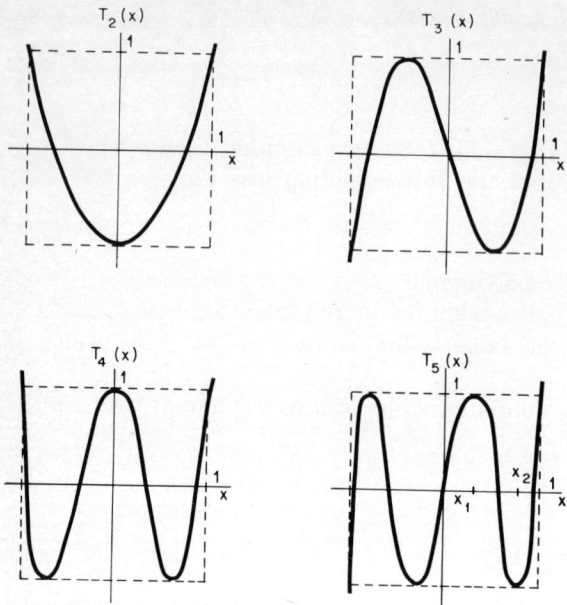

Fig. 3.3 Tschebycheff polynomial responses.

To understand why the loss curve in Fig. 3.2 is given by Eq. (3.3), consider the following points:

1. $[T_4(\omega/\omega_B)]^2$ is an even function of frequency, as is $A(\omega)$.
2. $[T_4(\omega/\omega_B)]^2$ has four second-order zeros in the region $-\omega_B < \omega < \omega_B$, so that $A(\omega)$ has four second-order zeros in that interval.
3. $[T_4(\omega/\omega_B)]^2$ is unity at three frequencies in the region $-\omega_B < \omega < \omega_B$ and at $\omega = \pm\omega_B$. If we define

$$\epsilon^2 = 10^{0.1\,A_{\max}} - 1$$

it follows that $A = A_{\max}$ at these frequencies.

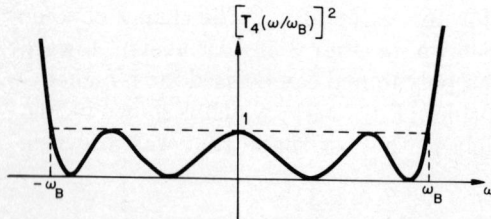

Fig. 3.4 The function $[T_4(\omega/\omega_B)]^2$ is similar to the loss response of a fourth-order Tschebycheff lowpass approximation.

These are the properties that $A(\omega)$ must possess in order to be a Tschebycheff lowpass approximation. Generalizing, we can conclude:

Theorem 3.3

The nth-order Tschebycheff lowpass approximation is described by

$$A(\omega) = 10 \log \left\{ 1 + \left[\epsilon T_n \left(\frac{\omega}{\omega_B} \right) \right]^2 \right\} \qquad (3.4)$$

where T_n is the nth-order Tschebycheff polynomial.

In the next section we shall find an equation for the nth-order Tschebycheff polynomial; however, before proceeding to it we shall make some observations about Eq. (3.4). The passband of the filter is supposed to approximate zero loss, so that $|H(j\omega)|$ should approximate unity. This is why we usually work with the characteristic function $K(s)$ instead of the input/output transfer function $H(s)$. These functions are related by

$$H(s)H(-s) = 1 + K(s)K(-s) \qquad (3.5)$$

Comparing (3.4) and (3.5) yields

$$|K(j\omega)|^2 = \epsilon^2 T_n^2 \left(\frac{\omega}{\omega_B} \right) \qquad (3.6)$$

That is, the characteristic function is simply a scaled version of a Tschebycheff polynomial.

3.5 THE TSCHEBYCHEFF POLYNOMIAL

The nth-order Tschebycheff polynomial $T_n(x)$ was defined to be the function that has the following properties:

1. T_n is even (odd) if n is even (odd).
2. T_n has all its zeros in the interval $-1 < x < 1$.
3. T_n oscillates between values of ± 1 in the interval $-1 \leq x \leq 1$.
4. $T_n(1) = +1$.

We shall find an expression for $T_n(x)$ by first examining the curve for $T_5(x)$ in Fig. 3.3 and then making some generalizations. The derivative of $T_5(x)$ is zero at the maximums and minimums; thus we can write

$$\frac{dT_5}{dx} = C_1(x - x_1)(x - x_2)(x + x_1)(x + x_2) \qquad (3.7)$$

or, in general,

Theorem 3.4

The derivative of $T_n(x)$ can be expressed in terms of its maximums and minimums x_i as

n even:
$$\frac{dT_n}{dx} = C_1 x \prod_{i=1}^{n/2-1} (x^2 - x_i^2) \qquad (3.8a)$$

n odd:
$$\frac{dT_n}{dx} = C_1 \prod_{i=1}^{(n-1)/2} (x^2 - x_i^2) \qquad (3.8b)$$

We shall now show that the derivative of $T_n(x)$ is related to another function. This is done by examining $T_5(x) + 1$ and $T_5(x) - 1$. From Fig. 3.3 we see that we can write

$$T_5 + 1 = C_2(x + 1)[(x + x_1)(x - x_2)]^2$$
$$T_5 - 1 = C_3(x - 1)[(x - x_1)(x + x_2)]^2$$

Multiplying these equations together yields

$$T_5^2 - 1 = C_4(x^2 - 1)[(x^2 - x_1^2)(x^2 - x_2^2)]^2$$

These results can be generalized to yield the next theorem.

Theorem 3.5

n even:
$$T_n^2 - 1 = C_4(x^2 - 1)x^2 \prod_{i=1}^{n/2-1} (x^2 - x_i^2)^2 \qquad (3.9a)$$

n odd:
$$T_n^2 - 1 = C_4(x^2 - 1) \prod_{i=1}^{(n-1)/2} (x^2 - x_i^2)^2 \qquad (3.9b)$$

Comparing the formulas in Theorems 3.4 and 3.5 yields the following important result:

Theorem 3.6

$$\left(\frac{dT_n}{dx}\right)^2 = M^2 \frac{T_n^2 - 1}{x^2 - 1} \qquad (3.10)$$

where M^2 is a constant.

We now have a differential equation that describes the nth-order Tschebycheff polynomial. Solving the differential equation yields T_n as given by the next theorem.

Theorem 3.7

The nth-order Tschebycheff polynomial is given by

$$T_n(x) = \cos(n \cos^{-1} x) \qquad (3.11)$$

Proof:

From Theorem 3.6,

$$\frac{dT_n}{dx} = M \left(\frac{T_n^2 - 1}{x^2 - 1} \right)^{1/2}$$

Thus,
$$\frac{dT_n}{\sqrt{1 - T_n^2}} = M \frac{dx}{\sqrt{1 - x^2}}$$
$$\cos^{-1} T_n = M \cos^{-1} x + C$$
$$T_n = \cos(M \cos^{-1} x + C)$$

As x goes from -1 to $+1$, $\cos^{-1} x$ goes from $-\pi$ to 0. Thus, since T_n should have n zeros in this region, $M \cos^{-1} x$ must go through $n\pi$ radians so that $M = n$. Finally, since $T_n(1) = 1$, we must have $C = 2\pi m$. Taking $m = 0$ yields the theorem.

Theorem 3.7 expresses $T_n(x)$ as a function of an inverse cosine. In this form it is not at all obvious that $T_n(x)$ is an nth-order polynomial. The following theorem not only proves that $T_n(x)$ as given by Eq. (3.11) is an nth-order polynomial, it also gives a simple way to find the polynomial.

Theorem 3.8

The nth-order Tschebycheff polynomial $T_n(x)$ can be found by using the following recursion relation with the initial values $T_0(x) = 1$ and $T_1(x) = x$:

$$T_{n+1}(x) = 2xT_n(x) - T_{n-1}(x) \tag{3.12}$$

Proof:

From Theorem 3.7 we have $T_{n+1} = \cos[(n+1) \cos^{-1} x]$. Thus if we define $z = \cos^{-1} x$ we can write

$$T_{n+1} = \cos(n+1)z$$
$$= \cos nz \cos z - \sin nz \sin z$$

Similarly, $\quad T_{n-1} = \cos nz \cos z + \sin nz \sin z$
Thus, $\quad T_{n+1} + T_{n-1} = 2 \cos z \cos nz$
$$= 2 \cos(\cos^{-1} x) \cos(n \cos^{-1} x)$$
$$= 2xT_n$$

This establishes the recursion relation in (3.12). The initial values $T_0(x) = 1$ and $T_1(x) = x$ follow from $T_n(x) = \cos(n \cos^{-1} x)$.

If one wanted to find $T_{18}(x)$ by applying (3.12) one would first have to find the 17 polynomials of lower degree. Hand computations would probably give rise to many mistakes, and even with a computer the process would take longer than necessary. It is instead quicker to use the expression[2]

$$T_n(x) = \frac{n}{2} \sum_{r=0}^{\text{IP}(n/2)} \frac{(-1)^r (n-r-1)!}{r!(n-2r)!} (2x)^{n-2r} \qquad (3.13)$$

where IP stands for "integer part of."

If one uses Tschebycheff polynomials very often, even the above expression might be tedious. In that case, Table 3.1 would be helpful.

3.6 THE NORMALIZED TSCHEBYCHEFF LOWPASS

The normalized Tschebycheff lowpass approximation has the following properties:

1. $H(s)$ is an nth-order polynomial.
2. $|H(j)|^2 = 2$.
3. $|H(j\omega)| \geq 1$.
4. The peak deviation of $|H(j\omega)|$ from unity is minimized over the interval $-\omega_B \leq \omega \leq \omega_B$ for any ω_B between 0 and 1.

In this section we investigate the losses of some normalized Tschebycheff approximations, using the formulas developed in the first part of the chapter. In particular, from Theorem 3.3 the loss of an nth-order (unnormalized) Tschebycheff lowpass filter is given by

$$A(\omega) = 10 \log \left\{ 1 + \left[\epsilon T_n \left(\frac{\omega}{\omega_B} \right) \right]^2 \right\} \qquad (3.14)$$

To use this for a normalized lowpass, we require that $A(1) = 10 \log 2$. Thus it follows that, for the normalized lowpass,

$$\epsilon T_n \left(\frac{1}{\omega_B} \right) = 1 \qquad (3.15)$$

Equation (3.15) indicates that ϵ and ω_B are related; given one, we can find the other. The passband ripple $A_{\max}$ is related to ϵ by

$$\epsilon^2 = 10^{0.1 A_{\max}} - 1 \qquad (3.16)$$

Combining (3.15) and (3.16) yields

$$10^{0.1 A_{\max}} = 1 + \left[T_n \left(\frac{1}{\omega_B} \right) \right]^{-2} \qquad (3.17)$$

This equation can be used to find ω_B for any value of $A_{\max}$ and n; some

TABLE 3.1 The Tschebycheff Polynomials
$T_n(x) = T_n x^n + T_{n-2} x^{n-2} + \cdots$

n	T_1	T_3	T_5	T_7	T_9	T_{11}	T_{13}	T_{15}	T_{17}
1	1								
3	-3	4							
5	5	-20	16						
7	-7	56	-112	64					
9	9	-120	432	-576	256				
11	-11	220	-1232	2816	-2816	1024			
13	13	-364	2912	-9984	16640	-13312	4096		
15	-15	560	-6048	28800	-70400	92160	-61440	16384	
17	17	-816	11424	-71808	239360	-452608	487424	-278528	65536

n	T_0	T_2	T_4	T_6	T_8	T_{10}	T_{12}	T_{14}	T_{16}	T_{18}
2	-1	2								
4	1	-8	8							
6	-1	18	-48	32						
8	1	-32	160	-256	128					
10	-1	50	-400	1120	-1280	512				
12	1	-72	840	-3584	6912	-6144	2048			
14	-1	98	-1568	9408	-26880	39424	-28672	8192		
16	1	-128	2688	-21504	84480	-180224	212992	-131072	32768	
18	-1	162	-4320	44352	-228096	658944	-1118208	1105920	-589824	131072

TABLE 3.2 A_{max} versus ω_B for a Normalized Tschebycheff Lowpass Filter

	ω_B, rad/s, for			
n	$A_{max} = 0.1$	$A_{max} = 0.5$	$A_{max} = 1$	$A_{max} = 10 \log 2$
3	0.720	0.857	0.913	1
4	0.824	0.915	0.950	1
5	0.881	0.944	0.967	1

results are given in Table 3.2. The table demonstrates that (for a fixed n) there is a tradeoff between ripple and passband width: If one wants a small ripple, then the passband must be narrow. If one wants both a small ripple and a wide passband, then n must be chosen large enough.

Table 3.2 can be used to obtain some insight into the passband behavior of a normalized Tschebycheff lowpass filter. Some insight into the stopband behavior can be obtained by studying Fig. 3.5, which demonstrates that the larger the degree n the greater the stopband attenuation. This is best appreciated by looking at the asymptotic behavior at high frequency. For a large loss, (3.14) becomes

$$A(\omega) \approx 10 \log \left[\epsilon T_n \left(\frac{\omega}{\omega_B} \right) \right]^2 \qquad (3.18)$$

But it follows from the recursion relation of (3.12) that, for large x,

$$T_n(x) \approx 2^{n-1} x^n$$

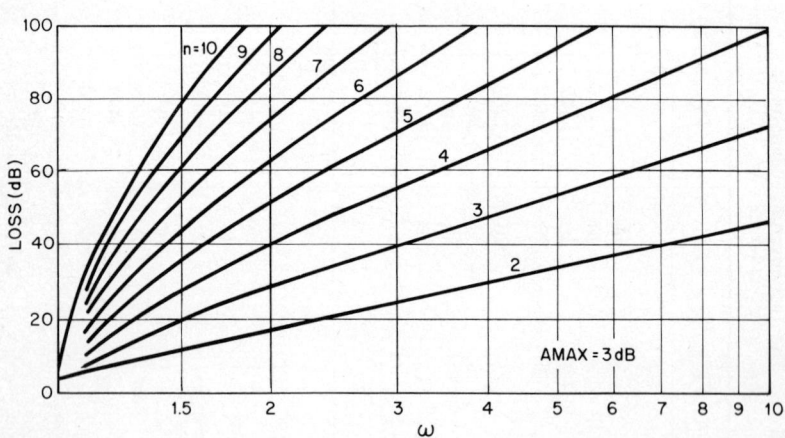

Fig. 3.5 Stopband loss of some normalized Tschebycheff lowpass approximations.

Thus (3.18) becomes

$$A(\omega) \approx 20 \log \epsilon + 20(n-1) \log 2 + 20n \log \frac{\omega}{\omega_B} \qquad (3.19)$$

This equation gives the asymptotic loss behavior of a Tschebycheff lowpass filter. If the filter is normalized so that $\omega = 1$ is the half-power point, then ϵ is determined by ω_B (see Table 3.2). However, for an unnormalized lowpass filter, ϵ and ω_B can be independently specified. Equation (3.19) indicates that decreasing the passband ripple ϵ decreases the stopband loss.

3.7 DETERMINATION OF DEGREE OF $T_n(x)$

Section 3.5 demonstrated that the nth-order Tschebycheff polynomial can be written as

$$T_n(x) = \cos(n \cos^{-1} x) \qquad (3.20)$$

From this equation, the following recursive relation was found:

$$T_{n+1}(x) = 2x T_n(x) - T_{n-1}(x) \qquad (3.21)$$

The recursive relation can be used to find an nth-order power-series expansion of $T_n(x)$.

The power-series expansion of $T_n(x)$ is valid for any x. On the other hand, one experiences difficulty for $x > 1$ if one tries to evaluate $T_n(x)$ by (3.20), as $\cos^{-1} x$ is then imaginary. When x is greater than unity, we will use the following theorem.

Theorem 3.9

The nth-order Tschebycheff polynomial can be expressed as

$$T_n(x) = \cosh(n \cosh^{-1} x) \qquad (3.22)$$

Proof

$$\cos jz = \frac{e^{j(jz)} + e^{-j(jz)}}{2} = \cosh z \triangleq x$$

so that

$$jz = \cos^{-1} x \qquad z = \cosh^{-1} x$$

which implies that

$$\cos^{-1} x = j \cosh^{-1} x$$

Thus,
$$T_n(x) = \cos(n \cos^{-1} x)$$
$$= \cos(nj \cosh^{-1} x) = \cosh(n \cosh^{-1} x)$$

Q.E.D.

Theorems 3.3 and 3.9 can be combined to yield the following important result:

$$A(\omega) = 10 \log \left\{ 1 + \left[\epsilon \cosh\left(n \cosh^{-1} \frac{\omega}{\omega_B} \right) \right]^2 \right\} \quad (3.23)$$

where $\epsilon^2 = 10^{0.1 A_{max}} - 1$.

This equation can be used to find n, the degree of the Tschebycheff filter. Assume that the filter requirements are as shown in Fig. 3.6. If the requirement is just met at $\dot\omega = \omega_H$, it follows from (3.23) that

$$A(\omega_H) = A_{min} = 10 \log \left\{ 1 + \left[\epsilon \cosh\left(n \cosh^{-1} \frac{\omega_H}{\omega_B} \right) \right]^2 \right\}$$

Substituting for ϵ^2 and solving for n yields*

$$n = \frac{\cosh^{-1} [(10^{0.1 A_{min}} - 1)/(10^{0.1 A_{max}} - 1)]^{1/2}}{\cosh^{-1} (\omega_H/\omega_B)} \quad (3.24)$$

Example 3.1

What degree Tschebycheff filter is necessary to meet the following lowpass requirements?

$$A_{max} = 0.1 \text{ dB} \qquad A_{min} = 30 \text{ dB} \qquad \frac{\omega_H}{\omega_B} = 1.3$$

Solution

Equation (3.24) yields $n = 7.9$. Since a filter requires that n be an integer, we must choose $n = 8$. At $\omega_H/\omega_B = 1.3$, Eq. (3.23) then yields $A(\omega_H) = 30.22$ dB. This is greater than $A_{min} = 30$ dB, as required.

3.8 A PROGRAM FOR THE LOSS OF TSCHEBYCHEFF LOWPASS FILTERS

If we are given filter requirements as in Fig. 3.6, it is an easy matter to write a program that will find the necessary degree of a Tschebycheff filter. The program in Fig. 3.7 not only finds the degree n, it also can find the loss of the filter at any frequency F.

* Note the similarity between this and the Butterworth formula (2.15).

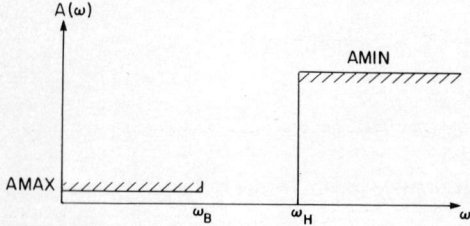

Fig. 3.6 Loss requirements for a lowpass filter.

The Tschebycheff Approximation

```
1; LØSS ØF TSCHEBYCHEFF LØWPASS FILTER
1.2 DEMAND AMAX,AMIN,FB,FH
1.3 EE=10^(.1*AMAX)-1,FF=10^(.1*AMIN)-1
1.31 X=SQRT(FF/EE),FX=LN(X+SQRT(X*X-1))
1.4 Y=FH/FB,FY=LN(Y+SQRT(Y*Y-1))
1.5 N=IP(FX/FY)+1
1.6 TYPE #,N,#
1.7 DEMAND F
1.8 TYPE #
1.9 DØ PART 6

6; CALCULATIØN ØF STØPBAND LØSS
6.1 TØ PART 10 IF F<FB
6.2 X=F/FB,FX=LN(X+SQRT(X*X-1))
6.3 TX=[EXP(N*FX)+EXP(-N*FX)]/2
6.4 A=10*LØG(1+EE*TX*TX)
6.5 TYPE F,A IN FØRM 1

10; CALCULATIØN ØF PASSBAND LØSS
10.2 X=F/FB,FX=ATN(SQRT(1-X*X),X)
10.3 TX=CØS(N*FX)
10.4 A=10*LØG(1+EE*TX*TX)
10.5 TYPE F,A IN FØRM 1
```

Fig. 3.7 Program for calculating the loss of a Tschebycheff lowpass filter.

```
DØ PART 1
    AMAX=.1
    AMIN=30
    FB=1
    FH=1.3

N=          8

F=1.3

   1.300+00      3.022+01
>
DØ PART 6 FØR F=0:.5:3
   0.000+00      1.000-01
   5.000-01      2.522-02
   1.000+00      1.000-01
   1.500+00      4.453+01
   2.000+00      6.916+01
   2.500+00      8.652+01
   3.000+00      1.001+02
>
```

Fig. 3.8 Sample use of program in Fig. 3.7.

Typing DO PART 1 causes the program to ask for the filter requirements A_{max}, A_{min}, FB, and FH. Steps 1.3 to 1.5 then solve for the degree n by evaluating

$$n = \frac{\cosh^{-1}[(10^{0.1 A_{min}} - 1)/(10^{0.1 A_{max}} - 1)]^{1/2}}{\cosh^{-1}(\text{FH/FB})}$$

The inverse hyperbolic cosines are found by using

$$\cosh^{-1} x = \ln(x + \sqrt{x^2 - 1})$$

Once the program has found the necessary degree, it asks (Step 1.7) for F, the frequency at which it should calculate the loss. If F is in the stopband, then Part 6 finds the loss by evaluating

$$A = 10 \log\{1 + [\epsilon T_n(x)]^2\} \qquad \text{Step 6.4}$$

where $T_n(x)$ is found by evaluating

$$T_n(x) = \cosh(n \cosh^{-1} x) \qquad \text{Steps 6.2, 6.3}$$

If F is instead in the passband, then $T_n(x)$ is found by*

$$T_n(x) = \cos(n \cos^{-1} x) \qquad \text{Steps 10.2, 10.3}$$

A sample use of the program is given in Fig. 3.8 for the requirements of Example 3.1. Note that if one wants to find the losses at many frequencies, one need only type DO PART 6 FOR F = $\cdots$.

---

* The inverse cosine is found by using $\cos^{-1} x = \tan^{-1}(\sqrt{1 - x^2}/x)$.

3.9 AN OPTIMUM PROPERTY OF TSCHEBYCHEFF FILTERS

Example 3.1 investigated the eighth-degree Tschebycheff lowpass filter whose loss is sketched in Fig. 3.9. The filter has a very steep transition region—in fact, it is as steep as the transition region of the twenty-first-degree Butterworth filter in Example 2.1.

$H(s)$ for the curve in Fig. 3.9 is an eighth-degree polynomial. This polynomial is such that the maximum passband loss is 0.1 dB. We shall now show that if any other eighth-degree polynomial (e.g., a Butterworth function) has less than 0.1 dB of passband loss, then its stopband performance is worse than that of the eighth-order Tschebycheff. To show this we assume the opposite: assume an eighth-order polynomial $P(s)$ exists that has a loss shape as indicated by the dashed line in Fig. 3.9. This dashed curve is less than 0.1 dB in the passband and is above the Tschebycheff curve in the stopband. However, note that the two curves intersect 18 times. This implies that $H(s) - P(s)$ must have nine zeros. But this is impossible because $H(s)$ and $P(s)$ are only of eighth degree. Generalizing, we can conclude:

Theorem 3.10

Let $H(s)$ and $P(s)$ be nth-degree polynomials. Furthermore, let $H(s)$ represent the transfer function of a Tschebycheff filter with maximum passband loss $A_{\max}$. If the passband loss of $P(s)$ is less than $A_{\max}$, then its stopband performance must be poorer than that of the Tschebycheff filter.

3.10 $H(s)$ FOR THE TSCHEBYCHEFF APPROXIMATION

The previous section demonstrated how (given $A_{\max}$, $A_{\min}$, FB, and FH) to find the necessary degree of a Tschebycheff filter. It also showed

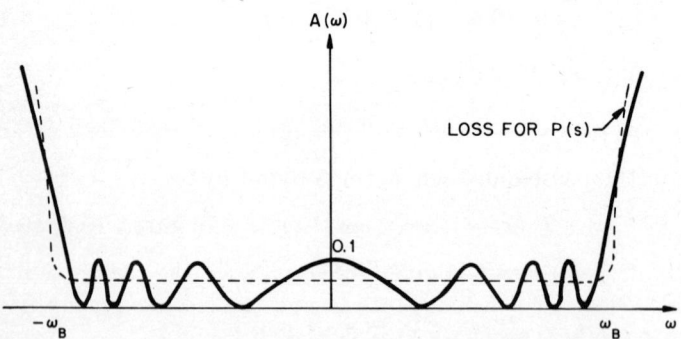

Fig. 3.9 Loss of an eighth-degree Tschebycheff lowpass filter.

how to find the loss $A(\omega)$. To synthesize a filter, we need to know the transfer function $H(s)$, which is related to $A(\omega)$ by

$$A(\omega) = 10 \log |H(j\omega)|^2 \qquad (3.25)$$

This section demonstrates how to find the polynomial $H(s)$.

For a Tschebycheff lowpass filter,

$$A(\omega) = 10 \log \left\{ 1 + \left[\epsilon T_n \left(\frac{\omega}{\omega_B} \right) \right]^2 \right\} \qquad (3.26)$$

In the material that follows, ω_B, the passband edge, will be normalized to unity. This frequency scaling will simplify the equations. For the normalized case, (3.26) becomes

$$A(\omega) = 10 \log \{ 1 + [\epsilon T_n(\omega)]^2 \} \qquad (3.27)$$

It follows from (3.25) and (3.27) that

$$H(s)H(-s) = 1 + \left[\epsilon T_n \left(\frac{s}{j} \right) \right]^2 \qquad (3.28)$$

To obtain $H(s)$, we find the roots of (3.28) and assign those in the left half-plane to $H(s)$. The roots s_k are the solutions to

$$T_n \left(\frac{s_k}{j} \right) = \pm \frac{j}{\epsilon} = \cos \left(n \cos^{-1} \frac{s_k}{j} \right) \qquad (3.29)$$

Defining

$$n \cos^{-1} \frac{s_k}{j} = x + jy \qquad (3.30)$$

we thus have

$$\pm \frac{j}{\epsilon} = \cos(x + jy) = \cos x \cos jy - \sin x \sin jy$$

$$= \cos x \cosh y - j \sin x \sinh y$$

Equating real and imaginary parts gives us

$$\cos x \cosh y = 0 \qquad \sin x \sinh y = \mp \frac{1}{\epsilon}$$

Since $\cosh y > 0$, the first equation implies

$$x = \frac{\pi}{2}(1 + 2k) \qquad k = 0, \pm 1, \pm 2, \ldots \qquad (3.31)$$

38 Approximation Methods for Electronic Filter Design

This, coupled with the second equation, yields

$$y = \mp \sinh^{-1} \frac{1}{\epsilon} \tag{3.32}$$

Combining equations (3.30), (3.31), and (3.32) gives

$$n \cos^{-1} \frac{s_k}{j} = \frac{\pi}{2}(1 + 2k) \mp j \sinh^{-1} \frac{1}{\epsilon}$$

$$\frac{s_k}{j} = \omega_k - j\sigma_k = \cos\left(\frac{\pi}{2}\frac{1+2k}{n} \mp \frac{j}{n}\sinh^{-1}\frac{1}{\epsilon}\right)$$

which implies

$$\begin{aligned}\sigma_k &= \pm \sin\frac{\pi}{2}\frac{1+2k}{n}\sinh\left(\frac{1}{n}\sinh^{-1}\frac{1}{\epsilon}\right) \\ \omega_k &= \cos\frac{\pi}{2}\frac{1+2k}{n}\cosh\left(\frac{1}{n}\sinh^{-1}\frac{1}{\epsilon}\right)\end{aligned} \tag{3.33}$$

By allocating the left half-plane roots of (3.33) to $H(s)$, we can find the input/output transfer function. Before doing this and simplifying the results, we shall first show that the roots in (3.33) lie on an ellipse. Squaring the expressions in (3.33) yields

$$\left(\sin\frac{\pi}{2}\frac{1+2k}{n}\right)^2 = \frac{\sigma_k^2}{\{\sinh[(1/n)\sinh^{-1}(1/\epsilon)]\}^2}$$

$$\left(\cos\frac{\pi}{2}\frac{1+2k}{n}\right)^2 = \frac{\omega_k^2}{\{\cosh[(1/n)\sinh^{-1}(1/\epsilon)]\}^2}$$

If we add these equations together, we obtain

$$1 = \left\{\frac{\sigma_k}{\sinh[(1/n)\sinh^{-1}(1/\epsilon)]}\right\}^2 + \left\{\frac{\omega_k}{\cosh[(1/n)\sinh^{-1}(1/\epsilon)]}\right\}^2 \tag{3.34}$$

This is the equation of an ellipse. That is, the roots of the Tschebycheff lowpass filter are located on an ellipse in the s plane. Recall that the Butterworth roots were located on a circle. It can be shown[3] that the Butterworth circle and Tschebycheff ellipse are related, but this will not be done here.

Before we digressed and studied the Tschebycheff ellipse, we derived (3.33), which gives the Tschebycheff roots. These expressions can be simplified by using the following formula, which comes from Problem 3.7:

$$e^{\sinh^{-1}\omega} = \omega + \sqrt{\omega^2 + 1} \tag{3.35}$$

Thus,

$$2 \sinh\left(\frac{1}{n} \sinh^{-1} \frac{1}{\epsilon}\right) = (e^{\sinh^{-1}(1/\epsilon)})^{1/n} - (e^{\sinh^{-1}(1/\epsilon)})^{-1/n}$$

$$= \left(\frac{1}{\epsilon} + \sqrt{\frac{1}{\epsilon^2} + 1}\right)^{1/n} - \left(\frac{1}{\epsilon} + \sqrt{\frac{1}{\epsilon^2} + 1}\right)^{-1/n}$$

(3.36)

Similarly,

$$2 \cosh\left(\frac{1}{n} \sinh^{-1} \frac{1}{\epsilon}\right) = \left(\frac{1}{\epsilon} + \sqrt{\frac{1}{\epsilon^2} + 1}\right)^{1/n} + \left(\frac{1}{\epsilon} + \sqrt{\frac{1}{\epsilon^2} + 1}\right)^{-1/n} \quad (3.37)$$

Substituting (3.36) and (3.37) into (3.33) yields the following left half-plane roots:

$$\sigma_k = -\frac{1}{2} \sin\left(\frac{\pi}{2} \frac{1+2k}{n}\right) \left[\left(\frac{1}{\epsilon} + \sqrt{\frac{1}{\epsilon^2} + 1}\right)^{1/n} - \left(\frac{1}{\epsilon} + \sqrt{\frac{1}{\epsilon^2} + 1}\right)^{-1/n}\right]$$

$$\omega_k = +\frac{1}{2} \cos\left(\frac{\pi}{2} \frac{1+2k}{n}\right) \left[\left(\frac{1}{\epsilon} + \sqrt{\frac{1}{\epsilon^2} + 1}\right)^{1/n} + \left(\frac{1}{\epsilon} + \sqrt{\frac{1}{\epsilon^2} + 1}\right)^{-1/n}\right]$$

(3.38)

These formulas allow us to calculate the quality of the zeros of $H(s)$ for the Tschebycheff lowpass filter. As mentioned in Chap. 2, quality is an important quantity, since it indicates how difficult it will be to realize the transfer function. For example, the higher the quality, the more sensitive will an active filter realization be to operational amplifier gain. The quality Q of a pair of complex roots at $-\sigma_k \pm j\omega_k$ is defined by

$$s^2 + 2\sigma_k s + \sigma_k^2 + \omega_k^2 = s^2 + \frac{\omega_z}{Q} s + \omega_z^2 \quad (3.39)$$

It thus follows that

$$Q = \frac{1}{2}\left(1 + \frac{\omega_k^2}{\sigma_k^2}\right)^{1/2} \quad (3.40)$$

Equations (3.38) and (3.40) can be used to find the quality of the roots of a Tschebycheff lowpass filter. The results are shown in Table 3.3 for two different values of passband ripple.*

We can make an important observation from Table 3.3. Assume that one designs a seventh-order Tschebycheff lowpass filter that has $A_{\max} = 3$ dB. From Table 3.3 the maximum quality is $Q_{\max} = 17.46$. If the seventh-degree filter more than meets the stopband requirements, but a

* Passband ripple is related to ϵ via $\epsilon^2 = 10^{0.1 A_{\max}} - 1$.

TABLE 3.3 Quality of Some Tschebycheff Roots*

\multicolumn{9}{c}{Quality for n, $A_{max} = 0.5$ dB}
2	3	4	5	6	7	8	9	10
0.86	1.71	0.71	1.18	0.68	1.09	0.68	1.06	0.67
	2.94	4.54	1.81	2.58	1.61	2.21	1.53
			6.51	8.84	3.47	4.48	2.89
					11.53	14.58	5.61
							17.99

\multicolumn{9}{c}{Quality for n, $A_{max} = 3$ dB}
1.30	3.07	1.08	2.14	1.04	1.98	1.03	1.93	1.03
	5.58	8.82	3.46	5.02	3.08	4.32	2.94
			12.78	17.46	6.83	8.87	5.70
					22.87	29.00	11.15
							35.85

* For n odd there is also a real root for which $Q = 0.5$.

sixth-degree filter does not, what can be done to reduce Q_{max}? To answer this question, consider a seventh-order filter that has $A_{max} = 0.5$ dB. From Table 3.3 this has $Q_{max} = 8.84$, which is less than that for $A_{max} = 3$ dB. Thus we can conclude that lowering the passband ripple lowers Q_{max}. How much can the passband ripple be reduced? This question can be answered by realizing that lowering A_{max} reduces the stopband loss; thus A_{max} can be reduced until the stopband requirement is no longer exceeded.

Comparing Table 3.3 with Table 2.1, we can see that a Tschebycheff filter requires much higher quality roots than does a Butterworth filter of similar degree. But, of course, the Tschebycheff filter has a much sharper transition region.

3.11 CONCLUDING REMARKS

The first type of approximation we discussed used Butterworth polynomials. This yielded maximally flat lowpass filters. These networks had the disadvantage of requiring a very high order polynomial for a sharp transition region. In this chapter we discussed the Tschebycheff approximation for $H(s)$. As in the Butterworth case, this is a polynomial, but, because of the equiripple nature, the transition region is much steeper.

As just noted, $H(s)$ for both the Butterworth and Tschebycheff cases was a polynomial; that is, all the attenuation poles were at infinity. A polynomial filter (passive) of degree n can be realized by using a total of

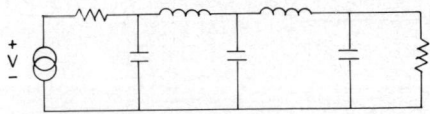

Fig. 3.10 Passive realization of a fifth-order-polynomial lowpass filter.

n inductors and capacitors. An example for $n = 5$ is given in Fig. 3.10. This has a particularly simple form because all attenuation poles are at infinity (at infinity the capacitors are short circuits and the inductors are open circuits). In the next chapter, we shall discuss a transfer function that is not a polynomial. It cannot be realized as simply as in Fig. 3.10, but it does have some advantages over the simpler polynomial filters.

There is a practical problem that arises if one attempts to realize an even-order Tschebycheff lowpass filter with a passive network. Even-order Tschebycheff lowpass filters have a zero-frequency loss which is equal to the passband ripple $A_{\max}$. However, it is shown in Sec. 16.3 that this implies that the source resistance cannot be equal to the load impedance. One way around this restriction is to use a frequency transformation (discussed in Chap. 6) which changes the loss at dc. However, it should also be pointed out here that active and digital filters can easily realize lowpass characteristics that have nonzero loss at dc.

REFERENCES

1. W. Cauer, *Synthesis of Linear Communication Networks*, McGraw-Hill Book Company, New York, 1958, app. 3.
2. W. W. Bell, *Special Functions for Scientists and Engineers*, D. Van Nostrand Company, Inc., Princeton, N.J., 1968, chap. 7, prob. 8.
3. E. A. Guillemin, *Synthesis of Passive Networks*, John Wiley & Sons, Inc., New York, 1957.

PROBLEMS

3.1   *a.* Evaluate $T_3(0.5)$ by using

$$T_n(x) = \cos(n \cos^{-1} x)$$

   b. Evaluate $T_3(2)$ by using

$$T_n(x) = \cosh(n \cosh^{-1} x)$$

   c. Check the results by evaluating

$$T_3(x) = -3x + 4x^3$$

3.2   *a.* If a fourth-order Tschebycheff lowpass filter has a passband ripple of 0.5 dB, find the loss at $f = 2f_B$.
   b. Repeat *a* for a passband ripple of 0.1 dB.

3.3 *a.* Find $dT_4(x)/dx$.
  b. Find $T_4(x) + 1$ and $T_4(x) - 1$.
  c. Check the results by demonstrating that

$$\left(\frac{dT_4}{dx}\right)^2 = M^2 \frac{T_4^2 - 1}{x^2 - 1}$$

3.4 Prove that the constant in Theorem 3.6 is $M = n$. Do this by using the fact that $T_n(x)$ behaves like cx^n for large x.

3.5 Find $T_4(x)$ by using (3.13).

3.6 What degree Tschebycheff filter is necessary to meet the following lowpass requirements:

$$A_{max} = 0.1 \text{ dB} \qquad A_{min} = 50 \text{ dB} \qquad \frac{\omega_H}{\omega_B} = 1.3$$

3.7 *a.* Show that

$$\frac{e^{\ln(x+\sqrt{x^2+1})} - e^{-\ln(x+\sqrt{x^2+1})}}{2} = x$$

  b. Use the result in *a* to demonstrate that

$$\sinh^{-1} x = \ln(x + \sqrt{x^2 + 1})$$

3.8 Show that $T_n^2(\omega) = \tfrac{1}{2}(T_{2n}(\omega) + 1)$.

CHAPTER FOUR

The Inverse Tschebycheff Filter

4.1 INTRODUCTION

The Butterworth input/output transfer function that we have discussed was a polynomial whose coefficients were chosen such that the transfer function was maximally flat at the origin. This transfer function had the disadvantage that high-order transfer functions were needed for sharp transition regions.

The inverse Tschebycheff filter is also maximally flat at the origin; however (unlike the Butterworth case), it is a ratio of polynomials. The finite poles of attenuation produce a much sharper transition region than is possible for a Butterworth filter of similar degree.

4.2 MANIPULATION OF TSCHEBYCHEFF CHARACTERISTICS

The characteristic function for a normalized Tschebycheff lowpass filter is shown in Fig. 4.1a, and its reciprocal is shown next to it.* The func-

---

* The constant c in the figure is related to $A_{\max}$. It will be discussed in more detail in later sections.

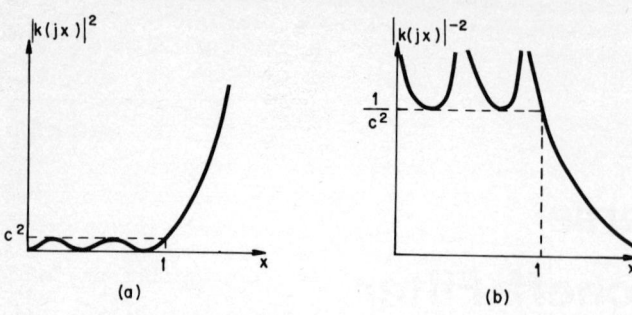

Fig. 4.1 Characteristic function for a normalized Tschebycheff lowpass filter and its reciprocal.

tion in Fig. 4.1b has the shape of a highpass characteristic function, but it can be transformed to a lowpass shape by making the substitution $x = a^2/\omega$. That is,

$$|K(j\omega)|^2 = \frac{1}{|k(jx)|^2} = \frac{1}{|k(ja^2/\omega)|^2} \qquad (4.1)$$

In Eq. (4.1), K is the characteristic function for the inverse Tschebycheff lowpass, and k is the characteristic function for the normalized Tschebycheff lowpass. Since k is related to the Tschebycheff polynomial, it follows from (4.1) that

$$|K(j\omega)|^2 = \frac{1}{|cT_n(a^2/\omega)|^2} \qquad (4.2)$$

The shape of this characteristic function (for $n = 5$) is shown in Fig. 4.2, which demonstrates that a fifth-order inverse Tschebycheff filter has two finite poles of attenuation. It also has two equal attenuation minimums in the stopband. In general, if n is even, an nth-order inverse Tschebycheff filter has $n/2$ attenuation poles and attenuation minimums. If n is odd, the number is instead $(n - 1)/2$.

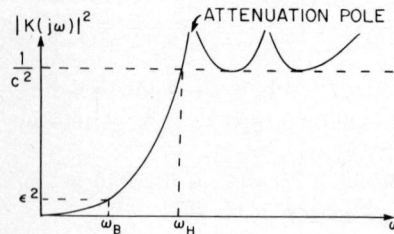

Fig. 4.2 Characteristic function of a fifth-order inverse Tschebycheff lowpass filter.

4.3 MAXIMALLY FLAT PROPERTY OF THE INVERSE TSCHEBYCHEFF FILTER

The characteristic function for the inverse Tschebycheff filter is

$$|K(j\omega)| = \frac{1}{|cT_n(a^2/\omega)|}$$

$T_n(x)$ is an nth-order polynomial of the variable x; thus, $T_n(a^2/\omega)$ is a ratio of polynomials. The denominator of $T_n(a^2/\omega)$ is simply a constant times ω^n. Since $|K(j\omega)|$ is inversely proportional to $|T_n(a^2/\omega)|$, it follows that the numerator of $|K(j\omega)|$ is simply a constant times ω^n. This implies that the first $n - 1$ derivatives of $|K(j\omega)|$ (evaluated as $\omega = 0$) are zero. That is, the inverse Tschebycheff approximation is maximally flat at the origin.

4.4 DETERMINATION OF INVERSE TSCHEBYCHEFF LOSS

From (4.2) the characteristic function for the inverse Tschebycheff low-pass filter is

$$|K(j\omega)| = \frac{1}{|cT_n(a^2/\omega)|}$$

Thus, the loss is given by

$$A(\omega) = 10 \log \left\{ 1 + \frac{1}{[cT_n(a^2/\omega)]^2} \right\} \qquad (4.3)$$

Recall that the stopband has equiminimums as shown in Fig. 4.2. These minimums occur where $T_n(a^2/\omega) = \pm 1$. Since one of these minimums is located at $\omega = \omega_H$, it follows that

$$a^2 = \omega_H \qquad (4.4)$$

By definition, the loss at $\omega = \omega_B$ is $A_{\max}$. Thus, combining (4.3) and (4.4) gives

$$A_{\max} = 10 \log \left\{ 1 + \frac{1}{[cT_n(\omega_H/\omega_B)]^2} \right\} \qquad (4.5)$$

If we define ϵ^2 so that

$$A_{\max} = 10 \log (1 + \epsilon^2)$$

it follows that

$$\epsilon^2 = \left[cT_n \left(\frac{\omega_H}{\omega_B} \right) \right]^{-2} \qquad (4.6)$$

Thus we can express the loss of an inverse Tschebycheff lowpass filter as

$$A(\omega) = 10 \log \left\{ 1 + \left[\epsilon \frac{T_n(\omega_H/\omega_B)}{T_n(\omega_H/\omega)} \right]^2 \right\} \quad (4.7)$$

where
$$\epsilon^2 = 10^{0.1 A_{\max}} - 1 \quad (4.8)$$

4.5 DETERMINATION OF DEGREE OF THE INVERSE TSCHEBYCHEFF FILTER

By definition, the loss at $\omega = \omega_H$ is $A(\omega_H) = A_{\min}$; from Eq. (4.7) we thus have

$$A_{\min} = 10 \log \left\{ 1 + \left[\epsilon T_n \left(\frac{\omega_H}{\omega_B} \right) \right]^2 \right\} \quad (4.9)$$

The above formula gives the minimum stopband loss of an inverse Tschebycheff filter. The surprising fact is that this formula is also valid for "regular" Tschebycheff filters, because the loss of a Tschebycheff filter is given by

$$A(\omega) = 10 \log \left\{ 1 + \left[\epsilon T_n \left(\frac{\omega}{\omega_B} \right) \right]^2 \right\} \quad (3.26)$$

Since the formula for $A_{\min}$ is identical for Tschebycheff and inverse Tschebycheff filters, it follows that if we are given requirements

$$A_{\max} \quad A_{\min} \quad \omega_H \quad \omega_B$$

then the necessary degree of an inverse Tschebycheff filter is the same as for a Tschebycheff filter. From (3.24) this is

$$n = \frac{\cosh^{-1} \left[(10^{0.1 A_{\min}} - 1)/(10^{0.1 A_{\max}} - 1) \right]^{1/2}}{\cosh^{-1} (\omega_H/\omega_B)} \quad (4.10)$$

Since n must be an integer, one of the requirements ($A_{\max}$, $A_{\min}$, ω_B, ω_H) will be met with a factor of safety. Usually $A_{\min}$ is the one we choose to exceed;* thus, we should like to know, "Given $A_{\max}$, n, ω_B, and ω_H, what is $A_{\min}$?" The answer to this is contained in Eqs. (4.8) and (4.9) which yield

$$A_{\min} = 10 \log \left\{ 1 + (10^{0.1 A_{\max}} - 1) \left[T_n \left(\frac{\omega_H}{\omega_B} \right) \right]^2 \right\} \quad (4.11)$$

This equation can be used to plot $A_{\min}$ (for a particular passband loss $A_{\max}$) as a function of ω_H/ω_B. Curves for $A_{\max} = 0.5$ dB are shown in

---

* If one were instead concerned about frequency shifts due to element changes, ω_B could be increased and ω_H decreased.

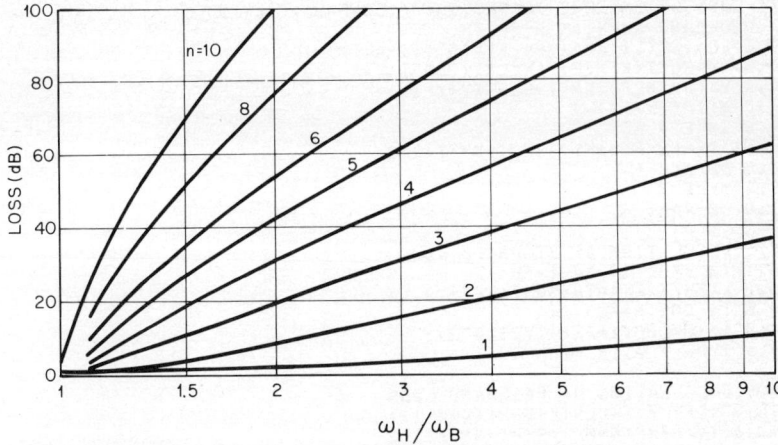

Fig. 4.3 Minimum stopband losses of inverse Tschebycheff lowpass filters of various orders. The curves are plotted for a passband loss of 0.5 dB.

Fig. 4.3.* If one wanted to, one could compile a set of these curves for various $A_{\max}$; however, in this computer age it makes much more sense to write a simple program.

4.6 A PROGRAM FOR THE LOSS OF INVERSE TSCHEBYCHEFF LOWPASS FILTERS

Assume that the filter requirements are specified in the usual manner as ω_B, ω_H, $A_{\max}$, and $A_{\min}$. This section describes the program in Fig. 4.4 which can be used to find the degree n of an inverse Tschebycheff filter that meets these requirements. The program can also be used to find the loss of the filter at any frequency f. It was written by analogy to the Tschebycheff lowpass filter program in Fig. 3.7 and can be used in a similar way.

Example 4.1

Use the program in Fig. 4.4 to find the lowest degree inverse Tschebycheff filter that meets the following requirements:

$$A_{\max} = 0.1 \text{ dB} \qquad A_{\min} = 30 \text{ dB} \qquad \frac{\omega_H}{\omega_B} = 1.3$$

Solution

These are the same requirements that were given in Example 3.1 for the Tschebycheff lowpass filter; thus $n = 8$. This is demonstrated in Fig. 4.5, which also

* This figure is valid for Tschebycheff filters as well as for inverse Tschebycheff filters.

```
1; LØSS ØF INVERSE TSCHEBYCHEFF LØWPASS FILTER
1.2 DEMAND AMAX,AMIN,FB,FH
1.3 EE=10^(.1*AMAX)-1,FF=10^(.1*AMIN)-1
1.31 X=SQRT(FF/EE),FX=LN(X+SQRT(X*X-1))
1.4 Y=FH/FB,FY=LN(Y+SQRT(Y*Y-1))
1.5 N=IP(FX/FY)+1
1.6 TYPE #,N,#
1.61 TY=[EXP(N*FY)+EXP(-N*FY)]/2
1.7 DEMAND F
1.8 TYPE #
1.9 DØ PART 6

6; CALCULATIØN ØF STØPBAND LØSS
6.1 TØ PART 10 IF F<FH
6.2 X=FH/F,FX=ATN(SQRT(1-X*X),X)
6.3 TX=CØS(N*FX)
6.4 A=10*LØG(1+EE*(TY/TX)^2)
6.5 TYPE F,A IN FØRM 1

10; CALCULATIØN ØF PASSBAND LØSS
10.2 X=FH/F,FX=LN(X+SQRT(X*X-1))
10.3 TX=[EXP(N*FX)+EXP(-N*FX)]/2
10.4 A=10*LØG(1+EE*(TY/TX)^2)
10.5 TYPE F,A IN FØRM 1
```

Fig. 4.4 Program for calculating the loss of an inverse Tschebycheff lowpass filter.

```
DØ PART 1
        AMAX=.1
        AMIN=30
         FB=1
         FH=1.3

         N=      8

         F=1.3

  1.300+00   3.022+01
>
  DØ PART 6 FØR F=.2:.2:2
  2.000-01   2.989-14
  4.000-01   2.662-09
  6.000-01   2.991-06
  8.000-01   7.060-04
  1.000+00   1.000-01
  1.200+00   8.679+00
  1.400+00   3.026+01
  1.600+00   4.180+01
  1.800+00   3.035+01
  2.000+00   3.202+01
```

Fig. 4.5 Sample use of program in Fig. 4.4.

finds the losses at certain frequencies. One of these frequencies is $\omega_H = 1.3$, at which $A_{min} = 30.22$ dB as it was for the Tschebycheff lowpass filter.

4.7 CONCLUDING REMARKS

The Butterworth lowpass filter is maximally flat at the origin. The input/output transfer function $H(s)$ is a polynomial; that is, it has no poles of attenuation. The inverse Tschebycheff lowpass filter is also maximally flat at the origin, but it does have poles of attenuation (see

Fig. 4.2). The attenuation poles give the inverse Tschebycheff filter a transition region much steeper than that of the Butterworth.

The Tschebycheff lowpass filter has an equiripple passband, and its input/output transfer function $H(s)$ is a polynomial. Even though it has no attenuation poles, its transition region is just as steep as that of the inverse Tschebycheff. Thus one might ask why we use the inverse Tschebycheff filter, which requires attenuation poles. This is a valid question, as the attenuation poles require additional elements in a physical realization. For example, Fig. 3.10 showed a network that could realize a fifth-degree Tschebycheff filter. A fifth-degree inverse Tschebycheff filter would require more elements, as indicated in Fig. 4.6.

The reason that inverse Tschebycheff filters are sometimes "better" than Tschebycheff filters is because of the maximally flat passband which results in better delay performance. Delay is a frequency-domain parameter that indicates how much a pulse will be distorted. This quantity is discussed in detail in Chap. 14; however, for now we shall just state that a constant delay is often desirable in the passband. With this as motivation, consider Fig. 4.7, which shows the delay performance of the Tschebycheff filter of Example 3.1 and the inverse Tschebycheff filter of Example 4.1. Both filters are described by

$$A_{\max} = 0.1 \qquad n = 8 \qquad \frac{\omega_H}{\omega_B} = 1.3 \qquad A_{\min} = 30.22$$

Figure 4.7 demonstrates that the inverse Tschebycheff filter has much better delay performance.

In this chapter we shall not discuss how to find the numerator of $H(s)$ for the inverse Tschebycheff filter approximation, as we did in previous chapters for Butterworth and Tschebycheff filters. This requires a root-finding subroutine and is postponed until Chap. 11. However, it might be worthwhile to borrow a result from that program: Table 4.1 shows the quality of the zeros of $H(s)$ for the Tschebycheff filter of Example 3.1 and the inverse Tschebycheff filter of Example 4.1. It indicates that for similar requirements the inverse Tschebycheff filter needs lower-Q roots.

It was pointed out at the end of Chap. 3 that an even-order Tschebycheff lowpass filter can present problems for passive network synthesis because it requires that the source impedance is not equal to the load. Even-order Tschebycheff lowpass filters have an even more fundamental

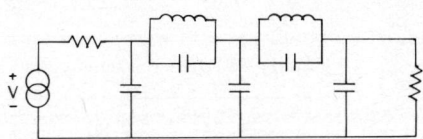

Fig. 4.6 Passive realization of a fifth-order inverse Tschebycheff lowpass filter.

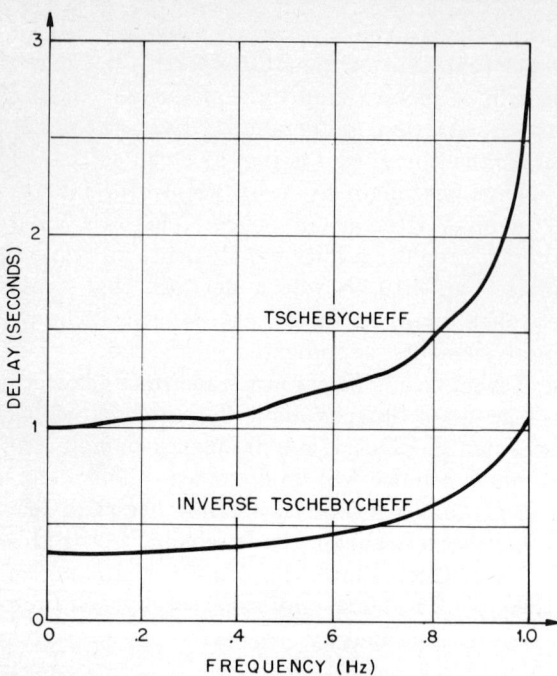

Fig. 4.7 Comparison of delays for a Tschebycheff and an inverse Tschebycheff filter.

drawback for passive filter realizations: they do not have an attenuation pole at infinity. That is, at high frequencies the loss approaches a constant value $A_{\min}$. As is discussed in Chap. 16, passive filters require an attenuation pole at infinity so that the zero-shifting technique can be used. Thus a frequency transformation (as discussed in Chap. 6) must be used to produce an attenuation pole at infinity. Again it should be pointed out that active and digital filters do not have this restriction—they do not need an attenuation pole at infinity.

TABLE 4.1 Comparison of the Quality of Tschebycheff Roots and Inverse Tschebycheff Roots

Quality of Tschebycheff Roots for Example 3.1			
0.59	1.18	2.45	8.08
Quality of Inverse Tschebycheff Roots for Example 4.1			
0.54	0.86	1.64	5.27

CHAPTER FIVE

Elliptic Filters

5.1 INTRODUCTION

We have discussed two different maximally flat filters, the Butterworth and inverse Tschebycheff. The Butterworth filter is a polynomial filter and thus has no finite poles of attenuation. Since the inverse Tschebycheff filter does have finite attenuation poles, it has a sharper transition region than the Butterworth.

Another polynomial filter that we have discussed is the Tschebycheff, which has a sharper transition region than the Butterworth because it does not concentrate all its "approximation power" at the origin, but instead applies it in an equiripple manner across the passband. Adding attenuation poles to a maximally flat polynomial filter can produce a stopband that has equal minimums (the inverse Tschebycheff). Similarly, adding attenuation poles to an equiripple-passband polynomial filter can produce a stopband that has equal minimums.

In this chapter we discuss filters that are equiripple in the passband and stopband. These filters are found by using elliptic functions and are thus referred to as *elliptic* filters; they are also known as *Cauer* or *Zolotarev* filters.* A typical elliptic characteristic is shown in Fig. 5.1.

* For that matter, they could also be called *Darlington* filters, as S. Darlington[1] did much original work.

52 Approximation Methods for Electronic Filter Design

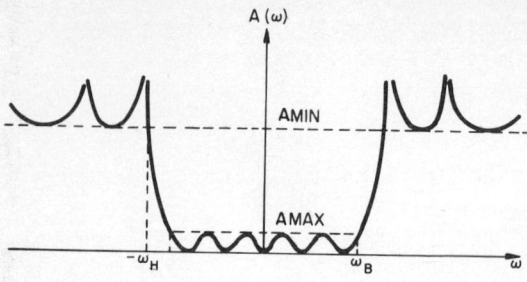

Fig. 5.1 Loss of a fifth-order elliptic filter.

Tschebycheff, inverse Tschebycheff, and Butterworth filters can all be considered to be special cases of elliptic filters. For example, in Fig. 5.1, if ω_H and $A_{\min}$ both approach infinity, then the stopband loss is monotonic whereas the passband is still equiripple; this represents a Tschebycheff characteristic. If, instead, ω_B and $A_{\max}$ approach zero (while ω_H and $A_{\min}$ are unchanged), then an inverse Tschebycheff characteristic results. Letting $A_{\max}$ and ω_B approach zero while $A_{\min}$ and ω_H approach infinity yields the Butterworth characteristic.

Elliptic filters have equal loss maximums in the passband and equal loss minimums in the stopband; thus, they are often said to be equiripple in the passband and stopband. They are a special case of filters that have an equiripple passband and an arbitrary stopband. As is demonstrated in later chapters, iterative computer programs can be written for this more general type of filter; the general programs can then be used to design elliptic filters. Thus, it is unnecessary for the filter designer to understand the elliptic theory presented in this chapter, but it is included for those who are fascinated by such filters. Since elliptic filters are constantly referred to in the literature, the reader may want to become familiar with the equations that describe them. This chapter contains all the background an electrical engineer needs in order to understand how elliptic function theory can be applied to yield elliptic filters. Those who are not interested in this mathematical treatment may proceed to Chap. 6 without loss of continuity.

5.2 INTRODUCTION TO TSCHEBYCHEFF RATIONAL FUNCTIONS

The Tschebycheff polynomial $T_n(x)$ was very helpful in our work with Tschebycheff filters. Similarly, the Tschebycheff rational function $R_n(x,L)$ will be used to find elliptic filters. This section sketches $R_n(x,L)$ for various degrees n and then discusses how it can be used to produce an elliptic loss shape.

Before we start our search for the Tschebycheff rational function, it will be helpful to review the Tschebycheff polynomial $T_n(x)$. $T_n(x)$ is an nth-degree polynomial that has the following properties:

1. T_n is even (odd) if n is even (odd).
2. T_n has all its zeros in the interval $-1 < x < 1$.
3. T_n oscillates between values of ± 1 in the interval $-1 \leq x \leq 1$.
4. $T_n(1) = +1$.

Sketches of $T_n(x)$ for various degrees were given in Fig. 3.3. These polynomials were used to make Tschebycheff filters by having

$$|H(j\omega)|^2 = 1 + \epsilon^2 T_n^2\left(\frac{\omega}{\omega_B}\right) \tag{5.1}$$

The Tschebycheff rational function $R_n(x,L)$ is very similar to the Tschebycheff polynomial $T_n(x)$; the major difference is that $R_n(x,L)$ is not a polynomial—it is a ratio of polynomials.

Definition 5.1

The Tschebycheff rational function $R_n(x,L)$ is a rational function of x that has the following properties:

1. R_n is even (odd) if n is even (odd).
2. R_n has all its n zeros in the interval $-1 < x < 1$ and all its n poles outside that interval.
3. R_n oscillates between values of ± 1 in the interval $-1 \leq x \leq 1$.
4. $R_n(1,L) = +1$.
5. $1/R_n$ oscillates between values of $\pm 1/L$ in the interval $|x| < x_L$.*

The following sections develop formulas for $R_n(x,L)$ that can be used to find the shape of $R_n(x,L)$ for any n and L. The shapes of some Tschebycheff rational functions are shown in Fig. 5.2. As a step toward understanding how the Tschebycheff rational function can be used for an elliptic approximation, consider $[R_5(\omega/\omega_B)]^2$, shown in Fig. 5.3. This function is very similar to the loss shape that was drawn in Fig. 5.1; in fact, we can write

$$A(\omega) = 10 \log \left\{ 1 + \left[\epsilon R_5\left(\frac{\omega}{\omega_B}, L\right) \right]^2 \right\} \tag{5.2}$$

To understand why the loss curve in Fig. 5.1 is given by Eq. (5.2), consider the following:

1. $[R_5(\omega/\omega_B, L)]^2$ is an even function of frequency as is $A(\omega)$.
2. R_5^2 has five second-order zeros in the region $-\omega_B < \omega < \omega_B$ so that $A(\omega)$ has five second-order zeros in that region.

---

* x_L is defined to be the first value of x at which $R_n(x,L) = L$. We will see later that L and x_L are related; that is, $L = f(x_L)$.

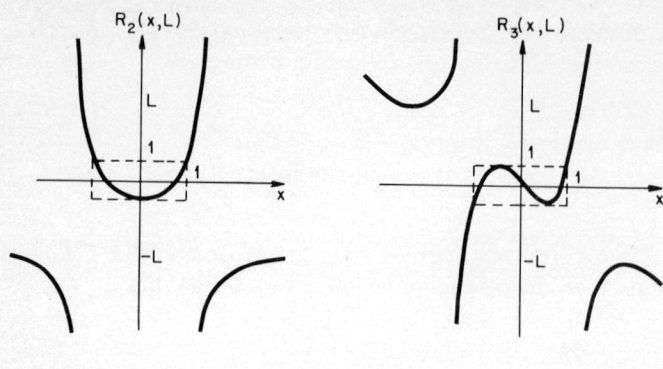

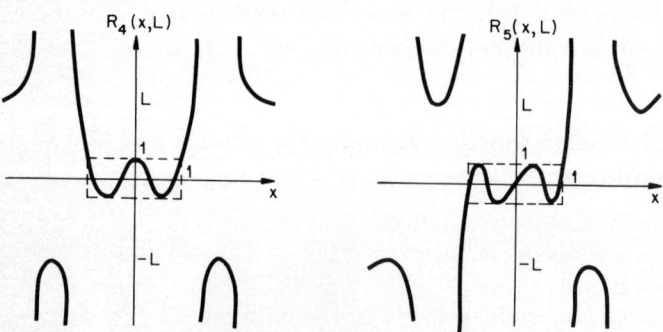

Fig. 5.2 Low-order Tschebycheff rational functions.

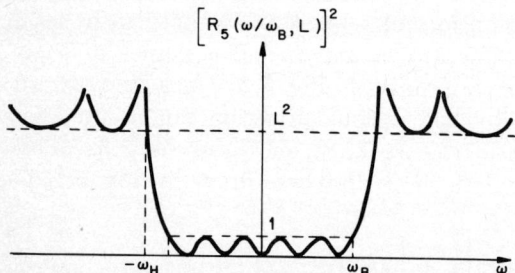

Fig. 5.3 The square of a Tschebycheff rational function is similar to the loss of an elliptic filter.

3. $[R_5(\omega/\omega_B,L)]^2$ is unity at four points in the region $-\omega_B < \omega < \omega_B$ and at $\omega = \pm\omega_B$. Defining

$$\epsilon^2 = 10^{0.1 A_{max}} - 1 \qquad (5.3)$$

it follows that $A = A_{max}$ at these frequencies.

4. $[R_5(\omega/\omega_B,L)]^2$ has four minimums of value L^2 in the region $|\omega| > \omega_H$.*
Defining

$$L^2 = \frac{10^{0.1 A_{min}} - 1}{10^{0.1 A_{max}} - 1} \quad (5.4)$$

it follows that $A = A_{min}$ at these frequencies.

The above properties describe the filter characteristic in Fig. 5.1; generalizing, we can conclude:

Theorem 5.1

The nth-order lowpass filter described by

$$A(\omega) = 10 \log \left\{ 1 + \left[\epsilon R_n\left(\frac{\omega}{\omega_B}, L\right) \right]^2 \right\} \quad (5.5)$$

where ϵ and L are given by (5.3) and (5.4), has the following properties:

1. An equiripple passband of amplitude A_{max}
2. An equiminimum stopband of amplitude A_{min}
3. A passband edge of ω_B

5.3 A BASIC FORM FOR $R_n(x,L)$

We saw in the previous section that the Tschebycheff rational function is very important in the design of elliptic filters. In fact, it follows from (5.5) that the characteristic function for an elliptic filter is given by

$$|K(j\omega)|^2 = \left[\epsilon R_n\left(\frac{\omega}{\omega_B}, L\right) \right]^2 \quad (5.6)$$

Since the Tschebycheff rational function is so important to elliptic filters, we shall investigate it in some detail. This section starts the investigation by presenting one form for the Tschebycheff rational function $R_n(x,L)$.

The Tschebycheff rational function has the property

$$R_n(x,L) = \frac{L}{R_n(x_L/x,L)} \quad (5.7)$$

To understand why this is a desirable property, recall that

$$|R_n(x,L)| \leq 1 \quad \text{for} \quad |x| \leq 1$$

But from (5.7) this implies that

$$|R_n(x,L)| \geq L \quad \text{for} \quad |x| > x_L$$

* ω_H is a function of the parameter L. It is also related to x_L since $x_L = \omega_H/\omega_B$.

Thus if $R_n(x,L)$ is described by (5.7), then an equiripple passband automatically yields an equiripple stopband.

Equation (5.7) implies that if $x = x_i$ is a pole of $R_n(x,L)$, then $x = x_L/x_i$ is a zero. Also, since R_n is an even or odd function, if x_i is a pole, then $-x_i$ is a pole too. Thus we can write:

n even:

$$R_n(x,L) = C_1 \prod_{i=1}^{n/2} \frac{x^2 - (x_L/x_i)^2}{x^2 - x_i^2}$$

n odd:
$\hspace{9cm}$ (5.8)

$$R_n(x,L) = C_2 x \prod_{i=1}^{(n-1)/2} \frac{x^2 - (x_L/x_i)^2}{x^2 - x_i^2}$$

It follows from (5.7) that

$$C_1 = L^{1/2} \prod_{i=1}^{n/2} \frac{x_i^2}{x_L} \qquad C_2 = \left(\frac{L}{x_L}\right)^{1/2} \prod_{i=1}^{(n-1)/2} \left(\frac{x_i}{x_L}\right)^2 \qquad (5.9)$$

The basic form for $R_n(x,L)$ which is given by (5.8) is very convenient. It implies that if we can learn how to pick the zeros x_L/x_i so that the passband is equiripple, then the stopband will automatically be equiripple. Thus, we can confine our attention to the passband, just as we did for the Tschebycheff approximation in Chap. 3. There, we studied* the passband behavior and found a differential equation that described the Tschebycheff polynomial. In the next section we shall study the passband behavior of the Tschebycheff rational function and find a differential equation that describes $R_n(x,L)$.

The solution of the differential equation for $R_n(x,L)$ will require a knowledge of elliptic integrals and elliptic functions. Because this area of mathematics is not familiar to most engineers, a detailed discussion is included in this chapter. For those not interested in such details, the results are summarized in the remaining part of this section. The equations are numbered as they appear in the rest of the chapter, so that the reader who wants greater detail may easily find the in-depth presentation.

The poles and zeros in Eq. (5.8) can be expressed in terms of the elliptic sine function, which is a function of two variables and thus can be written as sn (u,k). Many tables[3] exist that can be used to evaluate the elliptic sine function.

If one does not wish to use tables, the elliptic sine can be evaluated by using some simple recursive formula (e.g., see Part 3 of the program in Fig. 5.12). Thus, even though the Tschebycheff rational function is expressed in terms of the elliptic sine function, we should not consider it

* See Sec. 3.5.

to be a function that is too complicated to be of practical use. On the contrary, the elliptic sine function is only slightly harder to evaluate than the more familiar trigonometric sine function—we just need the proper tables or computer routines.

Theorem 5.5 establishes that the Tschebycheff rational function can be expressed as

n odd:

$$R_n(x,L) = C_1 x \prod_{\nu=1}^{(n-1)/2} \frac{x^2 - \text{sn}^2(2\nu K/n)}{x^2 - [x_L/\text{sn}(2\nu K/n)]^2} \quad (5.66a)$$

n even:

$$R_n(x,L) = C_2 \prod_{\nu=1}^{n/2} \frac{x^2 - \text{sn}^2[(2\nu-1)K/n]}{x^2 - \{x_L/\text{sn}[(2\nu-1)K/n]\}^2} \quad (5.66b)$$

In these equations, the modulus k of the elliptic sine function is $k = 1/x_L$. The modulus $k = x_L^{-1}$ also affects the parameter K, which is known as the complete elliptic integral.* That is, we can write

$$K = K(k) = K(x_L^{-1})$$

Like the elliptic sine function, the complete elliptic integral is tabulated in many tables,[7] or it can be evaluated by recursive relations.

Since we now have an expression for $R_n(x,L)$, we can evaluate the loss of an elliptic filter at any frequency by using

$$A(\omega) = 10 \log \left[1 + \epsilon^2 R_n^2 \left(\frac{\omega}{\omega_B}, L \right) \right]$$

A program that evaluates this expression is given in Fig. 5.12. The program evaluates the elliptic sine function and the complete elliptic integral by using some simple recursive relationships. The degree n of the elliptic filter is determined by the requirements $A_{\max}$, $A_{\min}$, FB, FH. The equation used to solve for n is

$$n = \frac{K(x_L^{-1})K'(L^{-1})}{K'(x_L^{-1})K(L^{-1})} \quad (5.49)$$

where

$$x_L = \frac{\text{FH}}{\text{FB}}$$

$$L^2 = \frac{10^{0.1 A_{\min}} - 1}{10^{0.1 A_{\max}} - 1} \quad (5.51)$$

A sample use of the elliptic filter program is demonstrated in Fig. 5.13. For the requirements $A_{\max} = 0.1$, $A_{\min} = 40$, FB = 20, FH = 26, a

* See Sec. 5.5 for the definitions of modulus, complete elliptic integral, etc.

sixth-degree filter was required. The program determined that the attenuation poles were located at

$$f_1 = 82.61 \quad f_2 = 33.29 \quad f_3 = 26.58$$

Chapter 10 will show how, given the attenuation poles, we can determine the natural modes [the zeros of $H(s)$].

The program in Fig. 5.12 can also be used to evaluate the loss at any frequency. This is demonstrated in Fig. 5.13 for some special frequencies:

1. Frequencies of maximum passband loss
2. Frequencies of minimum passband loss
3. Frequencies of minimum stopband loss

The program in Fig. 5.12 demonstrates that the elliptic functions are not difficult to evaluate. This fairly simple program evaluates the elliptic sine function, the complete elliptic integral, and the Tschebycheff rational function. Thus, while the derivations in the rest of this chapter may be lengthy, the application of the results is rather simple.

5.4 A DIFFERENTIAL EQUATION FOR $R_n(x,L)$*

The nth-order Tschebycheff rational function $R_n(x,L)$ was defined to be the function that has the following properties:

1. R_n is even (odd) if n is even (odd).
2. R_n has all its n zeros in the interval $-1 < x < 1$ and all its n poles outside that interval.
3. R_n oscillates between values of ± 1 in the interval $-1 \leq x \leq 1$.
4. $R_n(1,L) = +1$.
5. $1/R_n$ oscillates between values of $\pm 1/L$ in the interval $|x| > x_L$.

We shall obtain a differential equation for $R_n(x,L)$ by examining the curves in Fig. 5.2 and making some generalizations. First observe that dR_n/dx has $n - 1$ zeros that occur in the passband when $R_n = \pm 1$. The derivative also has $n - 1$ zeros that occur in the stopband where $R_n = \pm L$.† Thus,

(a) $(dR_n/dx)^2$ has $4(n - 1)$ zeros.‡ They are of second order and occur when $R_n = \pm 1, \pm L$ (not including $x = \pm 1, \pm x_L$).

Now, instead, consider the function $R_n + 1$. This has n zeros, and they occur at the passband minimums of R_n (a second-order zero is

---

* The approach used in this section is similar to Cauer's.[2]
† If n is even, one of these zeros occurs at $x = \infty$.
‡ A second-order zero is counted twice.

counted twice). Similarly, $R_n - 1$ has n zeros, and they occur at the passband maximums of R_n. Also, $R_n + L$ and $R_n - L$ have n zeros. Thus,

(b) $(R_n + 1)(R_n - 1)(R_n + L)(R_n - L) = (R_n^2 - 1)(R_n^2 - L^2)$ has $4n$ zeros. The zeros occur when $R_n = \pm 1, \pm L$. The zeros at $x = \pm 1, \pm x_L$ are of first order; the others are of second order.

Combining (a) and (b), we see that

$$\left(\frac{dR_n}{dx}\right)^2 \quad \text{and} \quad \frac{(R_n^2 - 1)(R_n^2 - L^2)}{(x^2 - 1)(x^2 - x_L^2)}$$

have the same zeros. Thus,

$$\left(\frac{dR_n}{dx}\right)^2 = M^2 \frac{(R_n^2 - 1)(R_n^2 - L^2)}{(x^2 - 1)(x^2 - x_L^2)}$$

which can be rewritten as

$$\frac{dR_n}{dx} = M \left[\frac{(1 - R_n^2)(L^2 - R_n^2)}{(1 - x^2)(x_L^2 - x^2)}\right]^{1/2}$$

Thus, we finally have

$$\frac{C \, dR_n}{[(1 - R_n^2)(L^2 - R_n^2)]^{1/2}} = \frac{dx}{[(1 - x^2)(x_L^2 - x^2)]^{1/2}} \triangleq du \quad (5.10)$$

The constant C is related to the constant M and will be determined later.

Equation 5.10 is a differential equation for the Tschebycheff rational function; the solution involves elliptic integrals. Since these will not be familiar to many readers, a digression is necessary. The next section defines the elliptic integral and discusses some of its properties. The following section introduces some Jacobian elliptic functions that are related to the elliptic integral. We shall discuss Jacobian elliptic functions because the solution to (5.10) can be expressed in terms of these functions.

5.5 THE ELLIPTIC INTEGRAL OF THE FIRST KIND

There are three elliptic integrals;[3-5] however, we will only be interested in the *first kind*, which is usually written as

$$u(\phi, k) = \int_0^\phi (1 - k^2 \sin^2 x)^{-1/2} \, dx \quad (5.11)$$

As indicated by this equation, the elliptic integral is a function of two variables, ϕ and k. The parameter k is called the *modulus* of the elliptic integral and is not allowed to be greater than unity, so that $u(\phi,k)$ is real for ϕ real. The parameter ϕ is referred to as the *amplitude* of the elliptic integral.

Perhaps the easiest way to become familiar with the elliptic integral is to plot it as a function of ϕ for various values of k. This can be done by consulting any one of a number of tables.[6,7] Some of these tables are given in terms of the *modular angle* θ instead of the modulus k. These quantities are related by the definition

$$k = \sin \theta \tag{5.12}$$

Figure 5.4 shows $u(\phi,k)$ for various values of the modulus k. The results have been normalized with respect to the *complete elliptic integral of the first kind*, which is defined as

$$u\left(\frac{\pi}{2}, k\right) \triangleq K = \int_0^{\pi/2} (1 - k^2 \sin^2 x)^{-\frac{1}{2}} \, dx \tag{5.13}$$

The complete elliptic integral K is plotted in Fig. 5.5.

The following properties of the elliptical integral follow directly from the definition in (5.11). Some of the properties are also evident from

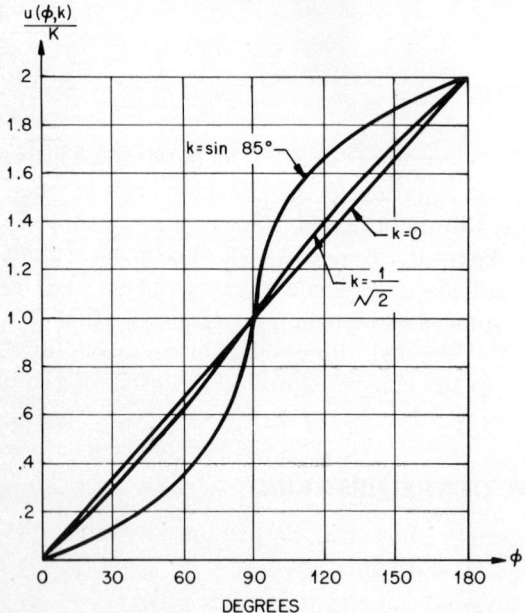

Fig. 5.4 Elliptic integral.

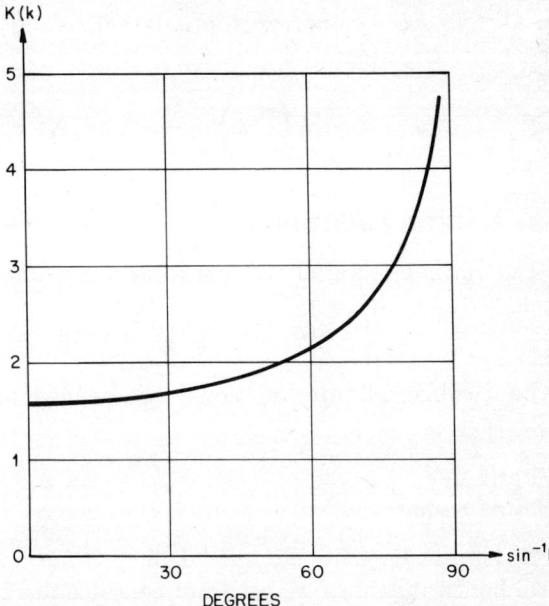

Fig. 5.5 Complete elliptic integral.

Fig. 5.4, whereas others are useful in extending the curves for different values of the amplitude ϕ.

a. $u(\phi,k) = -u(-\phi,k)$
b. $u(\pi,k) = 2K$
c. $u(\pi + \phi, k) = 2K + u(\phi,k)$
d. $du/d\phi = (1 - k^2 \sin^2 \phi)^{-1/2}$

Before terminating this brief discussion of the elliptic integral, we define two more quantities that we shall use in later sections. The *complementary modulus* k' is defined in terms of the modulus k as

$$k' = \sqrt{1 - k^2} \qquad (5.14)$$

Just as the complete elliptic integral K was defined in terms of k, so is the *complementary complete elliptic integral* K' defined in terms of k':

$$u\left(\frac{\pi}{2}, k'\right) \triangleq K' = \int_0^{\pi/2} [1 - (1 - k^2) \sin^2 x]^{-1/2} \, dx \qquad (5.15)$$

Section 5.13 contains a program that can be used to design elliptic lowpass filters. One important part of the program computes the elliptic integral of the first kind $u(\phi,k)$. Of course, the program can also be used to find the complete elliptic integral K or the complementary complete elliptic integral K', since

$$K = u\left(\frac{\pi}{2}, k\right) \quad \text{and} \quad K' = u\left(\frac{\pi}{2}, k'\right)$$

5.6 ELLIPTIC FUNCTIONS

The elliptic integral of the first kind was defined as

$$u(\phi, k) = \int_0^\phi (1 - k^2 \sin^2 x)^{-\frac{1}{2}} dx \qquad (5.11)$$

The Jacobian elliptic functions* are defined with respect to the above notation as

elliptic sine: $\quad$ sn $(u,k) = \sin \phi$
elliptic cosine: $\quad$ cn $(u,k) = \cos \phi \qquad (5.16)$

The elliptic tangent, etc., are defined in an analogous manner, but we will not use them so we need not consider them. However, there is one additional elliptic function that we will find useful: the difference function dn, which is the derivative of ϕ; that is,

$$\text{dn } (u,k) = \frac{d\phi}{du} \qquad (5.17)$$

The elliptic functions cn and dn can be expressed in terms of the elliptic sine. From (5.16) it follows that

$$\text{cn } (u,k) = [1 - \text{sn}^2 (u,k)]^{\frac{1}{2}} \qquad (5.18)$$

while from (5.11)

$$\text{dn } (u,k) = \frac{d\phi}{du} = (1 - k^2 \sin^2 \phi)^{\frac{1}{2}} = [1 - k^2 \text{sn}^2 (u,k)]^{\frac{1}{2}} \qquad (5.19)$$

As was the case for the elliptic integral of the first kind, perhaps the easiest way to become familiar with the elliptic functions is to plot them as a function of u for various values of k. Again one can use existing tables[6,8] to make a graph as in Fig. 5.6.

The following properties of the elliptic functions follow from the basic definitions in (5.11) and (5.16). The properties would be useful if one wanted to extend the curves in Fig. 5.6 to different frequencies. For example, property a could be applied to extend the curves to negative frequencies.

* Jacobi did much of the original work (starting about 1826) on elliptic functions, so these functions are often given his name.

a. $\text{sn}(u,k) = -\text{sn}(-u,k)$
   $\text{cn}(u,k) = \text{cn}(-u,k)$
b. $\text{sn}(0,k) = 0 \quad \text{sn}(K,k) = 1$
   $\text{cn}(0,k) = 1 \quad \text{cn}(K,k) = 0$
c. $\text{sn}(u+K, k) = \text{sn}(K-u, k)$
   $\text{cn}(u+K, k) = -\text{cn}(K-u, k)$
d. $\text{sn}(u+2K, k) = -\text{sn}(u,k)$
   $\text{cn}(u+2K, k) = -\text{cn}(u,k)$
e. $\text{sn}(u+4K, k) = \text{sn}(u,k)$
   $\text{cn}(u+4K, k) = \text{cn}(u,k)$

The last property indicates that the elliptic functions are periodic with period $4K$. These periodic functions resemble the normal trigonometric functions in many ways; in fact, since $u(\phi,0) = \phi$ it follows that

$$\text{sn}(u,0) = \sin u \qquad \text{cn}(u,0) = \cos u \qquad (5.20)$$

Other properties that have analogies in trigonometric functions are[5]

$$\text{sn}^2(u,k) + \text{cn}^2(u,k) = 1 \qquad (5.21)$$

$$\text{sn}(u+v) = \frac{\text{sn}\,u \cdot \text{cn}\,v \cdot \text{dn}\,v + \text{sn}\,v \cdot \text{cn}\,u \cdot \text{dn}\,u}{1 - k^2 \text{sn}^2 u \cdot \text{sn}^2 v} \qquad (5.22)$$

$$\text{cn}(u+v) = \frac{\text{cn}\,u \cdot \text{cn}\,v - \text{sn}\,u \cdot \text{sn}\,v \cdot \text{dn}\,u \cdot \text{dn}\,v}{1 - k^2 \text{sn}^2 u \cdot \text{sn}^2 v} \qquad (5.23)$$

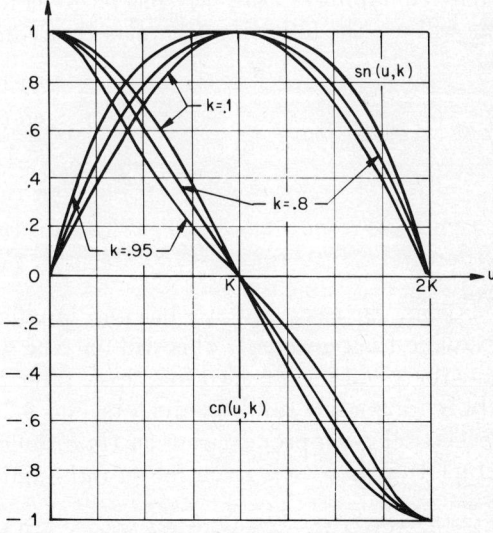

Fig. 5.6 Elliptic sine and elliptic cosine functions.

64 Approximation Methods for Electronic Filter Design

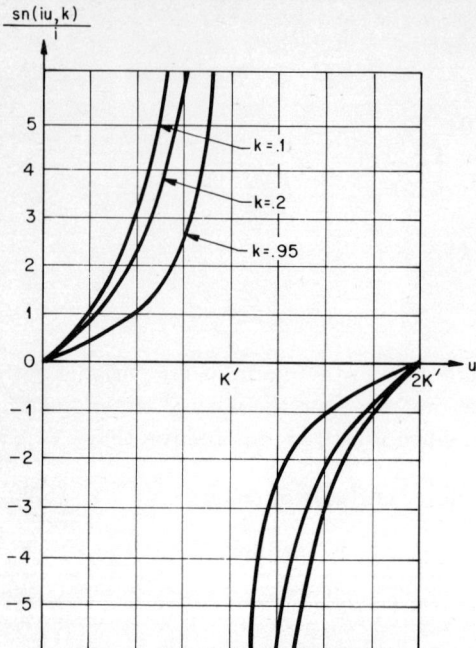

Fig. 5.7 Elliptic sine function for imaginary arguments.

The equations we have used so far have assumed that the arguments were real numbers; they can also be given for complex values. To do this we will use the following formulas that come from Chap. 8 of Greenhill:[9]

$$\text{sn}\,(iu,k) = i\,\frac{\text{sn}\,(u,k')}{\text{cn}\,(u,k')} \qquad (5.24)$$

$$\text{cn}\,(iu,k) = \frac{1}{\text{cn}\,(u,k')} \qquad (5.25)$$

$$\text{dn}\,(iu,k) = \frac{\text{dn}\,(u,k')}{\text{cn}\,(u,k')} \qquad (5.26)$$

Again, an easy way to become familiar with these functions is to examine their graphs. This will only be done for sn (iu,k), as the other functions will not be used in elliptic filter theory. To plot sn (iu,k), we apply (5.24); the result is given in Fig. 5.7.

The following properties of sn (iu,k) follow from (5.24) and the similar properties that were given for sn (u,k) and cn (u,k):

a. sn $(iu,k) = -\text{sn}\,(-iu,k)$
b. sn (iK',k) is infinite.

c. $\operatorname{sn}(i[u + K'], k) = -\operatorname{sn}(i[K' - u], k)$
d. $\operatorname{sn}(i[u + 2K'], k) = \operatorname{sn}(iu,k)$
e. $\operatorname{sn}(iu,1) = i \tan u$

Property d indicates that sn (iu,k) is periodic with a period of $2K'$.

Now that we have studied sn (u,k) and sn (iu,k) for u a real number, we are ready to delve into the complex argument $u = y + iz$. From the addition theorem of (5.22) and the relations in Eqs. (5.24) to (5.26), it follows that

$$\operatorname{sn}(u,k) = \operatorname{sn}(y + iz, k)$$
$$= \frac{\operatorname{sn}(y,k) \operatorname{dn}(z,k') + i \operatorname{cn}(y,k) \operatorname{dn}(y,k) \operatorname{sn}(z,k') \operatorname{cn}(z,k')}{\operatorname{cn}^2(z,k') + k^2 \operatorname{sn}^2(y,k) \operatorname{sn}^2(z,k')} \quad (5.27)$$

Equation (5.27) describes the behavior of sn (u,k), where u is any complex number. We are already familiar with the behavior of sn (u,k) when u is either pure real or pure imaginary (see Figs. 5.6 and 5.7). Recall the following properties:

$$\operatorname{sn}(y + 4K, k) = \operatorname{sn}(y,k)$$
$$\operatorname{sn}(i[z + 2K'], k) = \operatorname{sn}(iz,k)$$

These can be generalized to

$$\operatorname{sn}(u + 4K, k) = \operatorname{sn}(u,k)$$
$$\operatorname{sn}(u + 2iK', k) = \operatorname{sn}(u,k) \quad (5.28)$$

Thus sn (u,k) is a doubly periodic function: it has real period $4K$ and imaginary period $2K'$. This behavior is neatly summarized by the *periodic rectangle* in Fig. 5.8.

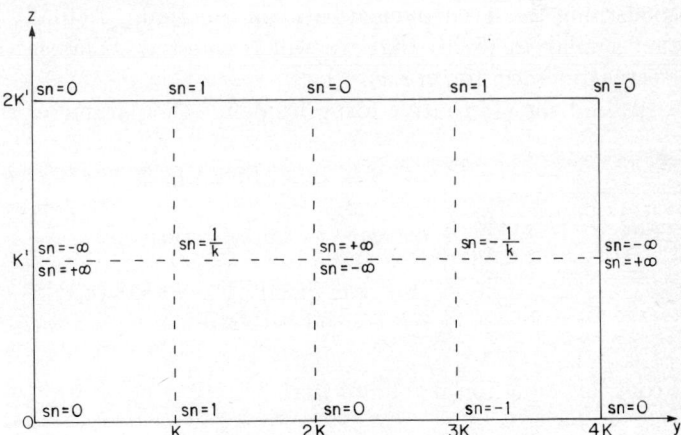

Fig. 5.8 Periodic rectangle for the elliptic sine function.

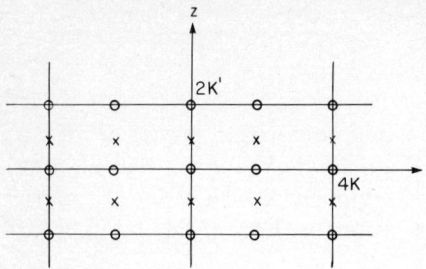

Fig. 5.9 Pole-zero locations for the elliptic sine function.

As implied by (5.28), if the behavior of sn (u,k) is known for all u inside the periodic rectangle, then it is known for the entire u plane. Thus we can picture the periodic rectangle as repeating, as shown in Fig. 5.9. This figure also indicates the periodic locations of the poles and zeros of sn (u,k).

In conclusion, this section has introduced the elliptic functions. Since the material in the sections that follow will use the elliptic sine most, we discussed sn (u,k) in greatest detail.

5.7 AN ALTERNATIVE FORM FOR THE ELLIPTIC INTEGRAL

The elliptic integral of the first kind was defined as

$$u(\phi,k) = \int_0^\phi (1 - k^2 \sin^2 x)^{-\frac{1}{2}} \, dx \qquad (5.29)$$

We discussed the elliptic integral because it is encountered in the theory concerning the Tschebycheff rational function. Actually, it is an equation similar to (5.29) that we will encounter. This section derives an alternative form for $u(\phi,k)$.

To find the alternative form, we define the parameter z as

$$z = \text{sn } (u,k) = \sin \phi \qquad (5.30)$$

Thus,
$$\frac{dz}{du} = \cos \phi \, \frac{d\phi}{du} = \text{cn } (u,k) \, \text{dn } (u,k)$$
$$= [1 - \text{sn}^2 (u,k)]^{\frac{1}{2}} [1 - k^2 \, \text{sn}^2 (u,k)]^{\frac{1}{2}}$$
$$= [(1 - z^2)(1 - k^2 z^2)]^{\frac{1}{2}}$$

From this equation it follows that

$$du = [(1 - z^2)(1 - k^2 z^2)]^{-\frac{1}{2}} \, dz \qquad (5.31)$$

Suppose we integrate this equation from z_1 to z_2. From (5.30), this is equivalent to integrating from $\sin \phi_1$ to $\sin \phi_2$; thus (5.31) leads to

$$u(\phi_2, k) - u(\phi_1, k) = \int_{\sin \phi_1}^{\sin \phi_2} [(1 - z^2)(1 - k^2 z^2)]^{-\frac{1}{2}} dz \quad (5.32)$$

As an example of the application of this formula, choose $\phi_2 = \phi$ and $\phi_1 = 0$. The equation then becomes

$$u(\phi, k) = \int_0^{\sin \phi} [(1 - z^2)(1 - k^2 z^2)]^{-\frac{1}{2}} dz \quad (5.33)$$

The expression in (5.33) is an alternative way of expressing the elliptic integral. This can be applied to find a z which is a solution to (5.31). Since $\phi = \operatorname{sn}(u, k)$, it follows from (5.33) that a solution is $z = \operatorname{sn}(u, k)$. A slight generalization of this result is stated in the next theorem.

Theorem 5.2

$$z = \operatorname{sn}(au + b, k) \quad (5.34)$$

is a solution to

$$a \, du = [(1 - z^2)(1 - k^2 z^2)]^{-\frac{1}{2}} dz \quad (5.35)$$

This differential equation is similar to (5.31), the variable u has simply been scaled by the parameter a, which accounts for the parameter a in (5.34). The constant b in (5.34) is determined by boundary conditions; for example, if we want $z = 0$ when $u = 0$, then $b = 0$.

The theorem is very easy to prove:

$$\frac{dz}{du} = \frac{d}{du} \operatorname{sn}(au + b, k) = a \operatorname{cn}(au + b, k) \operatorname{dn}(au + b, k)$$

$$= a[1 - \operatorname{sn}^2(au + b, k)]^{\frac{1}{2}}[1 - k^2 \operatorname{sn}^2(au + b, k)]^{\frac{1}{2}}$$

$$= a[(1 - z^2)(1 - k^2 z^2)]^{\frac{1}{2}}$$

from which (5.35) follows.

The alternative form of the elliptic integral in (5.33) can be applied to find an alternative form for the complete elliptic integral. We simply set $\phi = \pi/2$; thus

$$K(k) = u\left(\frac{\pi}{2}, k\right) = \int_0^1 [(1 - z^2)(1 - k^2 z^2)]^{-\frac{1}{2}} dz \quad (5.36)$$

We can also find an alternative form for the complementary complete elliptic integral. Problem 5.10 demonstrates that

$$K'(k) = \int_1^{1/k} [(z^2 - 1)(1 - k^2 z^2)]^{-\frac{1}{2}} dz \quad (5.37)$$

5.8 ELLIPTIC FUNCTIONS AND $R_n(x,L)$

The previous three sections have been devoted to the study of the elliptic integral and elliptic functions. We are finally prepared to relate the Tschebycheff rational function to elliptic functions. The Tschebycheff rational function $R_n(x,L)$ was described by the differential equations

$$du = [(1 - x^2)(x_L^2 - x^2)]^{-\frac{1}{2}} dx \qquad (5.38a)$$
$$= C[(1 - R_n^2)(L^2 - R_n^2)]^{-\frac{1}{2}} dR_n \qquad (5.38b)$$

Both equations can be expressed as

$$du = \frac{1}{a}[(1 - z^2)(1 - k^2 z^2)]^{-\frac{1}{2}} dz \qquad (5.39)$$

For example, (5.38a) can be put in this form by setting

$$x = z \qquad k = 1/x_L \qquad a = x_L$$

From Theorem 5.2, the solution to (5.39) is

$$z = \text{sn}\,(au + b, k) \qquad (5.40)$$

Applying this result to (5.38) yields

$$x = \text{sn}\,(x_L u + b_1, x_L^{-1}) \qquad R_n = \text{sn}\left(\frac{L}{C}u + b_2, L^{-1}\right) \qquad (5.41)$$

If we arbitrarily set the constant b_1 equal to zero, this implies that $x = 0$ when $u = 0$. But we also know that

$$R_n(x = 0, L) = \begin{cases} 0 & n \text{ odd} \\ (-1)^{n/2} & n \text{ even} \end{cases}$$
$$= \text{sn}\,(b_2, L^{-1})$$

This implies that

$$b_2 = \begin{cases} 0 & n \text{ odd} \\ (-1)^{n/2} K(L^{-1}) & n \text{ even} \end{cases} \qquad (5.42)$$

Combining (5.41) and (5.42) yields the following:

Theorem 5.3

The Tschebycheff rational function can be expressed as

$$R_n(x,L) = \begin{cases} \text{sn}\left(\frac{L}{C}u, L^{-1}\right) & n \text{ odd} \\ \text{sn}\left[\frac{L}{C}u + (-1)^{n/2} K(L^{-1}), L^{-1}\right] & n \text{ even} \end{cases} \qquad (5.43)$$

where u is the solution to

$$x = \text{sn}\,(x_L u, x_L^{-1}) \tag{5.44}$$

and C is a constant (to be determined in a later section).

This theorem expresses the Tschebycheff rational function $R_n(x,L)$ in terms of the elliptic sine function. However, the relation is not very convenient because the elliptic function in (5.43) does not depend explicitly on the variable x. Instead, for a specific value of x we must find the corresponding value of u by using (5.44). The next few sections will help us find a relation for $R_n(x,L)$ which is an explicit function of x.

5.9 THE PERIODIC RECTANGLE FOR $R_n(x,L)$

The elliptic sine function sn (u,k) is a doubly periodic function: it has a real period of $4K$ and an imaginary period of $2K'$. This behavior is conveniently summarized by the periodic rectangle in Fig. 5.8. Similarly, the Tschebycheff rational function is a doubly periodic function and has a periodic rectangle; this section investigates the periodic rectangle for $R_n(x,L)$. Much of the discussion centers on the case n odd; the discussion for n even would be similar, so only the results will be presented.

For n odd the Tschebycheff rational function is given by

$$R_n(x,L) = \text{sn}\left(\frac{L}{C}u, L^{-1}\right) \tag{5.43}$$

where
$$x = \text{sn}\,(x_L u, x_L^{-1}) \tag{5.44}$$

Since sn is periodic, R_n is periodic. If R_n were plotted as a function of Lu/C, then its real period would be $4K(L^{-1})$ where $K(L^{-1})$ is the complete elliptic integral evaluated for $k = L^{-1}$. However, it will be more convenient to consider R_n as a function of u. Its real period with respect to this parameter is $4(C/L)K(L^{-1})$. Similarly, its imaginary period is $2(C/L)K'(L^{-1})$. These facts lead to the periodic rectangle for R_n. The periodic rectangle in Fig. 5.10 is for R_n expressed as a function of u. To relate this to the variable x, we need a periodic rectangle for x. It follows from (5.44) that the real period of x is $4K(x_L^{-1})/x_L$, and the imaginary period is $2K'(x_L^{-1})/x_L$. Thus, the periodic rectangle for x is as drawn in Fig. 5.11.

The periodic rectangles in Figs. 5.10 and 5.11 are not independent of each other, as can be demonstrated by noting from (5.38a) that

$$u(x = 1) = \frac{1}{x_L}\int_0^1 [(1 - x^2)(1 - x_L^{-2}x^2)]^{-\frac{1}{2}}\,dx$$

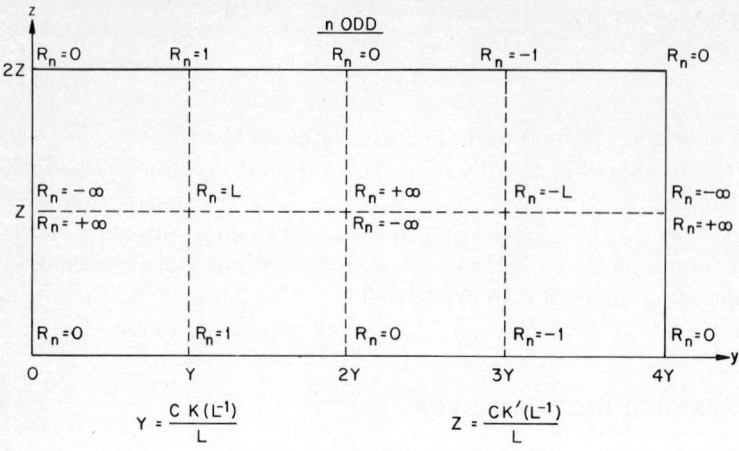
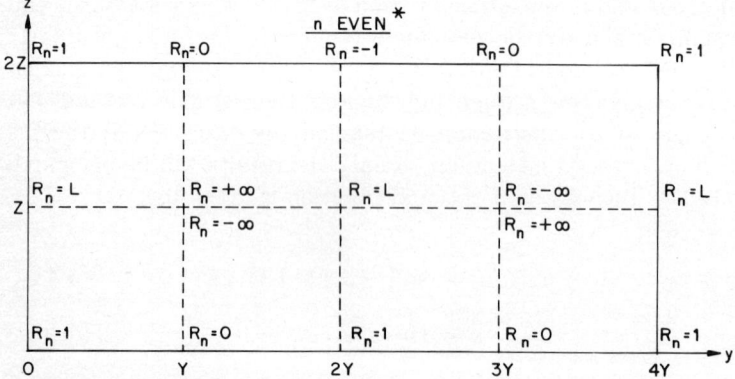

* THIS IS FOR n = 4, 8, 12, ··· IF n = 2, 6, 10 ··· THE PERIODIC RECTANGLE MUST BE SHIFTED BY y = 2Y

Fig. 5.10 Periodic rectangle for the Tschebycheff rational function.

But from (5.36) the right-hand side of this equation is equal to

$$\frac{K(x_L^{-1})}{x_L}$$

Thus, (5.36) and (5.38a) yield

$$u(x = 1) = \frac{K(x_L^{-1})}{x_L} \quad (5.45)$$

As x varies from 0 to 1, R_n varies between 0 and 1 or 0 and -1 exactly

n times (for example, see Fig. 5.2). Thus (5.38b) and (5.36) yield*

$$u(x = 1) = \frac{nC}{L} \int_0^1 [(1 - R_n^2)(1 - L^{-2}R_n^2)]^{-\frac{1}{2}} dR_n$$

$$= \frac{nCK(L^{-1})}{L} \qquad (5.46)$$

Comparing (5.45) and (5.46) gives us

$$\frac{K(x_L^{-1})}{x_L} = \frac{nCK(L^{-1})}{L} \qquad (5.47)$$

Equation (5.47) implies that if the periodic rectangles in Figs. 5.10 and 5.11 were drawn to the same scale, then the periodic rectangle for x would be n times wider than the periodic rectangle for R_n.

The heights of the rectangles are also related. The relation can be found by letting x increase from 1 to x_L. R_n then passes monotonically from 1 to L, and thus (5.37) and (5.38) yield

$$\frac{K'(x_L^{-1})}{x_L} = \frac{CK'(L^{-1})}{L} \qquad (5.48)$$

This equation implies that the periodic rectangles in Figs. 5.10 and 5.11 have the same height.

* To be mathematically precise, (5.38b) should contain a ± sign.

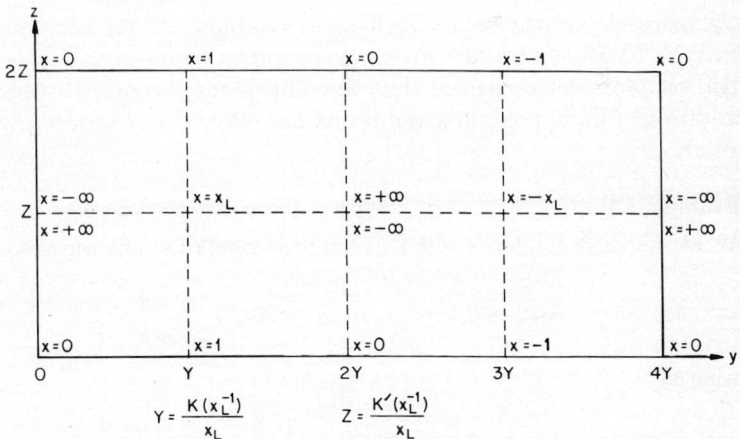

Fig. 5.11 Periodic rectangle for the "frequency" variable x.

5.10 DETERMINATION OF DEGREE OF ELLIPTIC FILTERS

The requirements for an elliptic lowpass filter can be stated in terms of $A_{\max}$, $A_{\min}$, ω_B, and ω_H. In this section we discuss how we can find the minimum degree n of an elliptic filter that meets these requirements.

We have, from the previous section,

$$\frac{K(x_L^{-1})}{x_L} = \frac{nCK(L^{-1})}{L} \quad \text{and} \quad \frac{K'(x_L^{-1})}{x_L} = \frac{CK'(L^{-1})}{L}$$

These equations imply that

$$n = \frac{K(x_L^{-1})K'(L^{-1})}{K'(x_L^{-1})K(L^{-1})} \qquad (5.49)$$

From the footnote at the bottom of page 55,

$$x_L = \frac{\omega_H}{\omega_B} \qquad (5.50)$$

The other parameter in (5.49) that needs to be evaluated is L. From (5.4), this is given by*

$$L^2 = \frac{10^{0.1 A_{\min}} - 1}{10^{0.1 A_{\max}} - 1} \qquad (5.51)$$

The following example demonstrates that these equations can be used to solve for the degree of an elliptic lowpass filter. The requirements in the example are the same as those in Example 3.1 for the Tschebycheff filter. The Tschebycheff filter was found to be of eighth degree, whereas this example demonstrates that the elliptic filter is of fifth degree. The addition of finite poles of attenuation has allowed us to use a lower-degree filter.

Example 5.1

What is the lowest-degree elliptic filter that meets the following requirements:

$$A_{\max} = 0.1 \text{ dB} \qquad A_{\min} = 30 \text{ dB} \qquad \frac{\omega_H}{\omega_B} = 1.3$$

Solution

$$K\left(\frac{1}{1.3}\right) = 1.9407 \qquad K'\left(\frac{1}{1.3}\right) = 1.7833$$

---

* G. Szentirmai[13] has given a nomograph for this.

From (5.51),
$$L = 207.1 \qquad K\left(\frac{1}{L}\right) = 1.5708 \qquad K'\left(\frac{1}{L}\right) = 6.7193$$

Thus,
$$n = \frac{1.9407 \times 6.7193}{1.7833 \times 1.5708} = 4.66$$

Since n must be an integer, a fifth-degree filter is needed to meet the requirements.

5.11 DETERMINATION OF L

The minimum stopband magnitude of $R_n(x,L)$ has been defined to be L. This value first occurs at $x = x_L$. In this section we derive an expression for L as a function of n and x_L. In the process of deriving the expression, we obtain a relation for $R_n(x,L)$ that will be used in following sections. The relation for $R_n(x,L)$ is given in the next theorem.

Theorem 5.4

If $x = \text{sn}(x_L u, x_L^{-1})$ then $R_n(x,L)$ can be expressed as
n odd:

$$R_n(x,L) = m \prod_{\nu=0}^{n-1} \text{sn}[x_L u + 2\nu n^{-1} K(x_L^{-1}), x_L^{-1}] \qquad (5.52a)$$

where
$$m^{-1} = \frac{cx_L}{L} \prod_{\nu=1}^{n-1} \text{sn}[2\nu n^{-1} K(x_L^{-1}), x_L^{-1}] \qquad (5.53a)$$

n even:

$$R_n(x,L) = m \prod_{\nu=0}^{n-1} \text{sn}[x_L u + (1+2\nu)n^{-1} K(x_L^{-1}), x_L^{-1}] \qquad (5.52b)$$

where
$$m^{-1} = (-1)^{n/2} \prod_{\nu=0}^{n-1} \text{sn}[(1+2\nu)n^{-1} K(x_L^{-1}), x_L^{-1}] \qquad (5.53b)$$

Proof

The proof will only be given for the case in which n is odd, as the other case is similar. For n odd we know from Theorem 5.3 that

$$R_n(x,L) = \text{sn}\left(\frac{L}{C} u, L^{-1}\right) \qquad (5.54)$$

where u is the solution to

$$x = \text{sn}(x_L u, x_L^{-1}) \qquad (5.55)$$

The elliptic sine function in (5.54) is zero at*

$$u = -\frac{2CK(L^{-1})\nu}{L} \qquad \nu = 0, 1, 2, \ldots$$

But from (5.47)

$$\frac{CK(L^{-1})}{L} = \frac{K(x_L^{-1})}{nx_L}$$

Thus, the zeros of $R_n(x,L)$ are located at

$$u = -\frac{2K(x_L^{-1})}{nx_L}\nu$$

These zeros agree with the right-hand side of (5.52a). Similarly, both (5.54) and (5.52a) imply that $R_n(x,L)$ has poles a distance of $CK'(L^{-1})/L$ "above" the zeros (see Fig. 5.10). Thus $R_n(x,L)$ as given by (5.52a) has the proper poles and zeros. Problem 5.13 demonstrates that the constant multiplier is also correct, so that (5.52a) is a valid expression for R_n.

Theorem 5.4 can be applied to yield an expression for the stopband height L. To do this, note from the periodic rectangle for R_n (for n odd) that

$$1 = R_n(x,L)\Big|_{u=CK(L^{-1})/L} \qquad (5.56a)$$

$$L = R_n(x,L)\Big|_{u=CK(L^{-1})/L+iCK'(L^{-1})/L} \qquad (5.56b)$$

Applying (5.47) and (5.48) gives us

$$1 = R_n(x,L)\Big|_{u=K(x_L^{-1})/nx_L} \qquad (5.57a)$$

$$L = R_n(x,L)\Big|_{u=K(x_L^{-1})/nx_L+iK'(x_L^{-1})/x_L} \qquad (5.57b)$$

Substituting the result from Theorem 5.4 and then solving for L yields

$$L^{-1} = \frac{\prod_{\nu=0}^{n-1} \operatorname{sn}\left[(1+2\nu)K(x_L^{-1})/n,\, x_L^{-1}\right]}{\prod_{\nu=0}^{n-1} \operatorname{sn}\left[(1+2\nu)K(x_L^{-1})/n + iK'(x_L^{-1}),\, x_L^{-1}\right]} \qquad (5.58)$$

This equation was derived for n odd, but it is also valid for n even. It

* This is indicated by the periodic rectangle in Fig. 5.10. The minus sign can be included, since R_n is an odd function.

can be simplified by using the following result from Problem 5.8:

$$\text{sn}\left[\frac{(1+2\nu)K(x_L^{-1})}{n} + iK'(x_L^{-1}), x_L^{-1}\right]$$

$$= \frac{x_L}{\text{sn}\left[(1+2\nu)K(x_L^{-1})/n, x_L^{-1}\right]} \quad (5.59)$$

Using this expression to rewrite (5.58) yields

$$L^{-1} = x_L^{-n} \prod_{\nu=0}^{n-1} \text{sn}^2\left[\frac{(1+2\nu)K(x_L^{-1})}{n}, x_L^{-1}\right]$$

Since $\text{sn}(2K - u) = \text{sn }u$, this can be rewritten as

$$L^{-1} = x_L^{-n} \prod_{\nu=0}^{IP(n/2)-1} \text{sn}^4\left[\frac{(1+2\nu)K(x_L^{-1})}{n}, x_L^{-1}\right] \quad (5.60)$$

where IP stands for "integer part of."

Example 5.2
For $n = 5$, $A_{\max} = 0.1$ dB, and $\omega_H/\omega_B = 1.3$, what is $A_{\min}$?

Solution
With $n = 5$ and $x_L = 1.3$, Eq. (5.60) yields $L = 340.6$. $A_{\min}$ can then be found from

$$\begin{aligned} A_{\min} &= A(\omega_H) = 10 \log(1 + \epsilon^2 L^2) \quad (5.61) \\ &= 10 \log[1 + (10^{0.1 A_{\max}} - 1)L^2] \\ &= 34.3 \text{ dB} \end{aligned}$$

5.12 A RATIONAL EXPRESSION FOR $R_n(x,L)$

In this section we shall finally obtain a rational expression for $R_n(x,L)$. The Tschebycheff rational function will be expressed in terms of its poles and zeros; i.e., it will be in the form

$$R_n(x,L) = C\frac{\Pi(x+z_i)}{\Pi(x+p_i)}$$

We will obtain the rational expression for $R_n(x,L)$ by returning to Theorem 5.4. Since $\text{sn}(2K - u) = \text{sn }u$, Eq. (5.52) can be rewritten as n odd:

$$R_n(x,L) = m \text{ sn}(x_L u) \prod_{\nu=1}^{(n-1)/2} \text{sn}(2\nu n^{-1}K + x_L u)\text{ sn}(2\nu n^{-1}K - x_L u)$$

$$(5.62a)$$

n even:

$$R_n(x,L) = m \prod_{\nu=0}^{(n-1)/2} \text{sn}\,[(2\nu+1)n^{-1}K + x_L u]\,\text{sn}\,[(2\nu+1)n^{-1}K - x_L u] \tag{5.62b}$$

The notation $K(x_L^{-1}) = K$ and $\text{sn}\,(x, x_L^{-1}) = \text{sn}\,(x)$ has been used in these equations and will be used in the rest of the chapter.

In Problem 5.14 it is shown that[12]

$$\text{sn}\,(u+v)\,\text{sn}\,(u-v) = \frac{\text{sn}^2 u - \text{sn}^2 v}{1 - k^2 \text{sn}^2 u \, \text{sn}^2 v} \tag{5.63}$$

Applying this result to (5.62a) yields

$$R_n(x,L) = m\,\text{sn}\,(x_L u) \prod_{\nu=1}^{(n-1)/2} \frac{\text{sn}^2\,(2uK/n) - \text{sn}^2\,(x_L u)}{1 - x_L^{-2}\,\text{sn}^2\,(2\nu K/n)\,\text{sn}^2\,(x_L u)} \tag{5.64}$$

But from (5.44), $x = \text{sn}\,(x_L u)$, so that (5.64) becomes

$$R_n(x,L) = mx \prod_{\nu=1}^{(n-1)/2} \frac{x^2 - \text{sn}^2\,(2\nu K/n)}{x^2 x_L^{-2}\,\text{sn}^2\,(2\nu K/n) - 1} \tag{5.65}$$

The above expression was for n odd; a similar expression can be found for n even. This leads to the following theorem:

Theorem 5.5

The Tschebycheff rational function can be expressed as n odd:

$$R_n(x,L) = C_1 x \prod_{\nu=1}^{(n-1)/2} \frac{x^2 - \text{sn}^2\,(2\nu K/n)}{x^2 - [x_L/\text{sn}\,(2\nu K/n)]^2} \tag{5.66a}$$

n even:

$$R_n(x,L) = C_2 \prod_{\nu=1}^{n/2} \frac{x^2 - \text{sn}^2\,[(2\nu-1)K/n]}{x^2 - \{x_L/\text{sn}\,[2\nu-1)K/n]\}^2} \tag{5.66b}$$

These equations are rational expressions for $R_n(x,L)$, and we have finally accomplished our task! We have expressed the Tschebycheff rational function in terms of its poles and zeros. While the poles and zeros appear to be given by complicated expressions, one should bear in mind that the elliptic sine function can be evaluated to yield a numerical result just as the more usual sine function can be evaluated.

From Theorem 5.5, the zeros of $R_n(x,L)$ are given by

n odd:
$$x_{z\nu} = \text{sn}\,\frac{2\nu K}{n} \tag{5.67a}$$

n even:
$$x_{z\nu} = \text{sn}\,\frac{(2\nu - 1)K}{n} \tag{5.67b}$$

The poles of $R_n(x,L)$ are given by

$$x_\nu = \frac{x_L}{x_{z\nu}} \tag{5.68}$$

The poles and zeros of $R_n(x,L)$ uniquely determine the function, except for a constant multiplier. The easiest way to find the constant multiplier is to use the fact that $R_n(1,L) = 1$. Thus,

$$C_1 = \prod_{\nu=1}^{(n-1)/2} \frac{1 - x_\nu^2}{1 - x_{z\nu}^2} \qquad C_2 = \prod_{\nu=1}^{n/2} \frac{1 - x_\nu^2}{1 - x_{z\nu}^2} \tag{5.69}$$

Besides the poles and zeros of $R_n(x,L)$, one might want to know where $R_n = \pm 1$ and $R_n = \pm L$, as these points determine the location of maximum passband loss and minimum stopband attenuation. For n odd, we can see from Fig. 5.10 that $R_n(x,L) = \pm 1$ when $u = (1 + 2\nu)CK(L^{-1})/L$. Applying (5.47), we see that this is equivalent to saying that $R_n(x,L) = \pm 1$ when $u = (1 + 2\nu)K(x_L^{-1})/(nx_L)$. But $x = \text{sn}\,(x_L u)$, so that*

n odd:
$$R_n(x,L) = \pm 1 \quad \text{when} \quad x = \text{sn}\,\frac{(1 + 2\nu)K(x_L^{-1})}{n} \tag{5.70a}$$

n even:
$$R_n(x,L) = \pm 1 \quad \text{when} \quad x = \text{sn}\,\frac{2\nu K(x_L^{-1})}{n} \tag{5.70b}$$

To find where $R_n(x,L) = \pm L$, recall that

$$R_n(x,L) = \frac{L}{R_n(x_L/x,L)}$$

Thus, if the values in (5.70) are identified as $x_{e\nu}$, it follows that

$$R_n(x,L) = \pm L \quad \text{when} \quad x = \frac{x_L}{x_{e\nu}}$$

---

* The case "n even" is established in a similar manner.

5.13 A PROGRAM FOR ELLIPTIC LOWPASS FILTERS

This section describes a program that can be used to find the minimum degree of an elliptic lowpass filter that meets the requirements A_{max}, A_{min}, ω_H, ω_B. The program also finds the poles of attenuation and can be used to find the loss at any frequency. Before discussing the program, we shall review the material discussed so far in this chapter. The review should be helpful, as the equations necessary for the program are scattered throughout the chapter.

Elliptic filters have equiripple passband and stopband. The theory for elliptic filters is based on the elliptic integral of the first kind,

$$u(\phi,k) = \int_0^\phi (1 - k^2 \sin^2 x)^{-\frac{1}{2}} dx$$

The elliptic sine function was defined as sn $(u,k) = \sin \phi$. This function was found to be very important in our discussion of the Tschebycheff rational function $R_n(x,L)$.

The Tschebycheff rational function is used to design elliptic filters in a manner analogous to the way in which the Tschebycheff polynomial $T_n(x)$ is used to design Tschebycheff filters. That is, the loss function $A(\omega)$ is related to $R_n(x,L)$ by

$$A(\omega) = 10 \log \left\{ 1 + \left[\epsilon R_n \left(\frac{\omega}{\omega_B}, L \right) \right]^2 \right\} \quad (5.71)$$

The program in Fig. 5.12 can be used to evaluate (5.71). This program will now be explained.

The program first demands the input requirements A_{max}, A_{min}, FB, FH, and then finds the minimum degree n that will meet these requirements. The degree n is found by using (5.49), which is

$$n = \frac{K(x_L^{-1})K'(L^{-1})}{K'(x_L^{-1})K(L^{-1})}$$

The values $K(x_L^{-1}) \triangleq$ KK and $K'(x_L^{-1}) \triangleq$ KK1 are found by using Part 5. Similarly, $K(L^{-1}) \triangleq$ KKL and $K'(L^{-1}) \triangleq$ KK1L can be found once L is determined. This is found in Step 1.11, which applies (5.51).

Once n has been determined (Step 1.44), $R_n(\omega/\omega_B,L)$ can be evaluated by using

$$R_n(x,L) = Cx^{1-NI} \prod_i \frac{x^2 - x_{zi}^2}{x^2 - x_{pi}^2}$$

where NI = 0 (or 1) if n is odd (or even). The poles and zeros of R_n

```
1; LØSS ØF ELLIPTIC LØWPASS FILTERS
1.1 DEMAND AMAX,AMIN,FB,FH
1.11 EE=10^(.1*AMAX)-1,L=SQRT((10^(.1*AMIN)-1)/EE),XL=FH/FB
1.13 DØ PART 5 FØR K=1/L
1.14 KKL=KK; KKL=K(1/L)
1.16 DØ PART 5 FØR K=SQRT(1-1/L/L)
1.17 KK1L=KK; KK1L=K (1/L)
1.41 DØ PART 5 FØR K=SQRT(1-1/XL/XL)
1.42 KK1=KK; KK1=K (1/XL)
1.43 DØ PART 5 FØR K=1/XL
1.44 N=IP(KK1L/KKL*KK/KK1)+1
1.45 TYPE #,N
1.46 NI=0,RN=1
1.47 NI=1 IF N=2*IP(N/2)
1.5 DØ PART 7 FØR I=1:1:N/2
1.6 DØ STEP 6.3 FØR X=1
1.7 C=1/RN

3; CALCULATIØN ØF SN(U,K)
3.1 Q=EXP(-$PI*KK1/KK)
3.2 V=$PI/2*U/KK,SN=0,J=0
3.3 W=Q^(J+.5),SN=SN+W*SIN((2*J+1)*V)/(1-W*W),J=J+1
3.4 TØ PART 3.3 IF W>10^-7
3.5 SN=SN*2*$PI/K/KK

5; CALCULATIØN ØF KK=CØMPLETE ELLIPTIC INTEGRAL
5.1 A[0]=ATN(K,SQRT(1-K^2)),Ø[0]=$PI/2,P=1,I=0
5.2 X=2/(1+SIN(A[I]))-1,Y=SIN(A[I])*SIN(Ø[I])
5.3 A[I+1]=ATN(SQRT(1-X*X),X)
5.4 Ø[I+1]=.5*(Ø[I]+ATN(Y,SQRT(1-Y*Y)))
5.5 E=1-A[I+1]*2/$PI,I=I+1
5.6 TØ PART 5.2 IF E>10^-7
5.7 P=P*(1+CØS(A[J])) FØR J=1:1:I
5.8 X=$PI/4+Ø[I]/2,KK=LN(SIN(X)/CØS(X))*P

6.1 X=F/FB
6.2 RN=C*X^(1-NI)
6.3 RN=RN*[X*X-XZ[I]^2]/[X*X-X[I]^2] FØR I=1:1:N/2
6.4 A=10*LØG(1+EE*RN*RN)
6.5 TYPE F,A IN FØRM 1

7; DETERMINATIØN ØF PØLES AND ZERØS
7.1 DØ PART 3 FØR U=(2*I-NI)*KK/N
7.2 XZ[I]=SN,X[I]=XL/SN
7.3 TYPE F[I] FØR F[I]=FB*X[I]

FØRM      1
##.### ###.####
```

Fig. 5.12 Program that evaluates the Tschebycheff rational function $R_n(x,x_L)$.

are found in Part 7; they are then used in Part 6 to calculate the loss function $A(\omega)$.

Example 5.3

Use of the program is demonstrated in Fig. 5.13. This finds the attenuation poles for a sixth-degree elliptic lowpass filter which has $A_\text{max} = 0.1$, $A_\text{min} = 40$, FB = 20, FH = 26. (The fact that A_min was specified as 40 dB determined that $n = 6$. An A_min of 35 dB would have given the same results.)

```
DØ PART 1
      AMAX=.1
      AMIN=40
      FB=20
      FH=26

      N=        6
      F[ 1]=    82.6050933
      F[ 2]=    33.2857993
      F[ 3]=    26.5772346
>


DØ PART 6 FØR F=0,11.66,18.18,20
   0.000     .1000
  11.660     .1000
  18.180     .1000
  20.000     .1000
>


DØ PART 6 FØR F=6.296,15.662,19.566
   6.296     .0000
  15.662     .0000
  19.566     .0000
>


DØ PART 6 FØR F=26,28.6,44.6
  26.000    46.8540
  28.600    46.8540
  44.600    46.8540
>
```

Fig. 5.13 Sample use of the elliptic lowpass filter program.

The program can be used to find the loss at any frequency. Those in Fig. 5.13 were chosen because they are special frequencies, as indicated below.

a. Frequencies of maximum passband loss. From (5.70) the normalized frequencies of maximum passband loss are

$$x_{e\nu} = \operatorname{sn} \frac{2\nu K}{n}$$

Thus we have

$$f_{e0} = 0$$

$$f_{e1} = 20 \operatorname{sn}\left(\frac{2K}{6}, \frac{1}{1.3}\right) = 11.66$$

$$f_{e2} = 20 \operatorname{sn}\left(\frac{4K}{6}, \frac{1}{1.3}\right) = 18.18$$

$$f_{e3} = 20 \operatorname{sn}\left(\frac{6K}{6}, \frac{1}{1.3}\right) = 20$$

Elliptic Filters

b. Frequencies of minimum passband loss. From (5.67) the normalized zeros of attenuation are located at

$$x_{z\nu} = \text{sn}\,\frac{(2\nu - 1)K}{6}$$

Thus the unnormalized zeros of attenuation are

$$f_{z1} = 20\,\text{sn}\left(\frac{K}{6}, \frac{1}{1.3}\right) = 6.296$$

$$f_{z2} = 20\,\text{sn}\left(\frac{3K}{6}, \frac{1}{1.3}\right) = 15.622$$

$$f_{z3} = 20\,\text{sn}\left(\frac{5K}{6}, \frac{1}{1.3}\right) = 19.566$$

c. Frequencies of minimum stopband loss. From the last equation in Sec. 5.12, these frequencies are where $x = x_L/x_{e\nu}$. But $x = f/f_B$, $x_L = f_H/f_B$, and $x_{e\nu} = f_{e\nu}/f_B$; thus,

$$f_{E1} = \frac{f_B f_H}{f_{e1}} = 44.60$$

$$f_{E2} = \frac{f_B f_H}{f_{e2}} = 28.60$$

$$f_{E3} = \frac{f_B f_H}{f_{e3}} = 26$$

5.14 CONCLUDING REMARKS

We have discussed several types of lowpass filters: Butterworth, Tschebycheff, inverse Tschebycheff, and elliptic. For a specific type of lowpass filter, there are three quantities that determine the necessary degree: the maximum passband loss, the minimum stopband attenuation, and the transition ratio $t_r = \omega_B/\omega_A$. Decreasing $A_{\max}$, increasing $A_{\min}$, or decreasing t_r will increase the degree. However, for the same requirements, one type of filter may be of lower degree than another.

Before we leave the subject of elliptic filters, it might be worthwhile to review why they will often be of lower degree than other types of filters. We shall do this by reconsidering the requirements in Example 5.1. We found in Chap. 2 that a Butterworth filter of degree 21 would satisfy these requirements. Such a high degree is necessary because the Butterworth approximation concentrates its approximating power at the origin, where it is maximally flat. We then discovered that if we let the passband be equiripple, as in a Tschebycheff filter, then an eighth-degree filter is adequate. A Tschebycheff filter produces a response that is better than necessary in the stopband; that is, the loss increases monotonically.

If the stopband loss is allowed to have equal minimums, then an even lower-degree filter is possible; thus, a fifth-degree elliptic filter satisfied the requirements.

Elliptic filters are equiripple in both the passband and stopband. The previous section described a program that can be used to evaluate the loss of an elliptic filter at any frequency. The program evaluates the function

$$A(\omega) = 10 \log \left\{ 1 + \left[\epsilon R_n \left(\frac{\omega}{\omega_B}, L \right) \right]^2 \right\} \quad (5.71)$$

In order to construct an actual filter, we need to know the input/output transfer function $H(s)$, which is related to $A(\omega)$ by

$$A(\omega) = 10 \log |H(j\omega)|^2 \quad (5.72)$$

Thus, we can write

$$|H(j\omega)|^2 = 1 + \left[\epsilon R_n \left(\frac{\omega}{\omega_B}, L \right) \right]^2 \quad (5.73)$$

It follows from (5.73) that the poles of $H(s)$ are also the poles of $R_n(\omega/\omega_B, L)$. Thus the program in the previous section can be used to find the poles of $H(s)$.

Example 5.4

Find the denominator of $H(s)$ for the elliptic filter described in Fig. 5.13.

Solution

Figure 5.13 shows that the attenuation poles are located at

$$F_1 = 82.61 \quad F_2 = 33.29 \quad F_3 = 26.58$$

Thus, the denominator of $H(s)$ is

$$q(s) = [s^2 + (2\pi \times 26.58)^2][s^2 + (2\pi \times 33.29)^2][s^2 + (2\pi \times 82.61)^2]$$

We have just seen that the program in Sec. 5.13 can be used to yield the denominator of $H(s)$. The denominator of $H(s)$ affects the stopband behavior most—it determines where the attenuation is infinite. Since this is so, we can say that the program in Sec. 5.13 tells us how to pick the attenuation poles so that the stopband has equal loss minimums.

We also want to know the numerator of $H(s)$. That is, how do we find the zeros of $H(s)$ so that the passband is equiripple? This is the subject of Chap. 10, which discusses how (for any arbitrary set of attenuation poles) to find the attenuation zeros so that the passband is equiripple. The chapter is not limited to lowpass filters, but also discusses bandpass filters. Before discussing bandpass filters, though, we shall discuss some frequency transformations. One of these transformations can change a lowpass characteristic to a bandpass. Thus, although the material presented in this chapter was concerned with lowpass filters, it can be extended to the bandpass case.

REFERENCES

1. S. Darlington, "Synthesis of Reactance 4-poles," *J. Math. & Phys.*, September 1939, pp. 257–353.
2. W. Cauer, *Synthesis of Linear Communication Networks*, McGraw-Hill Book Company, New York, 1958, app. 3.
3. W. Magnus and F. Oberhettinger, *Formulas and Theorems for the Functions of Mathematical Physics*, Chelsea Publishing Company, New York, 1949.
4. W. W. Bell, *Special Functions for Scientists and Engineers*, D. Van Nostrand Company, Inc., Princeton, N.J., 1968.
5. F. S. Woods, *Advanced Calculus*, Ginn and Company, Boston, 1934, pp. 365–386.
6. M. Abramowitz and I. Stegun, *Handbook of Mathematical Functions with Formulas, Graphs, and Mathematical Tables*, Natl. Bur. Stand. Appl. Math. Ser. 55, June 1964.
7. C. D. Hodgman, *C.R.C. Standard Mathematical Tables*, 12th ed., Chemical Rubber Publishing Company, Cleveland, 1960.
8. H. E. Fettis and J. C. Caslin, *Ten Place Tables of the Jacobian Elliptic Functions*, Aerosp. Res. Lab. ARL 65-180, September 1965.
9. A. G. Greenhill, *The Applications of Elliptic Functions*, Dover Publications, Inc., New York, 1959.
10. A. J. Grossman, "Synthesis of Tschebycheff Parameter Symmetrical Filters," *Proc. IRE*, April 1957, pp. 454–473.
11. H. J. Orchard, "Computation of Elliptic Functions of Rational Fractions of a Quarter Period," *IRE Trans. on Circuit Theory*, December 1958, pp. 352–355.
12. E. A. Guillemin, *Synthesis of Passive Networks*, John Wiley & Sons, Inc., New York, 1957.
13. G. Szentirmai, "Nomographs for Designing Elliptic-Function Filters," *Proc. IRE*, January 1960, pp. 113–114.
14. A. Papoulis, "On the Approximation Problem in Filter Design," *IRE Conv. Rec.*, vol. 5, pt. 2, 1957.

PROBLEMS

5.1 If an elliptic filter has a 0.1-dB passband ripple and a 30-dB minimum passband loss, what is the value of the parameter L?

5.2 *a.* Prove that

$$u\left(\frac{\pi}{2} + \phi, k\right) = 2K - u\left(\frac{\pi}{2} - \phi, k\right)$$

b. Show that if $u(\pi/2 + \phi, k) = K + \delta$, then $u(\pi/2 - \phi, k) = K - \delta$.

5.3 Show that $\operatorname{sn}(K + \delta, k) = \operatorname{sn}(K - \delta, k)$.

5.4 Show that
 a. $\operatorname{sn}(u + 2K, k) = -\operatorname{sn}(u,k)$
 b. $\operatorname{sn}(u + 4K, k) = \operatorname{sn}(u,k)$

5.5 Show that for $k = 0$ the expansion for sn $(u + v)$ reduces to

$$\sin (u + v) = \sin u \cos v + \cos u \sin v$$

5.6 Prove that sn $[i(u + K'), k] = -\text{sn}\,[i(K' - u), k]$.

5.7 The elliptic integral of the first kind was defined as

$$u(\phi,k) = \int_0^\phi (1 - k^2 \sin^2 x)^{-\frac{1}{2}}\, dx$$

Starting from this definition, show that

$$u(i\psi,k) = i \int_0^\psi (1 + k^2 \sinh^2 x)^{-\frac{1}{2}}\, dx$$

5.8 *a.* By applying Eq. (5.27) show that

$$\text{sn}\,(u + iK', k) = \frac{1}{k\, \text{sn}\,(u,k)}$$

b. The periodic rectangle of Fig. 5.8 indicates that

$$\text{sn}\,(K + iK', k) = \frac{1}{k}$$

Prove this relation.

5.9 Use the CRC tables to find $u(\phi,k)$, sn (u,k), and cn (u,k) for $\phi = 190°$, $k = 0.5$.

5.10 *a.* If $du = [(1 - z^2)(1 - k^2 z^2)]^{-\frac{1}{2}}\, dz$ is integrated from $\phi_1 = \pi/2$ to $\phi_2 = \sin^{-1}(1/k)$, show that

$$u\left(\sin^{-1}\frac{1}{k}, k\right) - u\left(\frac{\pi}{2}, k\right) = \int_1^{1/k} [(1 - z^2)(1 - k^2 z^2)]^{-\frac{1}{2}}\, dz$$

b. Use the result in part *a* to establish (5.37).

5.11 This problem will demonstrate that sn $(iu,k) = i$ sn $(u,k')/\text{cn}\,(u,k')$.

a. If $iu = \int_0^\phi (1 - k^2 \sin^2 x)^{-\frac{1}{2}}\, dx$

and $\sin x = i \tan y$, show that

$$iu = i \int_0^\psi (1 - k'^2 \sin^2 y)^{-\frac{1}{2}}\, dy$$

where $\sin \phi = i \tan \psi$.

b. Using the fact that

$$\text{sn}\,(iu,k) = \sin \phi = i\,\frac{\sin \psi}{\cos \psi}$$

show that

$$\text{sn}\,(iu,k) = i\,\frac{\text{sn}\,(u,k')}{\text{cn}\,(u,k')}$$

5.12 If the modulus k is greater than unity; show that the elliptic integral can be evaluated by

$$u(\phi,k) = \frac{1}{k} u\left(\psi, \frac{1}{k}\right)$$

where ϕ and ψ are related by $k \sin \phi = \sin \psi$.

Hint: In the definition of $u(\phi,k)$, use the substitution $k \sin x = \sin y$.

5.13 If
$$R_n = \operatorname{sn}\left(\frac{L}{C} u, L^{-1}\right)$$
$$= m \prod_{j=0}^{n-1} \operatorname{sn}\left[x_L u + 2jn^{-1}K(x_L^{-1}), x_L^{-1}\right]$$

show that

$$m^{-1} = \frac{Cx_L}{L} \prod_{j=1}^{n-1} \operatorname{sn}\left[2jn^{-1}K(x_L^{-1}), x_L^{-1}\right]$$

Hint: The derivatives of both sides, evaluated at $u = 0$, must be equal.

5.14 By using Eq. (5.22), show that

$$\operatorname{sn}(u+v) \operatorname{sn}(u-v) = (\operatorname{sn}^2 u - \operatorname{sn}^2 v)/(1 - k^2 \operatorname{sn}^2 u \operatorname{sn}^2 v)$$

5.15 Let $R_n[x, x_{Ln}]$ be the nth-order Tschebycheff rational function that has its stopband edge located at x_{Ln}. At the stopband edge $R_n = L_n(x_{Ln})$; that is, $R_n(y,y) = L_n(y)$.

  a. Show that the stopband edge of $R_3[R_2(x, x_{L2}), L_2(x_{L2})]$ is x_{L2}.

Hint: Evaluate this expression at $x = x_{L2}$, and then let y equal $L_2(x_{L2})$.

  b. Using the result in a, Papoulis[14] proceeds to show that

$$R_6(x, x_{L2}) = R_3\{R_2[x, L_2(x_{L2})], L_2(x_{L2})\}.$$

Check this result for $R_6(3,2) = 82{,}834$.

CHAPTER SIX

Frequency Transformations

6.1 INTRODUCTION

The filter functions that we have discussed have been for the lowpass case. Frequency transformations exist that can be used to change these characteristics to highpass, bandpass, or bandstop functions. Thus the work we have done for the Butterworth, Tschebycheff, and elliptic lowpass filters can be easily modified for other types of filters.

Instead of transforming a lowpass into a bandpass, we might wish to change one type of lowpass into another. For example, the even-order elliptic filters that we discussed had no attenuation poles at infinity; this is objectionable if one is designing passive filters. A frequency transformation that produces a pole at infinity will be discussed.

In the notation of this chapter S is the original variable, and s is the transformed variable. Similarly, the original frequency is $S = j\Omega$, whereas the transformed frequency is $s = j\omega$.

6.2 NORMALIZED-LOWPASS-TO-UNNORMALIZED-LOWPASS TRANSFORMATION

The specifications for a normalized lowpass filter can be given as in Fig. 6.1. That is, for $|\Omega| \leq 1$ the passband loss should be less than A_{max}, while for $|\Omega| \geq \Omega_H$ the stopband attenuation should be greater than A_{min}.

By definition, the normalized lowpass filter has the passband edge normalized to unity. The unnormalized lowpass filter instead has the passband edge at ω_B, as demonstrated in Fig. 6.2, which is just a frequency-scaled version of Fig. 6.1. That is, the normalized and unnormalized lowpass filters are related by the transformation

$$S = \frac{s}{\omega_B} \qquad (6.1)$$

This transforms the normalized passband edge $S = j$ to the unnormalized passband edge $s = j\omega_B$.

If we are given requirements as in Fig. 6.2, we would like to find the normalized lowpass requirements; that is, we want to find Ω_H. This can easily be done by applying (6.1) to yield

$$\Omega_H = \frac{\omega_H}{\omega_B} \qquad (6.2)$$

Example 6.1

If the passband and stopband edges of a lowpass filter are $f_B = 2$ kHz and $f_H = 10$ kHz, what is the normalized stopband edge Ω_H?

Solution

$$\Omega_H = \frac{\omega_H}{\omega_B} = \frac{f_H}{f_B} = 5$$

6.3 LOWPASS-TO-HIGHPASS TRANSFORMATION

The specifications for a highpass characteristic are shown symbolically in Fig. 6.3. This implies that for $|\omega| \leq \omega_L$ the stopband loss should be greater than $A_{\min}$, while for $|\omega| > \omega_A$ the passband loss should be less than $A_{\max}$.

The transformation that can be used for these requirements is

$$S = \frac{\omega_A}{s} \qquad (6.3)$$

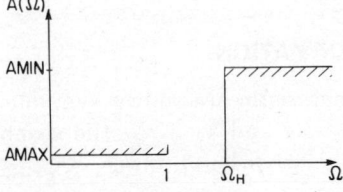

Fig. 6.1 Requirements for a normalized lowpass filter.

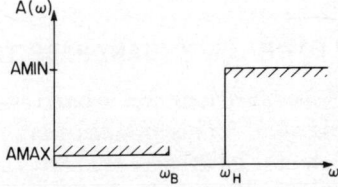

Fig. 6.2 Requirements for an unnormalized lowpass filter.

Because s is inversely proportional to S, the low-frequency and high-frequency behaviors are interchanged; thus the lowpass is transformed to a highpass. Given highpass requirements as in Fig. 6.3, we would like to find the normalized lowpass requirements. To find Ω_H, we notice that

$$A(\omega_L) = A_{\min} = A(\Omega_H)$$

Thus, applying (6.3) yields

$$\Omega_H = \frac{\omega_A}{\omega_L} \qquad (6.4)$$

To summarize, suppose that we are given the highpass requirements $A_{\max}$, $A_{\min}$, ω_L, ω_A. We can relate these to normalized lowpass require-

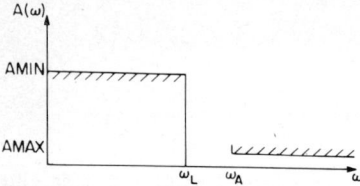

Fig. 6.3 Requirements for a highpass filter.

ments by making the transformation $S = \omega_A/s$. This yields the normalized lowpass $A_{\max}$, $A_{\min}$, $\Omega_B = 1$, $\Omega_H = \omega_A/\omega_L$. We can use these facts to modify lowpass filter programs to highpass filter programs.

Example 6.2

Find the attenuation poles for the lowest-degree elliptic filter that satisfies the following requirements: $A_{\max} = 0.1\,\text{dB}$, $A_{\min} = 40\,\text{dB}$, $f_L = 2\,\text{kHz}$, $f_A = 2.6\,\text{kHz}$.

Solution

We can transform this to a normalized lowpass described by $A_{\max} = 0.1$ dB, $A_{\min} = 40$ dB, $f_B = 1$, $f_H = 2.6/2 = 1.3$. The attenuation poles can be found for this lowpass filter, and then these poles can be transformed to highpass poles. This can be done by modifying the program in Fig. 5.12, which was for elliptic lowpass filters. The modifications for the highpass filter are shown at the top of Fig. 6.4, and the program is applied to this example at the bottom of Fig. 6.4.

6.4 LOWPASS-TO-BANDPASS TRANSFORMATION

The specifications for a bandpass characteristic are shown symbolically in Fig. 6.5. This implies that for $|\omega| \geq \omega_H$ and $|\omega| \leq \omega_L$ the stopband loss should be greater than $A_{\min}$. On the other hand, for $\omega_A \leq |\omega| \leq \omega_B$ the passband loss should be less than $A_{\max}$.

Compare Figs. 6.2, 6.3, and 6.5. The bandpass characteristic can be

```
10.1; MØDIFICATIØN ØF FIG. 5.12 FØR ELLIPTIC HIGHPASS FILTERS
10.2 DEMAND AMAX,AMIN,FA,FL
10.3 FB=1,FH=FA/FL
10.4 TYPE #,#,"    NØRMALIZED LØWPASS",#,AMAX,AMIN,FB,FH
10.5 DØ PART 1.11
10.6 F[I]=FA/F[I] FØR I=1:1:N/2
10.7 TYPE #,#,#,"    HIGHPASS LØSS PEAKS",#,#
10.8 TYPE F[I] FØR I=1:1:N/2

>

>DØ PART 10
        AMAX=.1
        AMIN=40
         FA=2600
         FL=2000

     NØRMALIZED LØWPASS

        AMAX=        .1
        AMIN=       40
          FB=        1
          FH=        1.3

          N=         6
        F[ 1]=       4.13025466
        F[ 2]=       1.66428997
        F[ 3]=       1.32886173


     HIGHPASS LØSS PEAKS

        F[ 1]=       629.501135
        F[ 2]=      1562.22777
        F[ 3]=      1956.56173
>
```

Fig. 6.4 Program for elliptic highpass filters.

thought of as a combination of a lowpass and a highpass characteristic, so it is not surprising that the bandpass transformation is a combination of a lowpass and a highpass transformation. That is,

$$S = \frac{s}{K_1} + \frac{K_2}{s} \qquad (6.5)$$

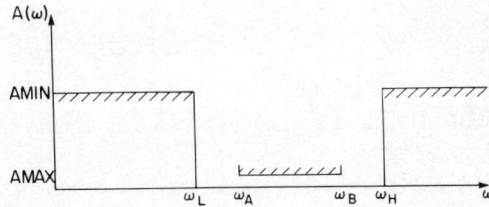

Fig. 6.5 Requirements for a bandpass filter.

90 Approximation Methods for Electronic Filter Design

To find the constants K_1 and K_2, we substitute $S = \pm j\Omega$ and $s = j\omega$. This yields

$$\omega = \pm \frac{K_1}{2}\left(\Omega \pm \sqrt{\Omega^2 + \frac{4K_2}{K_1}}\right) \tag{6.6}$$

Substituting $\Omega = 1$ to find ω_A and ω_B, and $\Omega = \Omega_H$ to find ω_L and ω_H, leads to

$$\begin{aligned}\omega_A &= \frac{K_1}{2}\left(-1 + \sqrt{1 + 4\frac{K_2}{K_1}}\right) \\ \omega_B &= \frac{K_1}{2}\left(1 + \sqrt{1 + 4\frac{K_2}{K_1}}\right) \\ \omega_L &= \frac{K_1}{2}\left(-\Omega_H + \sqrt{\Omega_H^2 + 4\frac{K_2}{K_1}}\right) \\ \omega_H &= \frac{K_1}{2}\left(\Omega_H + \sqrt{\Omega_H^2 + 4\frac{K_2}{K_1}}\right)\end{aligned} \tag{6.7}$$

If we multiply ω_A and ω_B together, it follows that $\omega_A\omega_B = K_1K_2$. Similarly, multiplying ω_L by ω_H yields $\omega_L\omega_H = K_1K_2$. Since $\omega_A\omega_B = \omega_L\omega_H$, the transformation of (6.5) results in a geometrically symmetrical filter.

Subtracting ω_A from ω_B in (6.7) yields $K_1 = \omega_B - \omega_A$. Since $\omega_A\omega_B = K_1K_2$, it follows that $K_2 = \omega_A\omega_B/(\omega_B - \omega_A)$. These results are summarized in the following theorem.

Theorem 6.1
The normalized lowpass-to-bandpass transformation

$$S = \frac{s^2 + \omega_A\omega_B}{(\omega_B - \omega_A)s} \tag{6.8}$$

transforms the normalized lowpass ($A_{\max}$, $A_{\min}$, $\Omega_B = 1$, Ω_H) to a geometrically symmetrical bandpass described by

$$\omega = \frac{\omega_B - \omega_A}{2}\left[\pm\Omega + \sqrt{\Omega^2 + \frac{4\omega_A\omega_B}{(\omega_B - \omega_A)^2}}\right] \tag{6.9}$$

The stopband edges (expressed in terms of the passband edges) are

$$\omega_{L,H} = \frac{\omega_B - \omega_A}{2}\left[\pm\Omega_H + \sqrt{\Omega_H^2 + \frac{4\omega_A\omega_B}{(\omega_B - \omega_A)^2}}\right] \tag{6.10}$$

If we are given a bandpass that has geometrically symmetrical requirements, we would like to find the normalized lowpass; that is, we would like to find Ω_H. This can be done by subtracting ω_L from ω_H. Equation (6.10) yields

$$\Omega_H = \frac{\omega_H - \omega_L}{\omega_B - \omega_A} \qquad (6.11)$$

This equation and Theorem 6.1 can be used to modify lowpass filter programs into bandpass filter programs.

Example 6.3
Find the attenuation poles for the lowest-degree elliptic filter that satisfies the following requirements: $A_{\max} = 0.1$ dB, $A_{\min} = 40$ dB,

$$f_L = \frac{\sqrt{25.69} - 1.3}{2} \qquad f_A = 2 \qquad f_B = 3 \qquad f_H = \frac{f_A f_B}{f_L}$$

Solution
Since $f_H f_L = f_A f_B$, the requirements are geometrically symmetrical, and we can use the previous results. These can be used to modify the program in Fig. 5.12, which is for elliptic lowpass filters. The modifications for the elliptic (geometrically symmetrical) bandpass filters are in Fig. 6.6. The important steps are:

Step 10.3 calculates FH for the normalized lowpass filter by using Eq. (6.11).

Steps 10.62 and 10.63 calculate the bandpass attenuation poles by using Eq. (6.6).

The program is applied to this example at the bottom of Fig. 6.6. Notice that the bandpass requirements FL, FA, FB were chosen such that FH = 1.3 for the normalized lowpass filter. Thus, the normalized lowpass is the same as in Fig. 6.4.

The results in Theorem 6.1 were for a geometrically symmetrical bandpass filter. If the bandpass requirements are not such that $\omega_A \omega_B = \omega_L \omega_H$, they can be changed so that the bandpass is geometrically symmetrical. There are two cases to be considered:

CASE 1 $\quad \omega_A \omega_B < \omega_L \omega_H$

One way we could solve this case is to choose a new upper stopband edge $\omega_H' < \omega_H$ such that $\omega_A \omega_B = \omega_L \omega_H'$. This is illustrated in Fig. 6.7. Another way we could solve this case is to choose a new upper passband edge $\omega_B' > \omega_B$ such that $\omega_A \omega_B' = \omega_L \omega_H$. This is also illustrated in Fig. 6.7.

Figure 6.7 demonstrates that for case 1 the passband can be extended to higher frequencies, or the upper stopband can be lowered. Which is desirable for a particular filter would depend on practical considerations.

92 Approximation Methods for Electronic Filter Design

```
        10.1; MODIFICATION OF FIG. 5.12 FOR ELLIPTIC BANDPASS FILTERS
        10.2 DEMAND AMAX,AMIN,FL,FA,FB
        10.24 FH=FA*FB/FL
        10.29 K1=FB-FA,K2=FA*FB/K1
        10.3 FH=(FH-FL)/(FB-FA),FB=1
        10.4 TYPE #,#,     NORMALIZED LOWPASS",#,AMAX,AMIN,FB,FH
        10.5 DO PART 1.11
        10.61 N2=IP(N/2)
        10.62 F[I+N2]=K1/2*(F[I]+SQRT(F[I]^2+4*K2/K1)) FOR I=1:1:N2
        10.63 F[I]=K1/2*(-F[I]+SQRT(F[I]^2+4*K2/K1)) FOR I=1:1:N2
        10.7 TYPE #,#,#,   BANDPASS LOSS PEAKS",#,#
        10.8 TYPE F[I] FOR I=1:1:2*N2
>



>DO PART 10
        AMAX=.1
        AMIN=40
           FL=(SQRT(25.69)-1.3)/2
           FA=2
           FB=3


    NORMALIZED LOWPASS

        AMAX=          .1
        AMIN=         40
          FB=          1
          FH=         1.3

           N=          6
        F[1]=       4.13025466
        F[2]=       1.66428997
        F[3]=       1.32886173


    BANDPASS LOSS PEAKS

        F[1]=       1.13873766
        F[2]=       1.75483497
        F[3]=       1.87357394
        F[4]=       5.26899232
        F[5]=       3.41912494
        F[6]=       3.20243567
>
```

Fig. 6.6 Program for geometrically symmetrical elliptic bandpass filters.

For example, we might expect the passband edge to be sensitive to element variations and thus want a factor of safety for the passband edge; for this problem we would design according to $\omega'_B = \omega_L \omega_H / \omega_A$.

Of course, a compromise can be made between $\omega'_B = \omega_L \omega_H / \omega_A$ and $\omega'_H = \omega_A \omega_B / \omega_L$. A reasonable compromise might be

$$\omega'_B \omega'_H = \omega_B \omega_H$$

Thus, using $\omega_A \omega_B' = \omega_L \omega_H'$ leads to

$$\omega_B' = \frac{(\omega_A \omega_B \omega_L \omega_H)^{1/2}}{\omega_A} \qquad \omega_H' = \frac{(\omega_A \omega_B \omega_L \omega_H)^{1/2}}{\omega_L} \qquad (6.12)$$

CASE 2 $\omega_A \omega_B > \omega_L \omega_H$

This case allows one to increase the lower stopband by picking $\omega_L' > \omega_L$ such that $\omega_A \omega_B = \omega_L' \omega_H$. Or the passband can be extended to lower frequencies by choosing $\omega_A' < \omega_A$ such that $\omega_A' \omega_B = \omega_L \omega_A$.

Alternatively, both stopband and passband can be improved by a choice such as

$$\omega_A' = \frac{(\omega_A \omega_B \omega_L \omega_H)^{1/2}}{\omega_B} \qquad \omega_L' = \frac{(\omega_A \omega_B \omega_L \omega_H)^{1/2}}{\omega_H} \qquad (6.13)$$

It has been demonstrated by cases 1 and 2 that if the original requirements are not geometrically symmetrical, they can be modified to be geometrically symmetrical.* An example of such a modification follows.

Example 6.4

Find the attenuation poles for the lowest-degree elliptic filter that satisfies the following requirements:

$A_{\max} = 0.25$ dB $A_{\min} = 35$ dB $f_L = 1$ $f_A = 1.1$ $f_B = 1.5$ $f_H = 1.6$

Make the filter geometrically symmetrical by using (6.12) or (6.13).

Solution

Since $f_A f_B > f_L f_H$, we will use (6.13). This yields $f_L' = 1.016$ and $f_A' = 1.083$. With these modified requirements we can now proceed as in Example 6.3. Alternatively, we can alter the program in Fig. 6.6 so that it automatically modifies the requirements. This is demonstrated in Fig. 6.8.

* Instead of modifying the requirements so that they are geometrically symmetrical, it is usually preferable to obtain transfer functions that are not symmetrical. This can result in simpler networks and will be covered in later chapters.

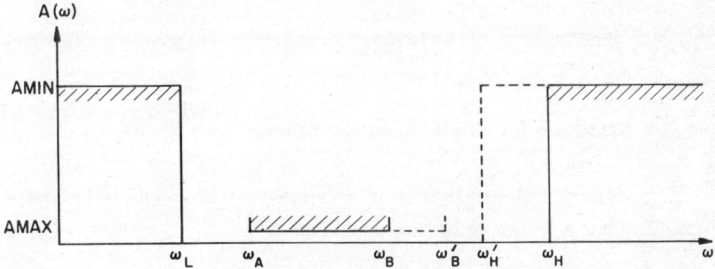

Fig. 6.7 Requirements for a nongeometrically symmetrical bandpass filter.

94 Approximation Methods for Electronic Filter Design

```
10.1; MODIFICATION OF FIG. 5.12 FOR ELLIPTIC BANDPASS FILTERS
10.2 DEMAND AMAX,AMIN,FL,FA,FB,FH
10.24 FB=SQRT(FA*FB*FL*FH)/FA,FH=FA*FB/FL IF FA*FB<FL*FH
10.25 FA=SQRT(FA*FB*FL*FH)/FB,FL=FA*FB/FH IF FA*FB>FL*FH
10.26 TYPE #,#,"   MODIFIED BANDPASS",#,FL,FA,FB,FH
10.29 K1=FB-FA,K2=FA*FB/K1
10.3 FH=(FH-FL)/(FB-FA),FB=1
10.4 TYPE #,#,"      NORMALIZED LOWPASS",#,AMAX,AMIN,FB,FH
10.5 DO PART 1.11
10.61 N2=IP(N/2)
10.62 F[I+N2]=K1/2*(F[I]+SQRT(F[I]^2+4*K2/K1)) FOR I=1:1:N2
10.63 F[I]=K1/2*(-F[I]+SQRT(F[I]^2+4*K2/K1)) FOR I=1:1:N2
10.7 TYPE #,#,#,"   BANDPASS LOSS PEAKS",#,#
10.8 TYPE F[I] FOR I=1:1:2*N2
>

>DO PART 10
        AMAX=.25
        AMIN=35
         FL=1
         FA=1.1
         FB=1.5
         FH=1.6

   MODIFIED BANDPASS

          FL=       1.0155048
          FA=       1.08320512
          FB=       1.5
          FH=       1.6

   NORMALIZED LOWPASS

        AMAX=           .25
        AMIN=         35
          FB=          1
          FH=          1.40235696

           N=          5
        F[1]=          2.14271929
        F[2]=          1.45269542


   BANDPASS LOSS PEAKS

        F[1]=           .90409344
        F[2]=          1.00739862
        F[3]=          1.79716787
        F[4]=          1.61287463
>
```

Fig. 6.8 Program for elliptic bandpass filters.

Theorem 6.1 described how a frequency $S = j\Omega$ is transformed to two frequencies $s = j\omega$, where ω is given by

$$\omega = \frac{\omega_B - \omega_A}{2}\left[\pm\Omega + \sqrt{\Omega^2 + \frac{4\omega_A\omega_B}{(\omega_B - \omega_A)^2}}\right]$$

This equation can be used to find the transformed attenuation poles, since they are located on the $j\omega$ axis. However, it cannot be used to determine the zeros of $H(s)$ because they are complex. To do this, we can use the following theorem.

Theorem 6.2

The normalized lowpass-to-bandpass transformation

$$S = \frac{s^2 + \omega_A \omega_B}{(\omega_B - \omega_A)s}$$

transforms the frequencies $S = -R \pm j\Omega$ (where R is a positive number) to

$$s = \frac{\omega_B - \omega_A}{2} \left(-R + \left[\frac{\sqrt{A^2 + (2R\Omega)^2} - A}{2} \right]^{1/2} \right.$$
$$\left. \pm j \left\{ \Omega - \left[\frac{\sqrt{A^2 + (2R\Omega)^2} + A}{2} \right]^{1/2} \right\} \right)$$

and (6.14)

$$s = \frac{\omega_B - \omega_A}{2} \left(-R - \left[\frac{\sqrt{A^2 + (2R\Omega)^2} - A}{2} \right]^{1/2} \right.$$
$$\left. \pm j \left\{ \Omega + \left[\frac{\sqrt{A^2 + (2R\Omega)^2} + A}{2} \right]^{1/2} \right\} \right)$$

where
$$A = \Omega^2 + \frac{4\omega_A \omega_B}{(\omega_B - \omega_A)^2} - R^2 \qquad (6.15)$$

Proof

This proof demonstrates that the transformation

$$S = \frac{s^2 + \omega_A \omega_B}{(\omega_B - \omega_A)s} \qquad (6.8)$$

transforms $S = -R + j\Omega$ into two of the frequencies given in Eq. (6.14). We could similarly show that $S = -R - j\Omega$ is transformed into the other two frequencies.

It follows from (6.8) that

$$s^2 - (\omega_B - \omega_A)Ss + \omega_A \omega_B = 0$$

Substituting $S = -R + j\Omega$ and solving the quadratic equation yield

$$s = \frac{\omega_B - \omega_A}{2} \left(-R + j\Omega \pm \sqrt{-A - 2jR\Omega} \right) \qquad (6.16)$$

where $A = \Omega^2 + 4\omega_A \omega_B / (\omega_B - \omega_A)^2 - R^2$.

96 Approximation Methods for Electronic Filter Design

It is shown in Problem 6.1 that if $y = -|y|$ then

$$\sqrt{-A + jy} = -\left(\frac{\sqrt{A^2 + y^2} - A}{2}\right)^{1/2} + j\left(\frac{\sqrt{A^2 + y^2} + A}{2}\right)^{1/2} \quad (6.17)$$

Substituting (6.17) into (6.16) with $y = -2R\Omega$ yields

$$s = \frac{\omega_B - \omega_A}{2}\left(-R + j\Omega \pm \left\{-\left[\frac{\sqrt{A^2 + (2R\Omega)^2} - A}{2}\right]^{1/2} + j\left[\frac{\sqrt{A^2 + (2R\Omega)^2} + A}{2}\right]^{1/2}\right\}\right)$$

Grouping the real and imaginary parts yields two of the equations in (6.14). The other two would result if we studied $S = -R - j\Omega$.

Theorem 6.3 can be used to investigate how the quality Q of lowpass roots is affected by the bandpass transformation. This is demonstrated in the next example.

Example 6.5

Assume that a normalized lowpass filter has the pair of roots $S = -1/\sqrt{2} \pm j/\sqrt{2}$.
 a. What are the undamped natural frequency and quality of the lowpass roots?
 b. If the lowpass roots are transformed to bandpass roots by

$$S = \frac{s^2 + 2}{s}$$

what are the bandpass roots?
 c. Find the undamped natural frequencies and qualities of the bandpass roots.

Solution

a. The undamped natural frequency Ω_0 and the quality Q are related to the roots S_1 and S_1^* by

$$S^2 + \frac{\Omega_0}{Q}S + \Omega_0^2 = (S - S_1)(S - S_1^*)$$

Since $S_1 = -1/\sqrt{2} + j/\sqrt{2}$, it follows that

$$\Omega_0 = 1 \qquad Q = \frac{1}{\sqrt{2}}$$

b. Applying Theorem 6.2 yields

$$s_{1,2} = -0.265 \pm j1.063$$
$$s_{3,4} = -0.442 \pm j1.771$$

c. By analogy to a,

$$s^2 + \frac{\omega_0}{q}s + \omega_0^2 = (s - s_1)(s - s_2)$$

Thus, for s_1, s_2

$$\omega_0 = 1.096 \qquad q = 2.065$$

and for s_3, s_4

$$\omega_0 = 1.825 \qquad q = 2.065$$

This example demonstrates two items that are true in general for the lowpass-to-bandpass transformation. First, the bandpass roots have a higher quality than the lowpass roots. Second, the two pairs of bandpass roots have the same quality.

6.5 LOWPASS-TO-BANDSTOP TRANSFORMATION

The specifications for a bandstop characteristic are shown symbolically in Fig. 6.9. This implies that for $|\omega| \leq \omega_A$ or $|\omega| \geq \omega_B$ the passband loss should be less than $A_{\max}$. For

$$\omega_L \leq |\omega| \leq \omega_H$$

the stopband loss should be greater than $A_{\min}$.

The normalized lowpass-to-bandpass transformation was

$$S = \frac{s^2 + \omega_A \omega_B}{(\omega_B - \omega_A)s} \tag{6.8}$$

A bandstop filter is a bandpass filter that has had its passband and stopband interchanged. Thus, a bandstop characteristic can be obtained by performing a bandpass transformation on a highpass filter instead of a lowpass filter (since a highpass and lowpass have their passbands and stopbands interchanged). The lowpass is transformed to a highpass by substituting $1/S$ for S. Thus, from (6.8), the normalized lowpass-to-bandstop transformation is

$$S = \frac{(\omega_B - \omega_A)s}{s^2 + \omega_A \omega_B} \tag{6.18}$$

Because the bandstop transformation is the same as the bandpass except S is replaced by $1/S$, the theorems for bandpass filters can be easily rewritten for bandstop filters. By analogy to Theorem 6.1 we have:

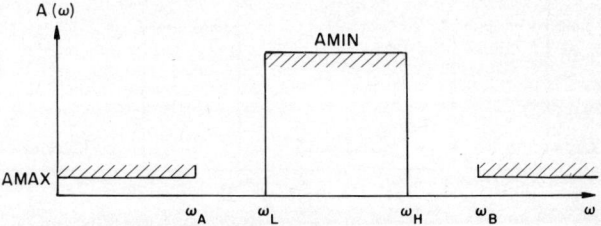

Fig. 6.9 Requirements for a bandstop filter.

Theorem 6.3

The normalized lowpass-to-bandstop transformation

$$S = \frac{(\omega_B - \omega_A)s}{s^2 + \omega_A \omega_B} \qquad (6.18)$$

transforms the normalized lowpass (A_{max}, A_{min}, $\Omega_B = 1$, Ω_H) to a geometrically symmetrical bandpass described by

$$\omega = \frac{\omega_B - \omega_A}{2}\left[\pm\frac{1}{\Omega} + \sqrt{\frac{1}{\Omega^2} + \frac{4\omega_A\omega_B}{(\omega_B - \omega_A)^2}}\right] \qquad (6.19)$$

The stopband edges (expressed in terms of the passband edges) are

$$\omega_{L,H} = \frac{\omega_B - \omega_A}{2}\left[\pm\frac{1}{\Omega_H} + \sqrt{\frac{1}{\Omega_H^2} + \frac{4\omega_A\omega_B}{(\omega_B - \omega_A)^2}}\right] \qquad (6.20)$$

Because of the direct analogy between the lowpass-to-bandpass transformation and the lowpass-to-bandstop transformation, the bandstop case will not be pursued further here.

6.6 LOWPASS-TO-MULTIPLE-BANDPASS TRANSFORMATION

The following discussion is given in terms of two passbands, but it can be generalized to many.* The bandpass case discussed in this section is shown in Fig. 6.10. This implies that for $\omega_{A1} \leq |\omega| \leq \omega_{B1}$ or $\omega_{A2} \leq |\omega| \leq \omega_{B2}$ the passband loss is less than A_{max}. Similarly, in the stopbands the attenuation should be greater than A_{min}.

A transformation that changes a normalized lowpass to a bandpass such as that shown in Fig. 6.10 is

$$S = \frac{(s^2 + \omega_1^2)(s^2 + \omega_2^2)}{Ks(s^2 + \omega_3^2)} \qquad (6.21)$$

* See Weinberg[1] or Cauer.[2]

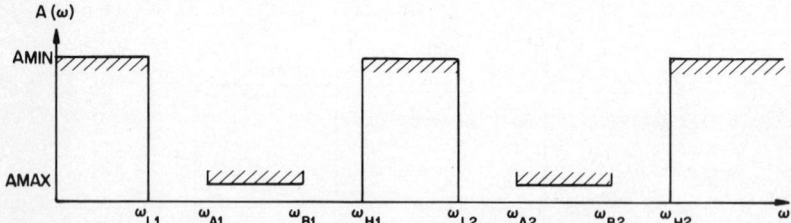

Fig. 6.10 Requirements for a multiple passband filter.

This transformation can be considered as a series of two transformations: first from a normalized lowpass to a bandpass, and then from the bandpass to a double bandpass. This is demonstrated by the equations

$$S = \frac{Z}{c_1} + \frac{c_2}{Z} \quad \text{and} \quad Z = \frac{s}{d_1} + \frac{d_2}{s} \qquad (6.22)$$

That is, eliminating Z between these two equations produces a transformation of the type in (6.21).

The transformation from S to Z produces two frequencies in the Z plane for every frequency in the S plane. Similarly, the transformation from Z to s produces two frequencies in the s plane for every frequency in the Z plane, so that one point in the S plane corresponds to four points in the s plane. For example, consider the normalized lowpass stopband edge $\pm\Omega_H$. The corresponding points in the s plane are $\pm\omega_{L1}$, $\pm\omega_{H1}$, $\pm\omega_{L2}$, $\pm\omega_{H2}$.

By using the concept of a series of two transformations, it is easy to show that the double passband filter must satisfy constraints analogous to those for the geometrically symmetric simple bandpass. However, this will not be pursued further here.

6.7 REACTANCE TRANSFORMATIONS

The transformations encountered so far in this chapter are known as *reactance* transformations. They transform a reactance in the S plane to another reactance in the s plane. For example, consider a capacitor in the S plane. The result of a lowpass-to-bandpass transformation is shown in Fig. 6.11. This leads to a conceptually easy design procedure for passive filters: design a prototype normalized lowpass, and then make element substitutions as in Fig. 6.11.

Frequency transformations other than reactance transformations do exist. For these, the new circuit can no longer be derived from the normalized lowpass by a reactance substitution. However, this does not imply that nonreactance transformations are impractical. On the contrary, they can transform a nonrealizable characteristic to a realizable one.

6.8 OTHER FREQUENCY TRANSFORMATIONS

The even-degree elliptic filters discussed in Chap. 5 had the property that for high frequencies the loss approached a (finite) constant value. That is, the even-degree elliptic filters did not have an attenuation pole at infinity. However, a practical passive ladder realization requires an

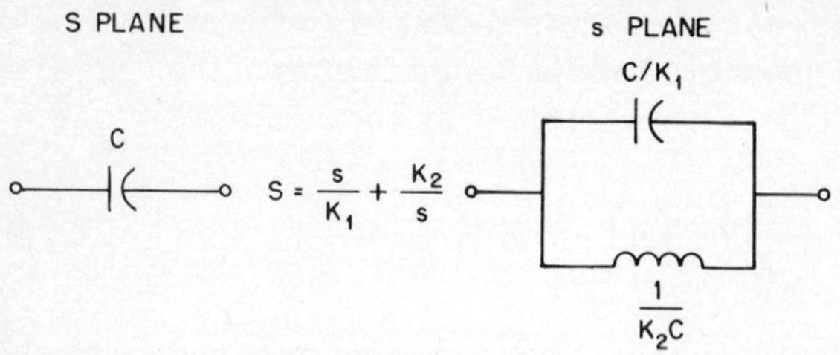

Fig. 6.11 A simple element substitution can produce a reactance transformation.

attenuation pole at infinity.[3] A transformation will now be presented that can produce a loss peak at infinity.

The general form of the transformation that we shall use is*

$$s^2 = K \frac{S^2 + \Omega_0^2}{S^2 + \Omega_c^2} \tag{6.23}$$

To see how this can produce an attenuation pole at $s = \infty$, consider Fig. 6.12a. This has its highest attenuation pole at $S = j\Omega_c$; but from (6.23), when $S = j\Omega_c$, we have $s = \infty$. Thus, the transformation in (6.23) shifts the attenuation pole at $S = j\Omega_c$ to $s = \infty$ as indicated in Fig. 6.12b.

We would like the transformation in (6.23) to leave the passband edge unaltered. That is, we want $s = j\Omega_B$ when $S = j\Omega_B$. This can be accomplished by requiring that

$$K = \frac{\Omega_c^2 - \Omega_B^2}{\Omega_B^2 - \Omega_0^2} \Omega_B^2 \tag{6.24}$$

In Eq. (6.24), Ω_B is the passband edge, whereas Ω_c is the location of the highest-attenuation pole in the original filter. From (6.23) it follows that the parameter Ω_0 determines the location of the origin in the s plane. These results are summarized in the following theorem.

Theorem 6.4

The transformation

$$s^2 = \left(\frac{\Omega_c^2 - \Omega_B^2}{\Omega_B^2 - \Omega_0^2} \Omega_B^2\right) \frac{S^2 + \Omega_0^2}{S^2 + \Omega_c^2} \tag{6.25}$$

1. Transforms $S = j\Omega_0$ to $s = 0$.
2. Transforms $S = j\Omega_B$ to $s = j\Omega_B$.
3. Transforms $S = j\Omega_c$ to $s = \infty$.

* This is a special case of a Möbius transformation, which is defined as a bilinear change of variables: $y = (ax + b)/(cx + d)$.

Example 6.6

In Example 5.3 we encountered the lowpass elliptic filter

$$A_{\max} = 0.1 \qquad A_{\min} = 40 \qquad F_B = 20 \qquad F_H = 26$$

This was a sixth-degree filter of the type shown in Fig. 6.12a. The attenuation poles were located at $F_1 = 82.61$, $F_2 = 33.29$, $F_3 = 26.58$. Use the transformation in Theorem 6.4 to produce an attenuation pole at infinity while keeping $A(\omega = 0) = A_{\max}$. (This filter is illustrated in Fig. 6.12b.)

Solution

Since we want the zero-frequency behavior to be unchanged, it follows that $\Omega_0 = 0$. Thus (6.25) yields

$$\omega^2 = (\Omega_c^2 - \Omega_B^2) \frac{\Omega^2}{\Omega_c^2 - \Omega^2}$$

which is equivalent to

$$f^2 = (F_c^2 - F_B^2) \frac{F^2}{F_c^2 - F^2}$$

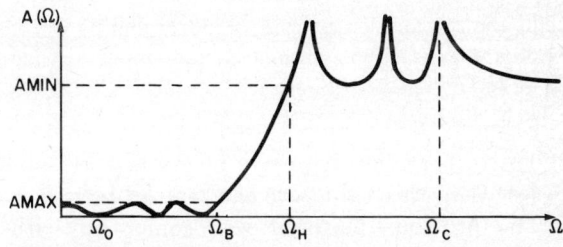

a) A SIXTH DEGREE ELLIPTIC FILTER

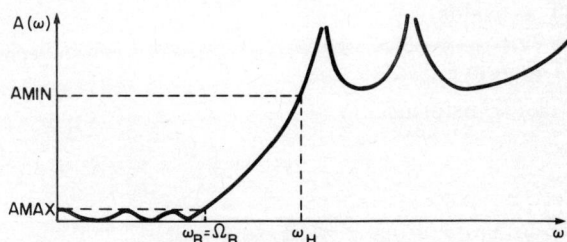

b) ILLUSTRATION OF TRANSFORMATION (6.23)

Fig. 6.12 A nonreactance transformation can shift an attenuation pole to infinity.

For this problem $F_B = 20$ and $F_c = F_1 = 82.61$. Thus the attenuation poles F_2 and F_3 are transformed to $f_2 = 35.29$ and $f_3 = 27.24$. Similarly, the stopband edge $F_H = 26$ is transformed to $f_H = 26.58$. In a practical design this would be compensated for by making the original F_H less than the design specifications.

In lowpass filter design it is often desirable to have the dc loss $A(0)$ less than the passband ripple A_{max}. For example, in passive network synthesis this allows a better ratio of source-to-load impedance. Example 6.6 had $A(0) = A_{max}$, but this could easily be modified.[4]

Example 6.7

Proceed as in Example 6.6, but this time have $A(\omega = 0) = 0$.

Solution

From Theorem 6.4,

$$f^2 = \left(\frac{F_c{}^2 - F_B{}^2}{F_B{}^2 - F_0{}^2} F_B{}^2 \right) \frac{F^2 - F_0{}^2}{F_c{}^2 - F^2}$$

From Example 5.3 we have

$$F_0 = 6.3 \qquad F_B = 20 \qquad F_c = 82.61$$

Thus the attenuation poles $F_2 = 33.29$ and $F_3 = 26.58$ are transformed to $f_2 = 36.52$ and $f_3 = 27.88$. Similarly, the stopband edge $f_H = 26$ is transformed to $f_H = 27.17$.

Example 6.7 demonstrated that we could make $A(\omega = 0) = 0$ by choosing Ω_0 as the frequency of the first attenuation zero. By choosing Ω_0 less than this value, we can instead have $0 < A(\omega = 0) < A_{max}$.

The previous material was concerned with filters that had finite attenuation peaks; i.e., there was some frequency Ω_c at which the prototype filter had a loss peak. The Tschebycheff filters discussed in Chap. 3 did not have finite loss peaks. What can we do if we want to have an even-degree Tschebycheff filter that has $A(0) < A_{max}$? We can still apply Theorem 6.4 if we let the loss peak located at Ω_c approach infinity. This yields

Theorem 6.5

The transformation

$$s^2 = \frac{\Omega_B{}^2}{\Omega_B{}^2 - \Omega_0{}^2} (S^2 + \Omega_0{}^2) \qquad (6.26)$$

1. Transforms $S = j\Omega_0$ to $s = 0$.
2. Transforms $S = j\Omega_B$ to $s = j\Omega_B$.

This theorem is very useful if we want to modify an even-degree Tschebycheff filter so that it has zero loss at the origin. To produce zero loss at the origin, we choose Ω_0 to be the location of the first Tschebycheff zero; thus, we want the first zero of $T_n(\Omega/\Omega_B)$. But the Tschebycheff polynomial can be expressed as

$$T_n(x) = \cos(n \cos^{-1} x)$$

The first zero of this expression is located at

$$x = \cos\frac{(n-1)\pi}{2n} = \sin\frac{\pi}{2n}$$

Thus, we should choose Ω_0 as

$$\Omega_0 = \Omega_B \sin\frac{\pi}{2n} \tag{6.27}$$

If we substitute (6.27) into (6.26), we have a transformation that will make an even-order Tschebycheff filter into an even-order polynomial filter that has a zero of attenuation at the origin. This transformation can be expressed as

$$\Omega^2 = \Omega_B^2 \sin^2\frac{\pi}{2n} + \omega^2\left(1 - \sin^2\frac{\pi}{2n}\right) \tag{6.28}$$

Example 6.8

If a Tschebycheff lowpass filter has a passband edge $\Omega_B = 1$ and a passband ripple of 3 dB, then its loss can be expressed as

$$A(\Omega) = 10 \log[1 + T_n^2(\Omega)]$$

For a fourth-degree Tschebycheff filter, transform the loss so that there is no attenuation at the origin. What is the loss at twice the passband edge?

Solution

From Table 3.1, $T_4(\Omega) = 1 - 8\Omega^2 + 8\Omega^4$. Thus, using the above equation for $A(\Omega)$ and (6.28) yields

$$A(\omega) = 10 \log[1 + (1 - 8\Omega^2 + 8\Omega^4)^2]$$

$$\Omega^2 = \sin^2\frac{\pi}{8} + \omega^2\left(1 - \sin^2\frac{\pi}{8}\right)$$

Evaluating this expression at $\omega = 2$ yields $\Omega^2 = 3.56$, so that $A(\omega = 2) = 37.4$ dB.

The behavior of the original Tschebycheff lowpass filter in Example 6.8 and the transformed filter is sketched in Fig. 6.13. The transformed characteristic in Fig. 6.13b is similar to a third-order Tschebycheff shape, but it is not the same. The transformed characteristic in Fig. 6.13b has an asymptotic slope of 80 dB/decade because it is a fourth-degree filter. On the other hand, a third-order Tschebycheff filter has an asymptotic slope of 60 dB/decade.

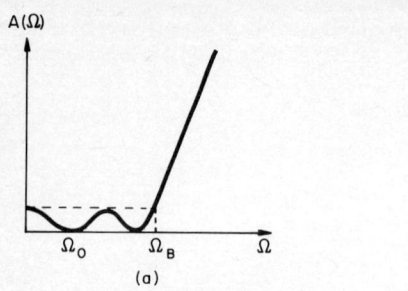
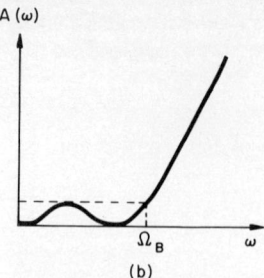

Fig. 6.13 (a) Fourth-order Tschebycheff lowpass filter; (b) transformed filter that is also fourth order.

The stopband performance of the transformed filter in Fig. 6.13b is poorer than the stopband performance of the Tschebycheff filter in Fig. 6.13a. This is because, as mentioned in Theorem 3.10, of all possible polynomial filters the Tschebycheff filter has the best stopband performance. As a demonstration of this fact we can compare the two filters at twice the passband edge. From Example 6.8 the transformed filter has $A(\omega = 2) = 37.4$ dB. The loss of the Tschebycheff filter is $A(\Omega = 2) = 39.7$ dB.

6.9 CONCLUSIONS

In the early sections of this chapter we discussed the frequently encountered transformations: normalized lowpass to unnormalized lowpass, highpass, or bandpass. We saw that if we were given the requirements for an unnormalized lowpass as

f_B passband edge
A_{max} maximum passband loss
f_H stopband edge
A_{min} minimum stopband loss

we could transform the requirements to a normalized lowpass filter described by

1   passband edge
A_{max} maximum passband loss
F_H stopband edge
A_{min} minimum stopband loss

The transformation was $S = s/\omega_B$; thus it follows that the normalized stopband edge was $F_H = f_H/f_B$. The normalized passband edge determines the complexity of the normalized lowpass filter, so that the ratio f_H/f_B determines the complexity of an unnormalized lowpass filter.

The requirements for a highpass filter can be given as

f_A	passband edge
$A_{\max}$	maximum passband loss
f_L	stopband edge
$A_{\min}$	minimum stopband loss

Instead of designing a highpass filter to meet these requirements, we can design a normalized lowpass filter. The degree of the lowpass filter is determined by the fact that its stopband edge is given by $F_H = f_A/f_L$. The transfer function of the normalized lowpass filter can be transformed to that of the highpass filter by using $S = \omega_A/s$.

The requirements for a bandpass filter can be given as

f_A, f_B	passband edges
$A_{\max}$	maximum passband loss
f_L, f_H	stopband edges
$A_{\min}$	minimum stopband loss

Instead of designing a bandpass filter to meet these requirements, we can again design a normalized lowpass filter. If we assume that the bandpass requirements are geometrically symmetrical, the degree of the normalized lowpass filter is determined by the fact that its stopband edge is given by $F_H = (f_H - f_L)/(f_B - f_A)$. The normalized lowpass filter can be transformed to a bandpass filter by

$$S = \frac{s^2 + \omega_A \omega_B}{(\omega_B - \omega_A)s}$$

It should be noted that this transformation doubles the degree; that is, every lowpass root is transformed to two bandpass roots. Also, the quality of the bandpass roots is higher than the quality of the lowpass roots. Thus, not only will the bandpass filter require more elements because its degree is higher, it will also require more ideal elements because its quality is higher.

We also discussed a lowpass-to-bandstop transformation and a lowpass-to-multiple-bandpass transformation. These were all reactance transformations; nonreactance transformations can also be very useful. One nonreactance transformation that is often helpful is

$$s^2 = K \frac{S^2 + \Omega_0^2}{S^2 + \Omega_c^2}$$

This can be used to shift to infinity an attenuation pole that was originally located at a finite frequency. It can also be used to shift an attenuation zero to the origin.

Before concluding this chapter on frequency transformations, we shall mention one more nonreactance transformation. This transformation is encountered in a procedure known as predistortion. Assume that we wish to synthesize the transfer function $H(S)$ by using passive elements: resistors, inductors, and capacitors. The inductors and capacitors will not be ideal but will have parasitic losses. One simple way to approximate the losses is to assume that all inductors and capacitors have the same dissipation factor; that is, assume that the impedance of an inductor can be expressed as $Z_L = (S + d)L$ and the admittance of a capacitor can be expressed as $Y_C = (S + d)C$, where d is the dissipation factor. The dissipation causes the roots of $H(S)$ to be shifted to the left by an amount d in the S plane. In the predistortion technique, one compensates for this shift in roots due to using lossy elements. The predistortion consists in using the nonreactance transformation $S = s - d$. Thus the transfer function $H(s)$ is very similar to $H(S)$; the only difference is that the roots of $H(S)$ have been shifted to the right by an amount d. If $H(s)$ is synthesized with lossy elements that have a dissipation factor d, the synthesized network will have the desired roots.

All the frequency transformations discussed here produced changes in the frequency scale, but they did not change the attenuation axis. For example, if the original stopband had equal minimums of loss, then the transformed filter had equal minimums. This can lead to a higher-degree filter than necessary, as filters often need not have equiminimum stopbands. Later chapters will describe how to design filters that have arbitrary stopbands. This will allow us to design bandpass filters directly; that is, we will not first design a lowpass filter and then transform the result. Thus, we will no longer be restricted to designing bandpass filters that are geometrically symmetrical.

REFERENCES

1. L. Weinberg, *Network Analysis and Synthesis*, McGraw-Hill Book Company, New York, 1962, sec. 11-9.
2. W. Cauer, *Synthesis of Linear Communication Networks*, McGraw-Hill Book Company, New York, 1958, pp. 302–315.
3. J. L. Herrero and G. Willoner, *Synthesis of Filters*, Prentice-Hall, Inc., Englewood Cliffs, N.J., 1966, chap. 6.
4. E. Christian and E. Eisenmann, "Broad-band Matching by Lowpass Transformations," *Proc. Fourth Allerton Conf.*, 1966, pp. 155–164.

PROBLEMS

6.1 Show that

    a. If $y = |y|$, then

$$(x + jy)^{1/2} = \left(\frac{\sqrt{x^2 + y^2} + x}{2}\right)^{1/2} + j\left(\frac{\sqrt{x^2 + y^2} - x}{2}\right)^{1/2}$$

b. If $y = -|y|$, then

$$(x + jy)^{1/2} = -\left(\frac{\sqrt{x^2 + y^2} + x}{2}\right)^{1/2} + j\left(\frac{\sqrt{x^2 + y^2} - x}{2}\right)^{1/2}$$

6.2 a. Show that the quality of the complex pair of roots $S_{1,2} = -R \pm j\Omega$ is $Q = \frac{1}{2}[1 + (\Omega/R)^2]^{1/2}$.

b. If a normalized lowpass-to-bandpass transformation is performed on the roots $S_{1,2}' = -R \pm j\Omega$, then two pairs of bandpass roots are produced. Their values are given by (6.14). Show that both pairs of roots have the same quality.

Hint: Part a indicates that this is equivalent to demonstrating that the magnitudes of the ratios of imaginary part to real part are the same for both pairs of roots.

6.3 Figure 6.6 contains a normalized lowpass filter that has an attenuation pole at $F(1) = 4.130255$. If the normalized lowpass is transformed to a bandpass described by $f_A = 2$ and $f_B = 3$, what does the pole at $F(1)$ transform to?

6.4 The passband edges of a bandpass filter are $f_A = 1.5$, $f_B = 2$. The stopband edges are $f_L = .75$, $f_H = 2.5$.

a. Redefine one of the requirements so that the new bandpass filter is geometrically symmetrical.

b. Use (6.12) or (6.13) to redefine two of the requirements so that the new bandpass filter is geometrically symmetrical.

6.5 The normalized second-order Butterworth function has the roots $S = -1/\sqrt{2} \pm j/\sqrt{2}$. If these roots are transformed by the bandstop transformation $S = s/(s^2 + 2)$,

a. What are the bandstop zeros?

b. What is the quality of the bandstop zeros?

6.6 Assume that a normalized lowpass filter is of degree n. If this is a polynomial filter, $H(S)$ has no finite poles, but it has n attenuation poles at $S = \infty$. If a normalized lowpass-to-bandpass transformation is performed, where are the attenuation poles of the bandpass filter?

6.7 Example 3.1 described an eighth-degree Tschebycheff lowpass filter that met the requirements

$$A_{max} = 0.1 \text{ dB} \qquad A_{min} = 30 \text{ dB} \qquad f_B = 2 \qquad f_H = 2.6$$

This was an even-order filter; thus $A(F = 0) = A_{max}$.

a. Find a frequency transformation such that $A(f = 0) = 0$ and $A(f = 2) = 0.1$ dB.

b. Does this meet the requirements? That is, find the loss at $f = 2.6$, and determine whether or not it is greater than 30 dB.

Hint: $T_8(x)$ has zeros at $x = 0.195, 0.556, 0.831,$ and 0.981.

CHAPTER SEVEN

The Transformed Variable

7.1 INTRODUCTION

In Chap. 6 we investigated a nonreactance transformation that can be written as

$$S = \left(\frac{as^2 + b}{s^2 + c}\right)^{1/2} \tag{7.1}$$

A special case of this transformation is discussed in this chapter. The transformation will be used so much in the following chapters that we shall assign it a name: the *transformed variable*. It will be denoted by Z.

The transformed variable will be used in later chapters for two reasons. First, it will simplify the formulation of many equations, and second, it will give improved computational accuracy. If the transformed variable is not used, the zeros of $H(s)$ tend to cluster near the passband edge, and they cannot be determined accurately. However, in terms of the transformed variable the zeros of $H(s)$ are spread apart and can be found much more accurately.

7.2 THE TRANSFORMED VARIABLE

We want a form of the transformation

$$S^2 = \frac{as^2 + b}{s^2 + c}$$

that spreads the passband $\omega_A \leq \omega \leq \omega_B$ over the frequency range $0 \leq \Omega \leq \infty$. The form we shall use is

$$Z^2 = \frac{s^2 + \omega_B^2}{s^2 + \omega_A^2} \tag{7.2}$$

This frequency transformation has been used by a number of people to improve computational accuracy in filter design. For example, Orchard and Temes used it in Ref. 1. Szentermai[2] used a similar transformation, the main difference being a constant multiplier ω_B/ω_A. His form will not be discussed here, as it cannot be used for lowpass filters. For lowpass filters ω_A is zero; thus, Eq. (7.2) simplifies to

$$Z^2 = 1 + \frac{\omega_B^2}{s^2}$$

The preceding comments lead to the following definition:

Definition 7.1

The transformed variable Z is related to the frequency s by

$$Z = \left(\frac{s^2 + \omega_B^2}{s^2 + \omega_A^2}\right)^{1/2} \tag{7.3}$$

In Eq. (7.2) there is an ambiguity of sign. For example, as $S \to \infty$ we see that $Z \to \pm 1$. The notation in (7.3) is meant to imply that we use the positive root; that is, $\text{Re}(Z) \geq 0$. The transformation is demonstrated in Fig. 7.1, which is valid for lowpass or bandpass filters; for bandpass filters, simply let $\omega_A = 0$.

Figure 7.1 indicates that the real frequencies $s = \sigma$ are transformed to $Z = X$, where $1 \leq X \leq \omega_B/\omega_A$, so that the real s axis is compressed to a portion of the real Z axis. This compression implies that for real s the transformation actually moves the roots closer together and thus can result in more numerical inaccuracy. However, this is not a problem because in any practical filter the real roots are widely separated; it is the roots near the passband edges that are close together.

Figure 7.1 also indicates that the stopband frequencies ($s = j\omega$, where $\omega > \omega_B$ or $\omega < \omega_A$) are transformed to part of the real Z axis. In particular, it should be noted that the lower passband edge ($s = j\omega_A$) is transformed to infinity, and the upper passband edge is transformed to the origin.

Finally, Fig. 7.1 demonstrates that the passband frequencies ($s = j\omega$, where $\omega_A < \omega < \omega_B$) are transformed to the entire imaginary Z axis, which implies that the transformed frequencies are more widely separated.

Example 7.1

If a bandpass filter has passband edges $\omega_A = 3$ and $\omega_B = 4$, then the transformed variable is given by $Z^2 = (s^2 + 16)/(s^2 + 9)$.
    a. Find the transformed frequency corresponding to $s = 2$.
    b. Find the transformed frequencies corresponding to $s = 2j$ and $s = 5j$.
    c. Find the transformed frequencies corresponding to $s = 3.01j$ and $s = 3.02j$.

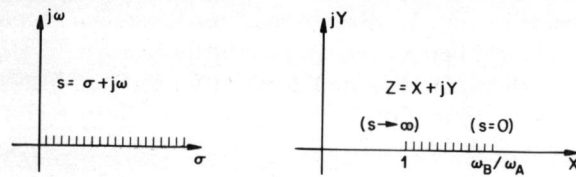

a) TRANSFORMATION OF RE(s)

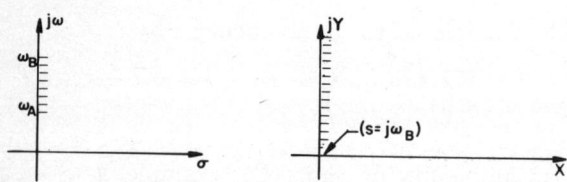

b) TRANSFORMATION OF PASSBAND

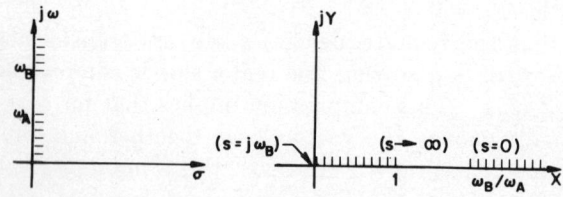

c) TRANSFORMATION OF STOPBANDS

Fig. 7.1 Transformation of different frequency regions s by the transformed variable Z.

Solution

Using $Z^2 = (s^2 + 16)/(s^2 + 9)$ yields the following:
a. $Z = 1.240$
b. $Z = 1.549, Z = 0.75$
c. $Z = 10.746j, Z = 7.559j$

In Example 7.1, the two passband frequencies $s = 3.01j$ and $s = 3.02j$ are transformed to $Z = 10.746j$ and $Z = 7.559j$. This illustrates that for frequencies near the passband edge there is a much greater separation in the Z plane than in the s plane. This is the main reason for using the transformed variable; it spreads out the passband frequencies and thus leads to improved numerical accuracy.

7.3 FUNCTIONS IN TERMS OF THE TRANSFORMED VARIABLE

The characteristic function is usually expressed as a ratio of polynomials:

$$K(s) = \frac{f(s)}{q(s)} \tag{7.4}$$

In this section we assume that $f(s)$ and $q(s)$ are of the form

$$f(s) = c_f \prod_{i=1}^{m/2} (s^2 + \omega_{fi}^2) \tag{7.5}$$

$$q(s) = s^{NZ} \prod_{i=1}^{N} (s^2 + \omega_i^2) \tag{7.6}$$

The function $q(s)$ determines where the characteristic function becomes infinite; thus, we see there are attenuation poles at $s = j\omega_i$. There are also NZ attenuation poles at the origin. The number of attenuation poles at infinity is

$$\text{NIN} = m - (\text{NZ} + 2N)$$

so that
$$m = \text{NZ} + \text{NIN} + 2N \tag{7.7}$$

Note from Eq. (7.5) that m must be an even number; thus, Eq. (7.7) implies that NZ + NIN must be even.

The function $f(s)$ determines where the characteristic function becomes zero; thus, there are attenuation zeros at $s = j\omega_{fi}$. We want to express $f(s)$ in terms of the transformed variable Z, where

$$Z^2 = \frac{s^2 + \omega_B^2}{s^2 + \omega_A^2} \tag{7.8}$$

Equation (7.8) implies that

$$s^2 = \frac{\omega_B{}^2 - Z^2\omega_A{}^2}{Z^2 - 1} \qquad (7.9)$$

Substituting into (7.5) yields

$$f(s) = c_f \prod_{i=1}^{m/2} \left(\frac{\omega_B{}^2 - Z^2\omega_A{}^2}{Z^2 - 1} + \omega_{fi}{}^2 \right)$$

This is a ratio of polynomials in Z; however, we would like the transformed function to be a polynomial. Thus, it is more convenient to use the functions

$$F(Z) \triangleq \frac{f(s)}{(s^2 + \omega_A{}^2)^{m/2}}$$

and

$$Q(Z) \triangleq \frac{q(s)}{(s^2 + \omega_A{}^2)^{m/2}} \qquad (7.10)$$

Theorems 7.1 and 7.2 give explicit relations for $F(Z)$ and $Q(Z)$. These relations demonstrate that $F(Z)$ and $Q(Z)$ are polynomials.

Theorem 7.1

If

$$f(s) = c_f \prod_{i=1}^{m/2} (s^2 + \omega_{fi}{}^2)$$

then

$$F(Z) \triangleq \frac{f(s)}{(s^2 + \omega_A{}^2)^{m/2}} = C_F \prod_{i=1}^{m/2} (Z^2 - Z_{fi}{}^2) \qquad (7.11)$$

where

$$C_F = c_f \prod_{i=1}^{m/2} \frac{\omega_{fi}{}^2 - \omega_A{}^2}{\omega_B{}^2 - \omega_A{}^2}$$

$$Z_{fi}{}^2 = \frac{\omega_{fi}{}^2 - \omega_B{}^2}{\omega_{fi}{}^2 - \omega_A{}^2} \qquad (7.12)$$

Equation (7.11) demonstrates why the transformed function $F(Z)$ was defined at it was: it resulted in $F(Z)$ being a polynomial. The constant multiplier C_F will be of little consequence in the material that follows—we will seldom use the relation given in Theorem 7.1 to evaluate it. However, the zeros Z_{fi} will be very important. Notice that Z_{fi} is simply the transformed variable Z evaluated at the frequency $s = j\omega_{fi}$, which is a zero of attenuation. Since the attenuation zeros are in the passband, this implies that $Z_{fi}{}^2$ is a negative number so that Z_{fi} is imaginary.

Proof

$$F(Z) = \frac{f(s)}{(s^2 + \omega_A^2)^{m/2}} = c_f \prod_{i=1}^{m/2} \frac{s^2 + \omega_{fi}^2}{s^2 + \omega_A^2}$$

$$= c_f \prod_{i=1}^{m/2} \frac{(\omega_B^2 - Z^2\omega_A^2)/(Z^2 - 1) + \omega_{fi}^2}{(\omega_B^2 - Z^2\omega_A^2)/(Z^2 - 1) + \omega_A^2}$$

$$= c_f \prod_{i=1}^{m/2} \frac{Z^2(\omega_{fi}^2 - \omega_A^2) + \omega_B^2 - \omega_{fi}^2}{\omega_B^2 - \omega_A^2}$$

$$= c_f \prod_{i=1}^{m/2} \left(\frac{\omega_{fi}^2 - \omega_A^2}{\omega_B^2 - \omega_A^2}\right)\left(Z^2 - \frac{\omega_{fi}^2 - \omega_B^2}{\omega_{fi}^2 - \omega_A^2}\right)$$

from which Theorem 7.1 follows.

We have just found an equation for $F(Z)$, which is a transformed version of the function $f(s)$; a similar equation can be given for $Q(Z)$, which is a transformed version of $q(s)$. Because $q(s)$ contains some zeros at the origin, the transformed function $Q(Z)$ differs in form from the transformed function $F(Z)$. Similarly, the attenuation poles at infinity make the forms of $F(Z)$ and $Q(Z)$ differ. This is illustrated in the next theorem, which is stated without proof.

Theorem 7.2

If
$$q(s) = s^{NZ} \prod_{i=1}^{N} (s^2 + \omega_i^2)$$

then

$$Q(Z) \triangleq \frac{q(s)}{(s^2 + \omega_A^2)^{m/2}}$$
$$= C_Q \left[1 - \left(\frac{\omega_A Z}{\omega_B}\right)^2\right]^{NZ/2} (Z^2 - 1)^{NIN/2} \prod_{i=1}^{N} (Z^2 - Z_i^2) \quad (7.13)$$

where
$$C_Q = \frac{\omega_B^{NZ}}{(\omega_B^2 - \omega_A^2)^{(NZ+NIN)/2}} \prod_{i=1}^{N} \frac{\omega_i^2 - \omega_A^2}{\omega_B^2 - \omega_A^2}$$

$$Z_i^2 = \frac{\omega_i^2 - \omega_B^2}{\omega_i^2 - \omega_A^2} \quad (7.14)$$

$$m = NZ + NIN + 2N$$

Equation (7.13) demonstrates that $Q(Z)$ can be expressed as a polynomial in Z. As was the case with $F(Z)$, the constant multiplier of $Q(Z)$ will be of little importance; however, the zeros Z_i will be very important. Because the attenuation poles ω_i are in the stopband, the transformed attenuation poles Z_i are real numbers.

7.4 $F(Z)$ AND $Q(Z)$ FOR LOWPASS FILTERS

In the previous section we defined $F(Z)$ and $Q(Z)$ as

$$F(Z) = \frac{f(s)}{(s^2 + \omega_A{}^2)^{m/2}} \qquad Q(Z) = \frac{q(s)}{(s^2 + \omega_A{}^2)^{m/2}}$$

where $m = NZ + NIN + 2N$. These functions were expressed in terms of the transformed variable Z by using the definition $Z^2 = (s^2 + \omega_B{}^2)/(s^2 + \omega_A{}^2)$. One advantage of this definition of the transformed variable is that it applies for lowpass filters as well as bandpass filters. For lowpass filters the lower passband edge ω_A is located at the origin; thus, by analogy to Definition 7.1, we have:

Definition 7.2

For lowpass filters, the transformed variable Z is related to the frequency s by

$$Z = \frac{(s^2 + \omega_B{}^2)^{1/2}}{s} \qquad (7.15)$$

Definition 7.2 is a special case of Definition 7.1; that is, the lower passband edge ω_A has been set equal to zero. Because the lowpass transformed variable is a limiting case of the bandpass transformed variable, we can easily find $F(Z)$ for lowpass filters. However, there are now two cases to consider, because the parameter m can be even or odd.

Theorem 7.3

a. m even: If

$$f(s) = c_f \prod_{i=1}^{m/2} (s^2 + \omega_{fi}{}^2)$$

then for a lowpass filter

$$F(Z) \triangleq \frac{f(s)}{s^m} = C_F \prod_{i=1}^{m/2} (Z^2 - Z_{fi}{}^2) \qquad (7.16a)$$

b. m odd: If

$$f(s) = c_f s \prod_{i=1}^{(m-1)/2} (s^2 + \omega_{fi}{}^2)$$

then for a lowpass filter

$$F(Z) \triangleq \frac{f(s)}{s^m} = C_F \prod_{i=1}^{(m-1)/2} (Z^2 - Z_{fi}^2) \qquad (7.16b)$$

where* $C_F = c_f \prod_{i=1}^{\text{IP}(m/2)} \left(\frac{\omega_{fi}}{\omega_B}\right)^2$

$Z_{fi}^2 = 1 - (\omega_B/\omega_{fi})^2$

This theorem will not be proved here because it is so similar to Theorem 7.1. Again, the important item to note is that $F(Z)$ is a polynomial. The zeros of this polynomial are the transformed values of $s = j\omega_{fi}$.

To find $Q(Z)$ for lowpass filters, we shall use the fact that there can be no attenuation poles in the passband, so that NZ must be zero. If we set NZ and ω_A equal to zero, then Theorem 7.2 becomes:

Theorem 7.4

If

$$q(s) = \prod_{i=1}^{N} (s^2 + \omega_i^2)$$

then for a lowpass filter

$$Q(Z) \triangleq \frac{q(s)}{s^m} = C_Q (Z^2 - 1)^{\text{NIN}/2} \prod_{i=1}^{N} (Z^2 - Z_i^2) \qquad (7.17)$$

where $C_Q = (\omega_B)^{-\text{NIN}} \prod_{i=1}^{N} \left(\frac{\omega_i}{\omega_B}\right)^2$

$Z_i^2 = 1 - (\omega_B/\omega_i)^2$

$m = \text{NIN} + 2N$

Theorems 7.1 to 7.4 give equations that can be used to calculate the transformed functions $F(Z)$ and $Q(Z)$ for bandpass or lowpass filters. The following example demonstrates their application to lowpass filters.

Example 7.2

Find $F(Z)$ and $Q(Z)$ for a normalized lowpass filter that has the characteristic function

$$K(s) = \frac{s(s^2 + 0.8)(s^2 + 0.95)}{(s^2 + 1.1)(s^2 + 1.5)}$$

* IP stands for "integer part of."

Solution

For a normalized lowpass filter, the passband edge is $\omega_B = 1$; thus, applying Theorem 7.3 yields

$$F(Z) = (0.8)(0.95)(Z^2 - Z_1^2)(Z^2 - Z_2^2)$$

where $Z_1^2 = 1 - 1/0.8$
$Z_2^2 = 1 - 1/0.95$

This yields

$$F(Z) = 0.76(Z^2 + 0.25)(Z^2 + 0.0526)$$

Similarly, applying Theorem 7.4 yields

$$Q(Z) = 1.65(Z^2 - 1)^{0.5}(Z^2 - 0.0909)(Z^2 - 0.3333)$$

7.5 THE INVERSE TRANSFORMATION

We have been writing the characteristic function as

$$K(s) = \frac{f(s)}{q(s)}$$

Instead of using $f(s)$ and $q(s)$ directly, we shall use the transformed functions $F(Z)$ and $Q(Z)$, as they will yield better numerical accuracy.

Once we have determined the characteristic function $K(s)$ for a specific filter, we shall want to find the input/output transfer function $H(s)$, which can be written as

$$H(s) = \frac{e(s)}{q(s)}$$

The function $q(s)$ is assumed to be known, but we must find $e(s)$. Since $H(s)H(-s) = 1 + K(s)K(-s)$, it follows that $e(s)$ is related to $f(s)$ and $q(s)$ via

$$e(s)e(-s) = f(s)f(-s) + q(s)q(-s) \qquad (7.18)$$

Dividing both sides of this equation by $(s^2 + \omega_A^2)^m$ yields

$$\frac{e(s)e(-s)}{(s^2 + \omega_A^2)^m} = \frac{f(s)f(-s)}{(s^2 + \omega_A^2)^m} + \frac{q(s)q(-s)}{(s^2 + \omega_A^2)^m} \qquad (7.19)$$

The functions $f(s)$ and $q(s)$ will be even or odd functions of s. For the sake of discussion, assume they are both even; then (7.19) becomes

$$\frac{e(s)e(-s)}{(s^2 + \omega_A^2)^m} = \frac{f^2(s)}{(s^2 + \omega_A^2)^m} + \frac{q^2(s)}{(s^2 + \omega_A^2)^m}$$
$$= F^2(Z) + Q^2(Z) \qquad (7.20)$$

The right-hand side of this equation is a polynomial in Z. Given this polynomial, we want to know how to find $e(s)$. Also, we want $e(s)$ to have only left half-plane roots. This section discusses how to find this inverse transformation.

Definition 7.3

The transformed function $E(Z)$ is defined as

$$E(Z) = \frac{e(s)}{(s^2 + \omega_A^2)^{m/2}} \tag{7.21a}$$

Similarly the transformed function $E^*(Z)$ is defined as

$$E^*(Z) = \frac{e(-s)}{(s^2 + \omega_A^2)^{m/2}} \tag{7.21b}$$

It follows from these definitions that

$$E(Z)E^*(Z) = \frac{e(s)e(-s)}{(s^2 + \omega_A^2)^m} \tag{7.22}$$

By assumption, this polynomial in Z is known; thus, the problem can be stated as: given $E(Z)E^*(Z)$, find the inverse transform $e(s)e(-s)$, and assign the left half-plane roots to $e(s)$.

It follows from (7.20) that $E(Z)E^*(Z)$ is an even function of Z, and its degree is the same as the degree of $F^2(Z)$. From Theorem 7.1, the degree of $F^2(Z)$ is $2m$; thus we can write $E(Z)E^*(Z)$ as

$$E(Z)E^*(Z) = e_0 + e_2 Z^2 + e_4 Z^4 + \cdots + e_{2m} Z^{2m} \tag{7.23}$$

This equation can be rewritten in one of two ways depending on whether m is even or odd.

m even:
$$E(Z)E^*(Z) = C \prod_{i=1}^{m/2} (Z^4 + p_i Z^2 + q_i) \tag{7.24a}$$

m odd:
$$E(Z)E^*(Z) = C(Z^2 - a) \prod_{i=1}^{(m-1)/2} (Z^4 + p_i Z^2 + q_i) \tag{7.24b}$$

We shall find the inverse transformation of Eq. (7.24b); the inverse transformation of Eq. (7.24a) will be obvious from this.

It follows from (7.24b) that

$$E(Z)E^*(Z) = C\left(\frac{s^2 + \omega_B^2}{s^2 + \omega_A^2} - a\right) \prod_{i=1}^{(m-1)/2} \left[\left(\frac{s^2 + \omega_B^2}{s^2 + \omega_A^2}\right)^2 + p_i \frac{s^2 + \omega_B^2}{s^2 + \omega_A^2} + q_i\right]$$

$$= \frac{e(s)e(-s)}{(s^2 + \omega_A^2)^m}$$

Thus we can express $e(s)e(-s)$ as

$$e(s)e(-s) = C[(s^2 + \omega_B^2) - a(s^2 + \omega_A^2)] \times \prod_{i=1}^{(m-1)/2} [(s^2 + \omega_B^2)^2 + p_i(s^2 + \omega_B^2)(s^2 + \omega_A^2) + q_i(s^2 + \omega_A^2)^2]$$

$$= -C[s^2(a-1) - (\omega_B^2 - a\omega_A^2)] \times \prod_{i=1}^{(m-1)/2} \left\{ (1 + p_i + q_i)s^4 + 2\left[\left(\frac{p_i}{2} + q_i\right)\omega_A^2 + \left(1 + \frac{p_i}{2}\right)\omega_B^2\right]s^2 + (\omega_B^4 + p_i\omega_A^2\omega_B^2 + q_i\omega_A^4) \right\}$$

This is summarized in the following theorem.

Theorem 7.5

If
$$E(Z)E^*(Z) = C(Z^2 - a) \prod_{i=1}^{(m-1)/2} (Z^4 + p_i Z^2 + q_i) \quad (7.24b)$$

then
$$e(s)e(-s) = -K^2(s^2 - \sigma_0^2) \prod_{i=1}^{(m-1)/2} (s^4 + 2b_i s^2 + a_i^2) \quad (7.25)$$

where
$$\sigma_0^2 = \frac{\omega_B^2 - a\omega_A^2}{a-1}$$

$$b_i = \frac{(p_i/2 + q_i)\omega_A^2 + (1 + p_i/2)\omega_B^2}{1 + p_i + q_i} \quad (7.26)$$

$$a_i^2 = \frac{\omega_B^4 + p_i\omega_A^2\omega_B^2 + q_i\omega_A^4}{1 + p_i + q_i}$$

In Theorem 7.5 the constant K is an unimportant constant multiplier. It could be given in terms of the other parameters, but that will not be necessary for the applications in later chapters. The important fact about Theorem 7.5 is that it allows one to find the roots of $e(s)$ because they must be in the left half-plane. That is, it follows from (7.25) that

$$e(s) = K(s + \sigma_0) \prod_{i=1}^{(m-1)/2} \left[s^2 + 2\left(\frac{a_i - b_i}{2}\right)^{1/2} s + a_i \right] \quad (7.27)$$

This equation is valid if m is odd. For m even,

$$e(s) = K \prod_{i=1}^{m/2} \left[s^2 + 2\left(\frac{a_i - b_i}{2}\right)^{1/2} s + a_i \right] \quad (7.28)$$

Example 7.3

A normalized lowpass filter has the characteristic function $K(s) = (s^2 + 0.9)/(s^2 + 1.1)$.

a. Find $F(Z)$ and $Q(Z)$.
b. Find $E(Z)E^*(Z)$.
c. Except for the constant multiplier, find $e(s)$.

Solution

a. By applying Theorems 7.3 and 7.4, we can find

$$F(Z) = 0.9(Z^2 + 0.1111) \qquad Q(Z) = 1.1(Z^2 - 0.0909)$$

b. Since $f(s)$ and $q(s)$ are even functions, it follows that

$$E(Z)E^*(Z) = F^2(Z) + Q^2(Z) = 2.02(Z^4 - 0.0198Z^2 + 0.0099)$$

c. Comparing the answer to part b with Eq. (7.24b), we see that

$$p_1 = -0.0198 \qquad q_1 = 0.0099$$

Thus setting $\omega_A = 0$ and $\omega_B = 1$ in (7.26) yields

$$b_1 = 1 \qquad a_1^2 = 1.01$$

Equation (7.28) then gives the following answer:

$$e(s) = K(s^2 + 0.0999s + 1.005)$$

7.6 CONCLUSIONS

The transformed variable is related to the frequency s by $Z^2 = (s^2 + \omega_B^2)/(s^2 + \omega_A^2)$. Theorems 7.1 to 7.4 indicate how certain polynomials may be written in terms of the transformed variable. The polynomial $Q(Z)$ is determined by the location of the attenuation poles. We shall see in the next chapter that if we know the attenuation poles for an equiripple filter, then we can determine the loss of the filter without ever solving for the zeros.

As implied in the preceding paragraph, if a filter is equiripple, then the attenuation poles uniquely determine the attenuation zeros. That is, given the passband ripple $A_{\max}$, the passband edges ω_A and ω_B, and the polynomial $Q(Z)$, we can find an analytic expression for $F(Z)$. The expression for the attenuation zeros is given in Chap. 9 and is of a simple form because of the use of the transformed variable.

Not only does the transformed variable simplify equations, it can also improve numerical accuracy. This will be very useful when we solve for the input/output transfer function $H(s)$. The function $H(s)$ will be determined from the characteristic function $K(s)$. Given $K(s)$, one must form a related polynomial and find its roots in order to solve for $H(s)$. Because the roots tend to cluster near the passband edges, numerical inaccuracies can result in the root-finding procedures. In order to avoid inaccuracies, some filter designers have used more than 30 digits in their calculations. The transformed frequency offers a way around this problem.

The transformed variable spreads the passband frequencies and thus leads to improved numerical accuracy. This means that the input/output transfer function $H(s)$ can be found accurately, and that is the purpose of this book: to present methods that accurately find a transfer function that meets certain specified requirements. After one has determined the transfer function, there are many alternative procedures for synthesizing it. If one decides to use a passive synthesis procedure, it may be best to calculate the elements in terms of the transformed variable, since use of the transformed variable allows one to determine the element values more accurately.[3]

REFERENCES

1. H. J. Orchard and G. C. Temes, "Filter Design using Transformed Variables," *IEEE Trans. on Circuit Theory*, December 1968, pp. 385–408.
2. G. Szentirmai, "A Filter Synthesis Program," in *System Analysis by Digital Computer*, edited by F. F. Kuo and J. F. Kaiser, John Wiley & Sons, Inc., New York, 1966.
3. J. A. C. Bingham, "A New Method of Solving the Accuracy Problem in Filter Design," *IEEE Trans. on Circuit Theory*, September 1964, pp. 327–341.

PROBLEMS

7.1 If a bandpass filter has passband edges $\omega_A = 5$ and $\omega_B = 6$, what is the transformed variable that corresponds to $s = -0.1 + 5.1j$?

7.2 The passband edges of a bandpass filter are $\omega_A = 4$ and $\omega_B = 5$. If the characteristic function is given by

$$K(s) = \frac{f(s)}{q(s)} = \frac{2(s^2 + 20)(s^2 + 22)}{s(s^2 + 27)}$$

find $F(z)$ and $Q(z)$.

7.3 Show that, as stated in Theorem 7.2, the constant C_Q is given by

$$C_Q = \omega_B^{NZ}(\omega_B^2 - \omega_A^2)^{-(NZ+NIN)/2} \prod_{i=1}^{N} \frac{\omega_i^2 - \omega_A^2}{\omega_B^2 - \omega_A^2}$$

Hint: Use the fact that

$$Q(z = 0) = \frac{q(-j\omega_B)}{(\omega_A^2 - \omega_B^2)^{m/2}}$$

7.4 This problem, which is similar to one in a paper by J. A. C. Bingham,* illustrates how the transformed variable can improve computational accuracy.

* J. A. C. Bingham, "A New Method of Solving the Accuracy Problem in Filter Design," *IEEE Trans. on Circuit Theory*, September 1964, pp. 327–341.

Assume that a normalized lowpass filter has

$$q(s) = (s^2 + 1.010)(s^2 + 1.020) \tag{1}$$

a. Find $q(s = j)$.

b. The polynomial in (1) can be put into the form

$$q(s) = s^4 + as^2 + b \tag{2}$$

Assume that we find the coefficients a and b by using a computer that is accurate to only four significant figures. What are a and b?

c. Find $q(s = j)$ by evaluating (2). Compare this answer with the one found in part a.

d. Let the transformed function $Q(z)$ be defined as

$$Q(Z) = \frac{q(s)}{s^4}\bigg|_{s^2 = 1/(z^2-1)}$$

From this it follows that $q(s = j) = Q(z = 0)$. Find $q(s = j)$ by calculating $Q(z = 0)$. Do the calculations to four significant figures and compare with the previous answers.

CHAPTER EIGHT

Attenuation Poles for Equiripple Passband Filters

8.1 INTRODUCTION

Elliptic filters, as illustrated in Fig. 8.1a, have an equiripple passband and an equiripple stopband. Often, however, the filter requirements will not be given so that an equiripple stopband is the best solution. An example of this is in Fig. 8.1b.

The characteristic in Fig. 8.1b has two loss peaks (poles of attenuation) in the lower stopband and three peaks (not counting the one at infinity) in the upper stopband. The peaks were chosen so that the stopband is not equiripple; however, it is still possible to have an equiripple passband. That is, for a given set of attenuation poles, it is possible to choose the natural frequencies [zeros of $H(s)$] so that the passband is equiripple. We shall not prove this statement in this chapter; the proof is deferred until Chap. 9.

The material in this chapter is presented in terms of the transformed variable Z introduced in the previous chapter, not because of its accuracy, but because the mathematical expressions are much simpler in terms of Z than in terms of s.

8.2 POLES AND ZEROS

While this chapter does not prove that the natural frequencies can be found so as to yield an equiripple passband, a little insight as to why it is possible can be gained from studying Fig. 8.2, which shows the poles and zeros of $H(s)$ for the bandpass filter of Fig. 8.1b. The input/output transfer function is of the form

$$H(s) = C_H \frac{s^{10} + a_9 s^9 + \cdots + a_0}{s(s^2 + \omega_1^2)(s^2 + \omega_2^2)(s^2 + \omega_3^2)(s^2 + \omega_4^2)} \quad (8.1)$$

The poles on the imaginary axis provide the necessary attenuation in the stopband. That is, near a pole ω_i the loss will be very large. However, in the passband ($\omega_A \leq \omega \leq \omega_B$) the loss due to the poles is offset by the influence of the zeros. By choosing these zeros correctly, one can make the passband equiripple.

The transfer function in Eq. (8.1) is a special case of the more general equation

$$H(s) = C_H \frac{s^m + a_{m-1} s^{m-1} + \cdots + a_0}{s^{NZ} \prod_{i=1}^{N} (s^2 + \omega_i^2)} \quad (8.2)$$

This transfer function has attenuation poles at ω_i ($i = 1, 2, \ldots, N$).

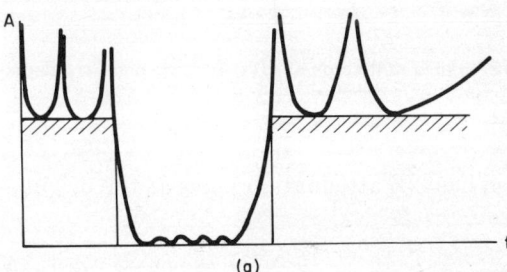

(a)

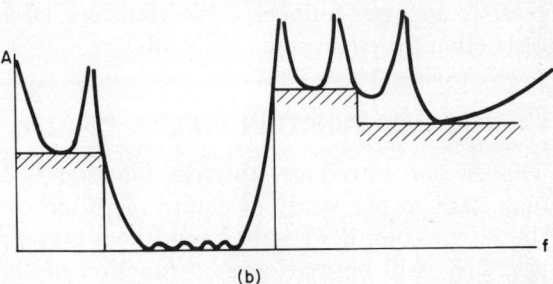

(b)

Fig. 8.1 An elliptic filter has an equiripple passband (a); however, a filter with an arbitrary stopband (b) can also have an equiripple passband.

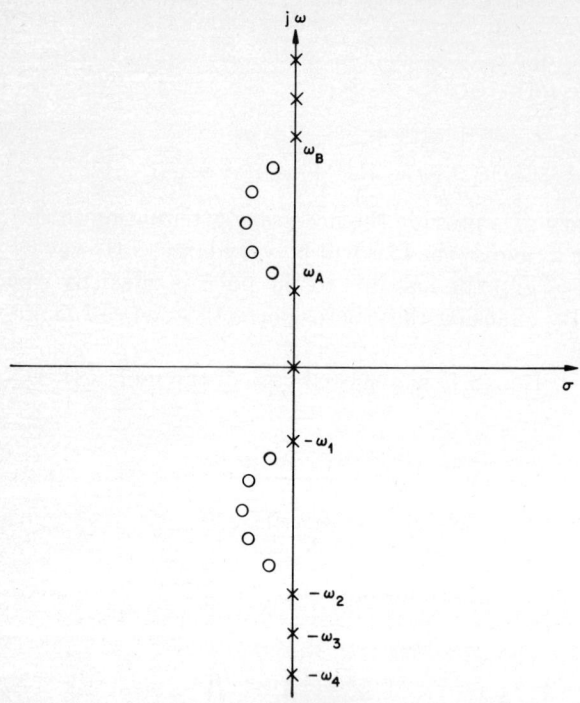

Fig. 8.2 Poles and zeros of $H(s)$ for a typical equiripple bandpass filter.

It also has NZ attenuation poles at the origin and NIN poles at infinity, where

$$\text{NIN} = m - (\text{NZ} + 2N) \tag{8.3}$$

In this chapter we assume that, for a bandpass filter, the sum NZ + NIN is an even number. We shall see later that this is a necessary restriction for equiripple passbands.

8.3 THE LOSS FUNCTION $L(Z)$

This section introduces the loss function $L(Z)$. This function is very important in our study of equiripple filters, as it allows us to calculate the attenuation if we only know the attenuation poles. The loss function $L(Z)$ will be written as a function of the transformed variable Z, which was defined as $Z^2 = (s^2 + \omega_B^2)/(s^2 + \omega_A^2)$. Thus, if there is an attenuation pole at $s = j\omega_i$, then it follows that $Z_i^2 = (\omega_i^2 - \omega_B^2)/(\omega_i^2 - \omega_A^2)$. This notation is used in the following definition.

Definition 8.1

If an equiripple filter has the passband $\omega_A \leq \omega \leq \omega_B$ and the following attenuation poles

1. NZ at the origin
2. N finite poles at ω_i ($i = 1, 2, \ldots, N$)
3. NIN at infinity

then the loss function is defined as

$$L(Z) = \left(\frac{Z + \omega_B/\omega_A}{Z - \omega_B/\omega_A}\right)^{NZ/2} \left(\frac{Z + 1}{Z - 1}\right)^{NIN/2} \prod_{i=1}^{N} \left(\frac{Z + Z_i}{Z - Z_i}\right) \quad (8.4)$$

where $Z_i^2 = (\omega_i^2 - \omega_B^2)/(\omega_i^2 - \omega_A^2)$.

In this definition, the terms in parenthesis represent the contributions to the loss function of the poles at the origin, infinity, and ω_i. In fact, if we notice that s equal to zero corresponds to Z equal to ω_B/ω_A, and s equal to infinity corresponds to Z equal to unity, it is easy to see that all three terms in parenthesis are of the same form. That is, they are written in the form $(Z + Z_i)/(Z - Z_i)$, where Z_i represents the pole of interest. Z_i is equal to ω_B/ω_A for a pole at the origin, and it is equal to unity for a pole at infinity.

Definition 8.1 was for bandpass filters; a simpler definition can be given for lowpass filters. The lowpass definition utilizes the fact that a lowpass filter can have no attenuation poles at the origin; that is, NZ = 0. Thus, for a lowpass filter, Definition 8.1 becomes

Definition 8.2

If an equiripple filter has the passband $0 \leq \omega \leq \omega_B$ and the following attenuation poles

1. N finite poles at ω_i ($i = 1, 2, \ldots, N$)
2. NIN at infinity

then the loss function is defined as

$$L(Z) = \left(\frac{Z + 1}{Z - 1}\right)^{NIN/2} \prod_{i=1}^{N} \left(\frac{Z + Z_i}{Z - Z_i}\right) \quad (8.5)$$

where $Z_i^2 = 1 - (\omega_B/\omega_i)^2$.

We shall establish, in the next chapter, that the attenuation $A(\omega)$ can be expressed in terms of the loss function. This is an important result because it implies that the attenuation poles uniquely determine the loss. The poles uniquely determine the loss because we are requiring the passband to be equiripple—we shall see in the next chapter that this uniquely specifies the natural frequencies in terms of the attenuation poles. The following theorem, which relates the stopband loss to the loss function $L(Z)$, is proved in the next chapter.

Theorem 8.1

In the stopband, the attenuation is given by

$$A(\omega) = 10 \log \left[1 + \frac{\epsilon^2}{4} \left(|L| + \frac{1}{|L|} \right)^2 \right] \tag{8.6}$$

where $|L|$ is evaluated at $Z = \sqrt{(\omega^2 - \omega_B{}^2)/(\omega^2 - \omega_A{}^2)}$.

Some insight into this relation can be gained from a result to be proved later: The loss function $L(Z)$ has a maximum of unity in the passband. In fact, $L(Z)$ is unity at the passband edges, so that (8.6) yields

$$A_{\max} = 10 \log (1 + \epsilon^2)$$

This is our usual expression for $A_{\max}$. The constant $\epsilon^2/4$ in (8.6) was chosen so that $A_{\max}$ has the right form.

8.4 THE TEMPLATE METHOD[1]

We have just mentioned that if we are given the poles of attenuation, we can determine the stopband performance (if we assume an equiripple passband). In an actual design problem we would not be given the attenuation poles—it would be our task to find them. These poles can be found by an iterative procedure:

1. Guess an initial set of poles.
2. Find the stopband loss.
3. Guess a new set of poles.
4. Find the stopband loss.
5. Repeat steps 3 and 4 as required.

In precomputer days, step 2 of the above iterative procedure was performed by using templates, physical shapes that could be moved to help expedite the calculation. This section discusses the template method, which is still widely referenced in the literature (although computers have decreased its use).

The template method can only be used to find the loss in the stopband; however, this is all we are interested in for the above iterative procedure.

In the stopband, the loss function $L(Z)$ is large, so that Theorem 8.1 yields

$$A(\omega) \approx 10 \log \left(1 + \frac{\epsilon^2}{4} |L|^2\right) \tag{8.7}$$

$$\approx 10 \log \frac{\epsilon^2}{4} + 10 \log |L|^2$$

We choose to write this as

$$A(\omega) \approx \text{IA}(\omega) - C$$

where
$$C = 10 \log \frac{4}{10^{0.1 A_{\max}} - 1} \tag{8.8}$$

$$\text{IA}(\omega) = 10 \log |L|^2 \tag{8.9}$$

$\text{IA}(\omega)$ is called the *image attenuation* and is very important in image parameter theory. However, we shall not discuss any of its properties, as they will not be needed here. The correction factor C relates the image attenuation IA to the actual attenuation A.

From (8.4) and (8.9), the image attenuation can be expressed as

$$\text{IA} = \frac{\text{NZ}}{2} 20 \log \left|\frac{Z + \omega_B/\omega_A}{Z - \omega_B/\omega_A}\right| + \frac{\text{NIN}}{2} 20 \log \left|\frac{Z + 1}{Z - 1}\right|$$
$$+ 20 \sum_{i=1}^{N} \log \left|\frac{Z/Z_i + 1}{Z/Z_i - 1}\right| \tag{8.10}$$

(For a lowpass, NZ = 0, and the first term can be ignored.) If one does not have access to a computer, Eq. (8.10) is easily evaluated with the help of the universal loss curve shown in Fig. 8.3.

Example 8.1
An equiripple bandpass filter is described by

$$f_A = 0.9 \quad f_B = 1/0.9 \quad \text{NZ} = 1$$
$$f_1 = 0.7 \quad f_2 = 1/0.7 \quad \text{NIN} = 1$$

Find the image attenuation at $f = 0.3$. What is the loss if $A_{\max} = 0.1$ dB?

Solution
Using $Z^2 = (f^2 - f_B^2)/(f^2 - f_A^2)$, the following values can be found:

$$f = 0.3 \quad f_1 = 0.7 \quad f_2 = \frac{1}{0.7}$$
$$Z = 1.2608 \quad Z_1 = 1.5254 \quad Z_2 = 0.8094$$

128   Approximation Methods for Electronic Filter Design

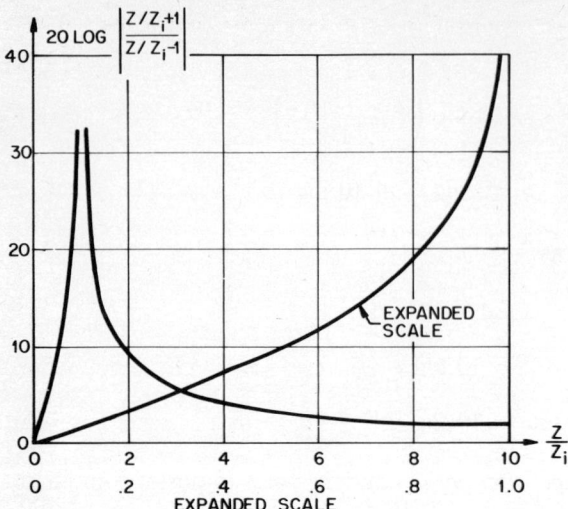

Fig. 8.3   Universal loss curve for the template method.

Using the universal loss curve, we find

$$20 \log \left| \frac{Z/Z_1 + 1}{Z/Z_1 - 1} \right| = 20 \log \left| \frac{0.827 + 1}{0.827 - 1} \right| = 20.4$$

Similarly, the universal loss curve can be used to find the other terms in (8.10) the result is

$$\text{IA} = \tfrac{1}{2}(40) + \tfrac{1}{2}(18.8) + 20.4 + 13.2 \approx 63 \text{ dB}$$

For $A_{\max} = 0.1$ dB, Eq. (8.8) yields $C = 22.35$ dB; thus, $A \approx 41$ dB.

In the above example, we calculated the image attenuation IA and then subtracted a correction factor C to yield the loss. The factor C was determined by the passband ripple as indicated in (8.8). If the passband ripple is changed, then C will change, but the image attenuation will remain the same; thus, changing the passband ripple will move the entire stopband loss up or down by a constant amount.* For example, suppose an elliptic filter is designed that has $A_{\max} = 0.1$ dB and $A_{\min} = 30$ dB. For $A_{\max} = 0.1$ dB, the correction factor is $C = 22.35$ dB. If the passband ripple is changed to $A_{\max} = 0.5$ dB, then the correction factor

---

* This statement is only true for regions in which the approximation of (8.7) is true. That is, it is only good for large amounts of stopband loss. However, as demonstrated in Problem 8.4, the error is quite small even for stopband losses of the order of 10 dB.

becomes $C = 15.16$ dB. This implies that the minimum stopband loss would increase to

$$A_{\min} = 30 + (22.35 - 15.16) = 37.19 \text{ dB}$$

8.5 INTUITION DEVELOPMENT

The template method is an iterative process: one guesses locations for the loss peaks, calculates the attenuation due to the peaks, and then moves the poles so as to improve the stopband performance. In order to move the poles to better locations, one should have an "intuitive feel" for the effect of a pole. This section should help to develop such an intuition.

The image attenuation introduced in the last section is a good tool for the development of our intuition. We shall use it to determine what happens when a single pole is placed at various frequencies. For a pole located at f_i, (8.10) becomes

$$\text{IA} = 20 \log \left| \frac{Z/Z_i + 1}{Z/Z_i - 1} \right| \quad \text{where} \quad Z^2 = \frac{f^2 - f_B{}^2}{f^2 - f_A{}^2}$$

Suppose, for example, we have $f_A = 0.9$, $f_B = 1/0.9$, and $f_i = 0.01$. The image attenuation for this pole is shown in Fig. 8.4. It approaches an infinite value as f approaches 0.01 and then decreases as f is increased toward the lower passband edge. Finally, as f is increased above the upper passband edge, the loss increases and approaches an asymptotic value.

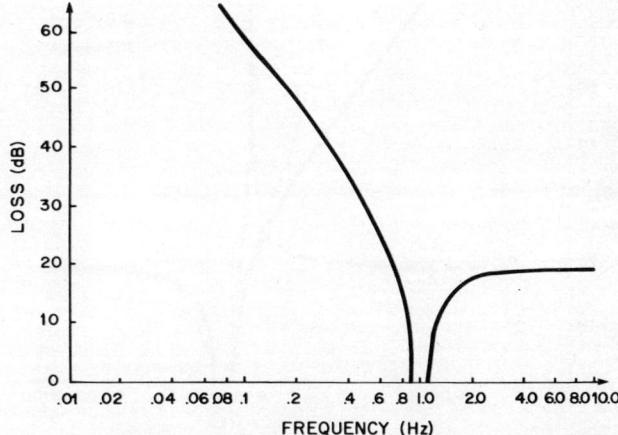

Fig. 8.4 Image attenuation for a bandpass filter that has an attenuation pole at $f = 0.01$.

If the peak frequency is varied, the loss shape is changed. This is demonstrated in Fig. 8.5. Curves 1, 2, and 3 give the image attenuation due to peaks at $f = 0.01, 0.3,$ and 0.7, respectively. Note that as the peak approaches the lower passband edge, the peak yields a sharper cutoff. However, we pay for this sharper cutoff in two ways:

1. The low-frequency loss is less.
2. The loss above the passband is less.

For a filter with more than one loss peak, the total image attenuation is found by adding the contributions due to the individual peaks. If only peaks below the passband are used, the upper stopband performance will be poor. Curves 4, 5, and 6 in Fig. 8.6 show attenuation due to peaks above the band at frequencies 1.5, 3, and infinity (NIN = 2), respectively. This figure illustrates the effect of varying loss peaks above the stopband.

The insight gained from Figs. 8.5 and 8.6 should be helpful in developing an intuitive feel for the effect of loss peaks. This could be of use if one wanted to use the template method of the previous section to find poles of attenuation for a specific filter design. It will also be of use in the next section.

8.6 A SIMPLE COMPUTER PROGRAM

In precomputer days the template method was important because a universal curve could be used to save computation time. However, the

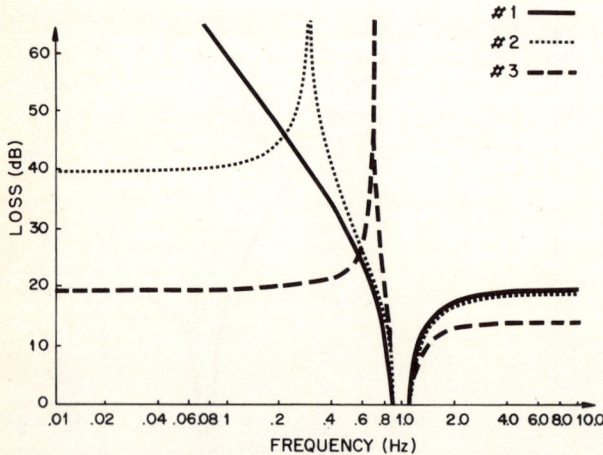

Fig. 8.5 Image attenuation due to attenuation poles at (1) $f = 0.01$, (2) $f = 0.3$, and (3) $f = 0.7$.

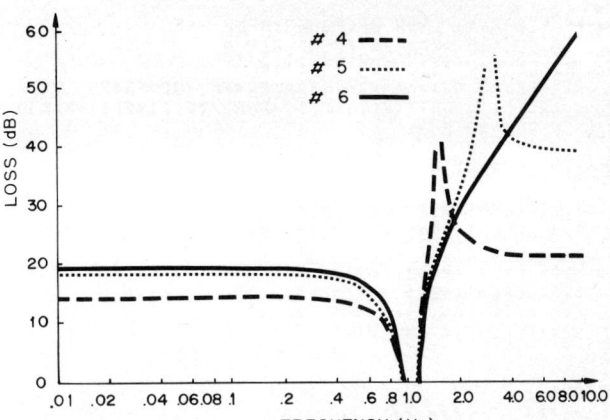

Fig. 8.6 Image attenuation due to attenuation poles at (4) $f = 1.5$, (5) $f = 3$, and (6) infinity.

template method is still tedious because of the necessity of transforming each frequency of interest to the Z plane. Also, the response at each frequency must be found by adding the contributions due to the individual poles.

The calculations are easily done on a computer. In fact, if one uses a computer, it takes little extra effort to use the exact expressions for loss instead of the approximate expressions of the template method. If one uses the exact expressions, the loss can be found for all frequencies outside the passband, and not just where the loss is large (e.g., in the template method, values below 10 dB are quite inaccurate). In fact, later chapters will extend this work so that the loss can also be found in the passband. However, this need not concern us here, as we are presently just interested in the stopband loss.

The stopband loss can easily be found by applying Theorem 8.1, which stated that

$$A(\omega) = 10 \log \left[1 + \frac{\epsilon^2}{4} \left(|\mathrm{L}| + \frac{1}{|\mathrm{L}|} \right)^2 \right]$$

where $\epsilon^2 = 10^{0.1 A_{\max}} - 1$

$$L = \left(\frac{Z + \omega_B/\omega_A}{Z - \omega_B/\omega_A} \right)^{NZ/2} \left(\frac{Z + 1}{Z - 1} \right)^{NIN/2} \prod_{i=1}^{N} \frac{Z + Z_i}{Z - Z_i}$$

$$Z^2 = \frac{\omega^2 - \omega_B{}^2}{\omega^2 - \omega_A{}^2}$$

A program based on these equations is shown in Fig. 8.7.

132 Approximation Methods for Electronic Filter Design

```
1; STØPBAND LØSS ØF EQUIRIPPLE BPF
1.2 DEMAND AMAX,FA,FB,NZ,NIN,N
1.3 DEMAND F[I] FØR I=1:1:N
1.4 EE=10^(.1*AMAX)-1,FAA=FA*FA,FBB=FB*FB
1.42 Z[I]=SQRT((F[I]*F[I]-FBB)/(F[I]*F[I]-FAA)) FØR I=1:1:N
1.5 DEMAND F
1.6 TYPE #
1.7 TØ PART 6

6; DETERMINATIØN ØF LØSS
6.4 Z=SQRT((F*F-FBB)/(F*F-FAA))
6.55 L=((Z+1)/ Z-1 )^(NIN/2)
6.56 L=L*((FA*Z/FB+1)/ FA*Z/FB-1 )^(NZ/2) IF NZ>0
6.57 L=L*(Z+Z[I])/(Z-Z[I]) FØR I=1:1:N
6.8 A=10*LØG(1+EE/4*(L+1/L)^2)
6.9 TYPE F,A IN FØRM 1


FØRM 1
#.###^^^   ###.##


>
```

Fig. 8.7 Program to calculate the stopband loss of equiripple bandpass filters.

Example 8.2

Use the program in Fig. 8.7 to find the loss of the following equiripple bandpass filter at $f = 0.3$:

$$A_{\max} = 0.1 \qquad F_A = 0.9 \qquad F_B = 1/0.9$$
$$\text{NZ} = 1 = \text{NIN} \qquad F_1 = 0.7 \qquad F_2 = 1/0.7$$

Solution

Note that this is the filter discussed in Example 8.1. Now if we want to find the loss at $f = 0.3$, all we need do is type DO PART 1 and answer the computer's questions, as illustrated in Fig. 8.8. To find the losses at other frequencies, simply type DO PART 6 FOR F = · · · · .

```
         AMAX=.1
           FA=.9
           FB=1/.9
           NZ=1
          NIN=1
            N=2
         F[1]=.7
         F[2]=1/.7
            F=.3
  3.000-01    40.49
>

  DØ PART 6 FØR F=.001,.1,.2,.4,.6
  1.000-03    90.36
  1.000-01    50.32
  2.000-01    44.18
  4.000-01    37.86
  6.000-01    36.12
>
```

Fig. 8.8 Sample use of the program in Fig. 8.7.

8.7 A COMPUTER TEMPLATE METHOD

The template method can be summarized as

1. Guess an initial set of poles.
2. Find the stopband loss.
3. Guess a new set of poles.
4. Find the stopband loss.
5. Repeat steps 3 and 4 as required.

The program described in the previous section makes step 2 simple. Thus, one need not convert the frequencies to the transformed variable before using a template; the program will do this. However, the simple program in Fig. 8.7 does not help one guess where the new poles should be. An involved computer program has been described by B. R. Smith and G. C. Temes[2] which places the poles to give an optimum stopband. The main features of this program are discussed in the rest of this chapter. This will allow us to write a fairly simple interactive computer program which is much shorter than the Smith-Temes program, for it assumes the user has a fair background in filter design and computer programming—as the reader of this book should now have. Thus, for example, the programmer is expected to supply the initial set of poles; the computer will not specify these.

The program described here does have one major advantage over previous programs. Unlike the original program, it computes the loss exactly. The original pole placer used the same type of approximations as were used in the template method. Because of these approximations, the original pole placers were accurate only when there was a substantial amount of loss in the stopband. However, the pole placer described in this chapter is accurate for any amount of loss (for example, 5 dB) in the stopband.

8.8 TERMINOLOGY USED FOR THE GENERAL PROBLEM

The filter requirements are assumed to be given in the form of Fig. 8.9. Assuming that the loss requirements are given as a series of step functions simplifies the problem.

The number of steps in the lower stopband is denoted as SA (SA = 4 in Fig. 8.9). These steps can be described by giving the frequency at the beginning of the step (FS_i) and the corresponding attenuation (A_i). By definition, the first-step frequency is $FS_0 = 0$. Also, the loss at the last-step frequency in the lower stopband is $A_{\text{SA}} = 0$.

Similarly, the number of steps in the upper stopband will be denoted as SB (SB = 3 in Fig. 8.9).

For loss requirements as in Fig. 8.9, we want to locate the attenuation poles so that the response is optimized. Before we can give our definition of optimum, we must introduce some more terms, with the help of Fig. 8.10.

The main concept introduced by Fig. 8.10 is that of an *arc*. There are three different types of arcs shown in the figure.

1. Arc$_1$ has two sections. One is the portion of the loss curve below the lowest pole at F_1. The other is the portion of the loss curve above the highest pole at F_4.
2. Internal arcs (e.g., arc$_2$ and arc$_5$) are bounded by two adjacent poles.
3. Stopband-edge arcs (arc$_3$ and arc$_4$) are bounded by a pole and a stopband edge.

We first consider a specific arc in Fig. 8.10: arc$_5$. We want to know how close this arc comes to the given requirement. We define the minimum difference for arc$_5$ as

$$D_{\min_5} = \min_{f_3 < f < f_4} [A(f) - A_R(f)]$$

where $A(f)$ describes arc$_5$ and $A_R(f)$ describes the given requirement. In Fig. 8.10, the minimum difference $D_{\min_5}$ occurs at a step frequency FS_7.

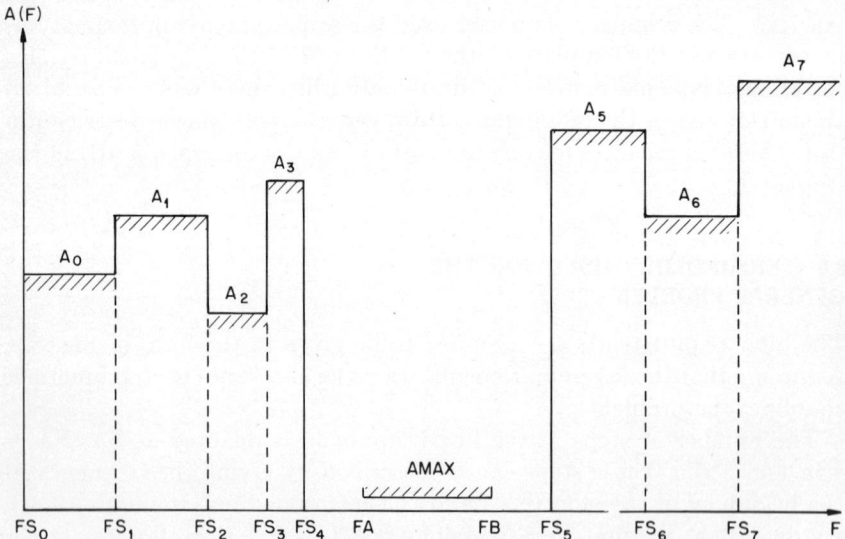

Fig. 8.9 Illustration of general attenuation requirements for bandpass filters.

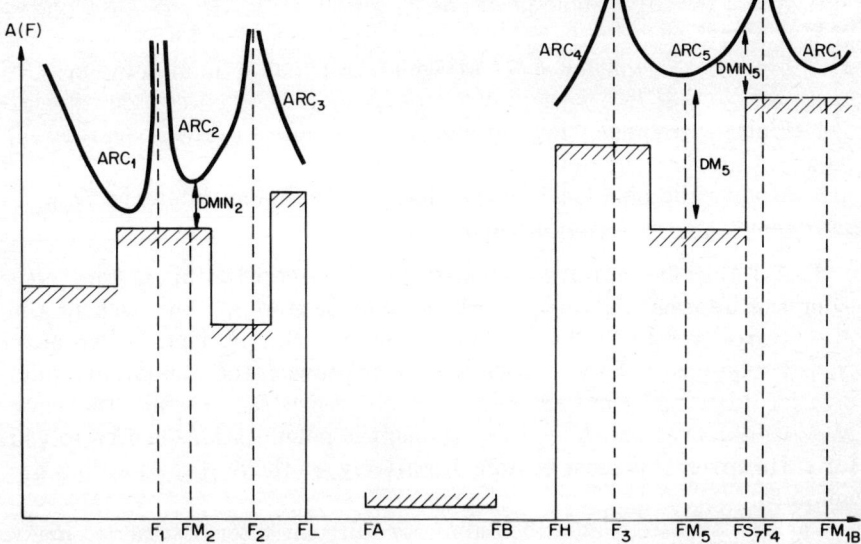

Fig. 8.10 Illustration of attenuation performance of a general bandpass filter.

Now consider a general arc: arc_i. The minimum difference $D_{\min_i}$ will occur at some frequency $F_{\min_i}$ where the attenuation is $A_{\min_i}$. The minimum frequency $F_{\min_i}$ might be a step frequency or it might be at the minimum of the arc (for example, $D_{\min_2}$ occurs at the minimum of arc_2).

We are now ready to describe what is meant by the optimum pole location.

Definition 8.3

The poles are optimally located if the minimum difference $D_{\min_i}$ is the same for each arc.

8.9 OUTLINE OF SOLUTION

The previous section described what is to be considered as the optimum pole location for a specific problem. This section outlines an iterative procedure that can be used to find the optimum solution. Later sections will investigate specific parts of the solution in greater detail. The procedure is as follows:

1. Guess NZ, the number of poles at the origin
   NIN, the number of poles at infinity
   NA, the number of poles in the lower stopband
   NB, the number of poles in the upper stopband

2. Guess the initial pole locations F_i for $i = 1, 2, \ldots, N$ (for $N = NA + NB$).

3. Find the frequency FM_i at which arc$_i$ has a minimum for $i = 1, 2, \ldots, N+1$.

4. Find the frequency $F_{\min_i}$ at which arc$_i$ is closest to the given requirement. Calculate the difference $D_{\min_i}$.

5. Modify the pole locations so that the minimum difference $D_{\min_i}$ is the same ($D_{\min_i} = D_{\min}$) for each arc.

Step 5 must be performed iteratively. To see why this is true, consider the information we have at the end of step 4. For each of the $N + 1$ arcs, we know the minimum distance $D_{\min_i}$. That is, we have $N + 1$ relations. We also have $N + 1$ unknowns: the N unknown final pole locations and the final minimum difference $D_{\min}$. Since there are $N + 1$ unknowns and $N + 1$ equations, the pole locations can be solved for. However, this must be done iteratively, as the $N + 1$ equations are nonlinear.

The first two steps described above require the filter designer to make some educated guesses. In step 1, if he overdesigns (chooses too many poles) the resulting approximation will give a large $D_{\min}$; that is, the design will meet the requirements with plenty of margin. On the other hand, if he chooses too few poles, the resulting approximation will give a negative $D_{\min}$; that is, the requirements will not be met. In either case, if the designer is unhappy with the results, he may simply take another guess at step 1.

The initial pole locations that are guessed at step 2 are not critical. However, the closer they are to optimum, the faster the final pole locations will be obtained.

8.10 DETERMINATION OF ARC MINIMUM

Step 3 of the previous section requires that we find the frequency at which an arc has a minimum. This section describes how the minimum can be found.

From (8.6) the stopband loss is given by

$$A = 10 \log \left[1 + \frac{\epsilon^2}{4} \left(|L| + \frac{1}{|L|} \right)^2 \right]$$

Thus, the loss has a local minimum when $|L| + 1/|L|$ has a local minimum. Problem 8.1 demonstrates that the minimums are zeros of the following function:

$$D(Z) = \frac{NZ}{2} \frac{f_B/f_A}{(f_B/f_A)^2 - Z^2} + \frac{NIN}{2} \frac{1}{1 - Z^2} + \sum_{i=1}^{N} \frac{Z_i}{Z_i^2 - Z^2} \quad (8.11)$$

This is a high-degree equation in Z^2; thus, the zeros must be found by some iterative process. The iterative solution of (8.11) requires initial guesses for the locations of the loss minima. There are three cases to be considered.

1. *Loss minimum between two finite (nonzero) loss peaks Z_j, Z_{j+1}.* The location of this minimum will be determined mainly by Z_j and Z_{j+1}. Thus, applying (8.11) gives

$$\frac{Z_j}{Z_j^2 - Z_m^2} + \frac{Z_{j+1}}{Z_{j+1}^2 - Z_m^2} \approx 0$$

This leads to

$$Z_m^2 \approx Z_j Z_{j+1} \qquad (8.12)$$

2. *Loss minimum between the origin ($\omega = 0$) and the peak nearest the origin (Z_1).* By analogy to case 1 the approximate location is

$$Z_m^2 \approx \frac{(NZ/2)Z_1 + f_B/f_A}{Z_1 + (NZ/2)(f_B/f_A)} \frac{f_B}{f_A} Z_1 \qquad (8.13)$$

3. *Loss minimum between the highest pole Z_N and infinity.* By analogy to case 1 the approximate location is

$$Z_m^2 \approx \frac{(NIN/2)Z_N + 1}{Z_N + NIN/2} Z_N \qquad (8.14)$$

One of the above equations can be used to obtain an initial guess for a solution to (8.11); this guess is denoted as Z_{m0}. The next guess is arbitrarily chosen as

$$Z_{m1}^2 = Z_{m0}^2 - 0.00001$$

Newton's iterative procedure can now be used to improve these guesses until the error is arbitrarily small. This procedure expresses the minimum at the ith iteration in terms of the two previous iterations as

$$Z_{mi}^2 = \frac{Z_{mi-2}^2 D(Z_{mi-1}) - Z_{mi-1}^2 D(Z_{mi-2})}{D(Z_{mi-1}) - D(Z_{mi-2})} \qquad (8.15)$$

This Newton iterative procedure is used in the pole-placer program (Fig. 8.11) for the determination of the minimum of an arc. The section of the program that does these calculations is Part 8. The initial guess for the loss minimum is $Z_0^2 \triangleq ZZ0$. This is found by Steps 8.1, 8.12, and 8.14, which are based on Eqs. (8.12), (8.13), and (8.14). The second guess for the loss minimum is $Z_1^2 = Z_0^2 - 0.00001$. Step 8.32 then uses Eq. (8.15) to find subsequent guesses.

```
1; BANDPASS PØLE PLACER
1.1 DEMAND SA,SB
1.12 DEMAND FS[I],A[I] FØR I=0:1:SA+SB
1.2 DEMAND AMAX,FA,FB,NZ,NIN,NA,NB
1.25 TYPE #,#,#
1.3 DEMAND F[I] FØR I=1:1:NA+NB
1.4 EE=10^(.1*AMAX)-1,FAA=FA*FA,FBB=FB*FB
1.41 N=NA+NB,S=SA+SB,FL=FS[SA],FH=FS[SA+1]
1.42 Z[I]=SQRT((F[I]*F[I]-FBB)/(F[I]*F[I]-FAA))  FØR I=1:1:N
1.43 DØ PART 6 FØR F=FL
1.44 AL=A
1.45 DØ PART 6 FØR F=FH
1.46 AH=A
1.6 TYPE #
1.7 TØ PART 2

2; CØNTRØL FØR LØWER STØPBAND
2.1 TØ PART 2.18 IF NA=0
2.12 DØ PART 6 FØR F=0 IF NZ=0
2.14 FM=0,AM=A,DM=A-A[0] IF NZ=0
2.15 ZA=FB/FA,ZB=Z[1]
2.16 DØ PART 8 IF NZ>0
2.18 FM=FL,AM=AL,DM=AL-A[SA-1] IF NA=
2.2 FX=0,FY=FL,I=1
2.22 FY=F[1] IF NA>0
2.24 DØ PART 30
2.3 TØ PART 3 IF NA=0
2.4 TØ PART 2.6 IF NA=1
2.5 I=I+1,ZA=Z[I-1],ZB=Z[I]
2.52 DØ PART 8
2.54 FX=F[I-1],FY=F[I]
2.56 DØ PART 30
2.58 TØ PART 2.5 IF I<NA
2.6 FM=FL,AM=AL,DM=AL-A[SA-1],FX=F[NA],FY=FL,I=NA+1
2.7 DØ PART 30
2.8 TØ PART 3

3; CØNTRØL FØR UPPER STØPBAND
3.1 FM=FH,AM=AH,DM=AH-A[SA+1]
3.2 FX=FH,FY=10*FS[S],I=NA+2
3.22 FY=F[NA+1] IF NB>0
3.24 DØ PART 30
3.3 TØ PART 3.8 IF NB=0
3.4 TØ PART 3.6 IF NB=1
3.5 ZA=Z[I-1],ZB=Z[I]
3.52 DØ PART 8
3.54 FX=F[I-1],FY=F[I]
3.56 DØ PART 30 FØR I=I+1
3.58 TØ PART 3.5 IF I<N+1
3.6 ZA=Z[N],ZB=1
3.64 DØ PART 8 IF NIN>0
3.66 DØ PART 6.55 FØR Z=.9999999 IF NIN=0
3.68 FM=10^20,AM=A,DM=A-A[S] IF NIN=0
3.69 FX=F[N],FY=10*FS[S],I=N+2
3.7 DØ PART 30
3.8 FMIN[I]=FMIN[I],AMIN[I]=AMIN[I],DMIN[I]=DMIN[I] IF DMIN[I]>DMIN[I]
3.82 TYPE #,#,#

6; DETERMINATIØN ØF LØSS
6.4 Z=SQRT((F*F-FBB)/(F*F-FAA))
6.55 L=((Z+1)/`Z-1´)^(NIN/2)
6.56 L=L*((FA*Z/FB+1)/`FA*Z/FB-1´)^(NZ/2) IF NZ>0
6.57 L=L*(Z+Z[I3])/(Z-Z[I3]) FØR I3=1:1:N
6.8 A=10*LØG(1+EE/4*(L+1/L)^2)
```

Fig. 8.11 Pole-placer program for equiripple bandpass filters. (*Continued on next page.*)

```
8; DETERMINATIØN ØF ARC MINIMUM   (FM=FREQ., AM=LØSS)
8.1  ZZØ=ZA*ZB,I1=0,ZZA=ZA*ZA,ZZB=ZB*ZB
8.12 ZZØ=(NZ/2*ZB+FB/FA)/(ZB+NZ/2*FB/FA)*ZZØ IF FB=FA*ZA
8.14 ZZØ=(NIN/2*ZA+1)/(ZA+NIN/2)*ZZØ IF ZB=1
8.16 DØ PART 8.9 FØR ZZ=ZZØ
8.2  DØ=D,ZZ1=ZZØ-.00001
8.3  DØ PART 8.9 FØR ZZ=ZZ1
8.32 D1=D,ZZ2=(ZZØ*D1-ZZ1*DØ)/(D1-DØ)
8.33 ZZ2=ZZA+.001 IF ZZ2<ZZA
8.34 ZZ2=ZZB-.001 IF ZZ2>ZZB
8.42 DØ=D1,ZZØ=ZZ1,ZZ1=ZZ2
8.5  TØ PART 8.3 IF  ZZ1-ZZØ >.0000001
8.6  F=SQRT((FBB-FAA*ZZ1)/(1-ZZ1))
8.7  DØ PART 6.55 FØR Z=SQRT(ZZ1)
8.82 AR=A[I1] IF FS[I1]<F
8.84 FM=F,AM=A,DM=A-AR,I1=I1+1
8.85 TØ PART 8.82 IF I1<S+1
8.9  D=NIN/2/(1-ZZ)+NZ/2*FB/FA/(FBB/FAA-ZZ)
8.91 D=D+Z[I2]/(Z[I2]*Z[I2]-ZZ) FØR I2=1:1:N

30; DETERMINATIØN ØF MINIMUM DIFFERENCE (DMIN)
30.1 J=1
30.2 TØ PART 30.6 IF FS[J]<FX ØR FS[J]>FY
30.3 AR=MAX(A[J-1],A[J])
30.4 DØ PART 6 FØR F=FS[J]
30.5 FM=F,AM=A,DM=A-AR IF DM>A-AR
30.6 J=J+1
30.7 TØ PART 30.2 IF J<S+1
30.8 FMIN[I]=FM,AMIN[I]=AM,DMIN[I]=DM
30.9 TYPE I,FM,AM,DM IN FØRM 1

49; DETERMINATIØN ØF NEW PØLES
49.1 N1=N+1,C1=LN(10)/40,I1=1
49.2 ZZ=(FMIN[I1]^2-FBB)/(FMIN[I1]^2-FAA),Z=SQRT(ZZ)
49.21 A[I1,I2]=Z/(ZZ-Z[I2]^2) FØR I2=1:1:N
49.215 I1=I1+1
49.22 TØ PART 49.2 IF I1<N1+1
49.3 A[I1,N1]=-C1,C[I1]=-C1*DMIN[I1] FØR I1=1:1:N1
49.4 TØ PART 50

50; CRØUT REDUCTIØN
50.1 R=1,J1=1,J2=N1
50.2 J=R+1
50.22 A[R,J]=A[R,J]-A[R,K]*A[K,J] FØR K=1:1:R-1
50.24 A[R,J]=A[R,J]/A[R,R],J=J+1
50.26 TØ PART 50.22 IF J<N1+1
50.27 C=R+1
50.28 A[I,C]=A[I,C]-A[I,K]*A[K,C] FØR K=1:1:C-1 FØR I=C:1:N1
50.29 R=R+1
50.3 TØ PART 50.2 IF R<N+1
50.4 C[J1]=C[J1]-A[J1,K]*C[K] FØR K=1:1:J1-1
50.42 C[J1]=C[J1]/A[J1,J1],J1=J1+1
50.43 TØ PART 50.4 IF J1<N1+1
50.5 X[J2]=C[J2]
50.52 X[J2]=X[J2]-A[J2,K]*X[K] FØR K=J2+1:1:N1
50.525 J2=J2-1
50.53 TØ PART 50.5 IF J2>0
50.55 TYPE #,#
50.6 DØ PART 50.8 FØR I=1:1:N
50.7 TØ PART 1.43
50.8 Z[I]=Z[I]+X[I],ZZ=Z[I]*Z[I]
50.82 F[I]=SQRT((FBB-FAA*ZZ)/(1-ZZ))
50.83 TYPE F[I]
```

Fig. 8.11 (Continued)   Pole-placer program for equiripple bandpass filters.

8.11 DETERMINATION OF THE MINIMUM DIFFERENCE $D_{\min_i}$

We have outlined a solution for finding the optimum pole location for a specific problem. Step 4 of this solution was, "Find the frequency $F_{\min_i}$ at which arc$_i$ is closest to the given requirement. Calculate the difference $D_{\min_i}$." This section describes how the program in Fig. 8.11 finds the minimum difference $D_{\min_i}$.

The minimum difference is found by first determining the relative minimum of the arc. Assume that this minimum occurs at FM_i and is a distance DM_i above the given requirement. We must next determine whether or not the requirement is closer to the arc at a breakpoint than at the relative minimum FM_i. For example, again consider arc$_5$ in Fig. 8.10. At the arc's minimum FM_5, the requirement is exceeded by DM_5 dB. However, the absolute minimum occurs at FS_7, where the requirement is exceeded by only $D_{\min_5}$ dB.

To understand the program in Fig. 8.11, first consider arc$_1$, which is in the lower stopband as shown in Fig. 8.10. If there are no poles in the lower stopband (NA = 0), then the relative minimum of arc$_1$ will be at the stopband edge ($f = f_L$). On the other hand, if there is a pole at $f = f_1$ but none at the origin, then the relative minimum of arc$_1$ will be at $f = 0$. The last case to consider is if arc$_1$ is bounded on the left by a pole at the origin and on the right by a pole at $f = f_1$. In this case, the relative minimum will be between the poles and can be found as described in Sec. 8.10. The absolute minimum can then be found as described above. The cases mentioned in this paragraph are analyzed in Steps 2.1 to 2.24 of the pole-placer program.

The remaining steps in Part 2 consider the other arcs in the lower stopband. The upper-stopband arcs are treated in Part 3.

8.12 DETERMINATION OF NEW POLES

Step 5 of the outlined solution was to modify the pole locations so that the minimum difference $D_{\min_i}$ is the same for each arc. In the discussion of the outlined solution, it was mentioned that the pole locations would have to be modified iteratively, as the equations are nonlinear. This section describes Part 49 of the poler-placer program, which can be used iteratively to determine the optimum pole locations.

Consider arc$_j$. By definition, the minimum difference $D_{\min_j}$ occurs at the frequency $F_{\min_j}$ where the loss is $A_{\min_j}$. This loss is a function of the pole locations Z_i ($i = 1, 2, \ldots, N$). If the poles are changed slightly, then the change in loss can be approximated as

$$\Delta A_{\min_j} \approx \sum_{i=1}^{N} \frac{\partial A_{\min_j}}{\partial Z_i} \Delta Z_i \qquad (8.16)$$

We want to determine the pole changes ΔZ_i so that the solution is optimum. To discover how to do this, consider a specific example: Assume that arc$_j$ has $D_{\min_j} = 5$ dB whereas the optimum solution has $D_{\min} = 7$ dB. The pole changes should then be chosen so that $\Delta A_{\min_j} = 7 - 5 = 2$ dB. Or, in general,

$$\Delta A_{\min_j} = D_{\min} - D_{\min_j} \qquad (8.17)$$

Combining (8.16) and (8.17) yields

$$\sum_{i=1}^{N} \frac{\partial A_{\min_j}}{\partial Z_i} \Delta Z_i - D_{\min} = -D_{\min_j} \qquad j = 1, 2, \ldots, N+1^* \qquad (8.18)$$

This is a set of $N + 1$ equations that can be solved for the $N + 1$ unknowns ΔZ_i and $D_{\min}$. To solve the set of equations we need $\partial A/\partial Z_i$. From Problem 8.2, this is

$$\frac{\partial A}{\partial Z_i} = \frac{(40/\ln 10)(|L|^2 - 1/|L|^2)}{4/\epsilon^2 + (|L| + 1/|L|)^2} Z \frac{1}{Z^2 - Z_i^2} \qquad (8.19)$$

where

$$Z^2(F_{\min}) = \frac{F_{\min_j}^2 - FB^2}{F_{\min_j}^2 - FA^2} \qquad (8.20)$$

Equation (8.19) is an exact relation; that is, no approximations were used in deriving it. However, if it were used in a program, it would add substantially to the computation time. Instead, it is usually sufficient to use the approximations†

$$|L|^2 \gg \frac{4}{\epsilon^2} \qquad \text{and} \qquad |L| \gg 1$$

Then (8.19) becomes

$$\frac{\partial A}{\partial Z_i} \approx \frac{40}{\ln 10} \frac{Z}{Z^2 - Z_i^2} \qquad (8.21)$$

Thus, (8.18) can be written as

$$\sum_{i=1}^{N} \frac{Z_{\min_j}}{Z_{\min_j}^2 - Z_i^2} \Delta Z_i - \frac{\ln 10}{40} D_{\min} = -\frac{\ln 10}{40} D_{\min_j} \qquad j = 1, 2, \ldots,$$

$$N + 1 \qquad (8.22)$$

* To see why there are $N + 1$ arcs, consider Fig. 8.10. Notice that the first and last arcs are both arc$_1$.

† The approximation is usually adequate because the ΔZ_i become smaller as the program proceeds, and minor inaccuracies in ΔZ_i will not stop convergence. If a case is encountered for which the program will not converge because of the approximation in (8.21), one could use the exact expression in (8.19). However, it would probably be faster to interrupt the program after it has typed a set of $F_{\min_j}$, $A_{\min}$, $D_{\min}$ values and guess new pole locations for the next iteration.

This equation can be rewritten as

$$\sum_{i=1}^{N+1} a_{ji} x_i = C_j \qquad j = 1, 2, \ldots, N+1 \tag{8.23}$$

where $\quad a_{ji} = \dfrac{Z_{\min_j}}{Z_{\min_j}^2 - Z_i^2}$

$\qquad\qquad x_i = \Delta Z_i \qquad i = 1, 2, \ldots, N$

$\qquad\qquad a_{j(N+1)} = -\dfrac{\ln 10}{40}$

$\qquad\qquad x_{N+1} = D_{\min}$

$\qquad\qquad C_j = -\dfrac{\ln 10}{40} D_{\min_j}$

The parameters a_{ji} and C_j are found in Part 49 of the pole-placer program. The program then solves the set of equations (8.23) by using the Crout reduction described in Ref. 3. The Crout reduction is fast and requires little storage.

8.13 SUMMARY OF POLE-PLACER PROGRAM

The pole-placer program in Fig. 8.11 can be started by typing DO PART 1. This will cause the program to demand the input data. Once the program has the initial pole locations, it finds the relative minimum of each of the arcs by doing Part 8.

After the relative minimum of arc_i has been determined, the breakpoints are checked to determine whether or not it is an absolute minimum. The program then prints (for each arc)

1. FMIN[I], the frequency at which the minimum occurs
2. AMIN, the attenuation at FMIN[I]
3. DMIN, the amount by which the requirement is exceeded at FMIN[I]

By examining $D_{\min}$ for each arc, the filter designer can decide whether or not he wants the pole locations moved so as to improve the loss performance. If he wants to change the pole locations, he can enter the new values and then type DO PART 1.4. The program will then calculate $D_{\min}$ for each of the new arcs.

Usually, the filter designer will want the program to determine the new pole locations. This is done by typing DO PART 49. The program then finds the new pole locations. A word of caution is in order here: If the initial poles were poorly chosen, these new poles may be at very different locations. They should be checked to make sure that each of the following is still true.

1. There are NA poles in the lower stopband.
2. There are NB poles in the upper stopband.
3. $F_i < F_{i+1}$ (that is, the poles are still ordered).

The most likely violation is that a pole has moved out of the stopband. For example, F[NA + 1] (the first pole in the upper stopband) may have been shifted below FH (the upper stopband edge). If this is the case, the program should be interrupted and F[NA + 1] set equal to a value slightly greater than FH. Then one can continue by typing DO PART 1.4.

Once the initial poles are properly chosen, the program can perturb the poles and find a better solution. The new pole locations can again be improved by typing DO PART 49, as is demonstrated in the next section. The pole locations can usually be improved until the optimum ones are obtained. However, if one stopband was allotted many more poles than the other (for example, NA ≫ NB), then it is possible that no pole locations will be found such that $D_{\min_i}$ is the same for each arc. Instead, the lower stopband might have $D_{\min_i}$ positive while the upper stopband has $D_{\min_i}$ negative. This of course indicates a poor initial allocation of poles to the upper and lower stopbands.

8.14 EXAMPLES FOR THE POLE-PLACER PROGRAM

This section contains some examples which illustrate the use of the program. As mentioned previously, the program is not meant to be used blindly by an inexperienced designer. Instead, some familiarity with the theory as presented in this chapter is required.

Example 8.3

This example finds the poles of attenuation for a geometrically symmetrical elliptical filter. The requirements, given below, are the same as for the geometrically symmetrical bandpass filter in Example 6.4; thus, the final pole locations should be the same.

Requirements:
   Passband edges: $f_A = 1.08320512$ $f_B = 1.5$
   Stopband edges: $f_L = 1.0155048$ $f_H = 1.6$
   Passband loss: $A_{\max} = 0.25$
   Stopband loss: $A_{\min} = 35$

These parameters are entered into the program as indicated in Fig. 8.12.
   The initial pole locations are $F_1 = 0.9, F_2 = 1, F_3 = 1.61, F_4 = 2$. For these pole locations, the minimums are found as indicated in Fig. 8.12. For example,

```
DØ PART 1
      SA=1
      SB=1
   FS[0]=0
    A[0]=35
   FS[1]=1.0155048
    A[1]=0
   FS[2]=1.6
    A[2]=35
    AMAX=.25
      FA=1.08325012
      FB=1.5
      NZ=1
     NIN=1
      NA=2
      NB=2


   F[1]=.9
   F[2]=1
   F[3]=1.61
   F[4]=2

FMIN[ 1]=6.969-01    AMIN= 45.56    DMIN= 10.56
FMIN[ 2]=9.705-01    AMIN= 46.39    DMIN= 11.39
FMIN[ 3]=1.016+00    AMIN= 38.66    DMIN=  3.66
FMIN[ 4]=1.600+00    AMIN= 43.09    DMIN=  8.09
FMIN[ 5]=1.675+00    AMIN= 38.37    DMIN=  3.37
FMIN[ 6]=2.727+00    AMIN= 49.79    DMIN= 14.79


>DØ PART 49


     F[1]=       .906029441
     F[2]=      1.00848015
     F[3]=      1.61303294
     F[4]=      1.7723765

FMIN[ 1]=7.010-01    AMIN= 42.93    DMIN=  7.93
FMIN[ 2]=9.792-01    AMIN= 43.04    DMIN=  8.04
FMIN[ 3]=1.016+00    AMIN= 44.54    DMIN=  9.54
FMIN[ 4]=1.600+00    AMIN= 43.84    DMIN=  8.84
FMIN[ 5]=1.658+00    AMIN= 44.54    DMIN=  9.54
FMIN[ 6]=2.276+00    AMIN= 42.38    DMIN=  7.38


>DØ PART 49


     F[1]=       .903776426
     F[2]=      1.0074392
     F[3]=      1.61290852
     F[4]=      1.79655578

FMIN[ 1]=6.988-01    AMIN= 43.40    DMIN=  8.40
FMIN[ 2]=9.779-01    AMIN= 43.35    DMIN=  8.35
FMIN[ 3]=1.016+00    AMIN= 43.42    DMIN=  8.42
FMIN[ 4]=1.600+00    AMIN= 43.37    DMIN=  8.37
FMIN[ 5]=1.662+00    AMIN= 43.42    DMIN=  8.42
FMIN[ 6]=2.322+00    AMIN= 43.37    DMIN=  8.37

>
```

Fig. 8.12 Sample use of bandpass pole-placer program for elliptical filter requirements.

the minimum of the fourth arc is at the stopband edge, where the attenuation is $A_{\min} = 43.09$.

Typing DO PART 49 causes the program to guess new pole locations. The filter characteristic for these new poles is much better than for the original poles. Using Part 49 one more time causes the arc minimums to be within a few hundreds of a decibel of each other, which is deemed satisfactory. It should be noted that the final pole locations agree with the results in Example 6.4.

Example 8.4

Example 8.3 was a geometrically symmetrical elliptic bandpass filter. As given initially in Example 6.4, the filter was not geometrically symmetrical. Instead,

$$f_L = 1 \quad f_A = 1.1 \quad f_B = 1.5 \quad f_H = 1.6$$

The pole locations for this nonsymmetrical filter were found by using the program in Fig. 8.11. The poles from Example 8.3 were used for the initial guess. The program was allowed to modify these poles until the result shown in Fig. 8.13 was obtained. Notice that the minimum distance is 11.56 dB, compared to 8.35 dB in Example 8.3. In this example we obtain greater stopband loss because we do not here have the constraint that the filter be geometrically symmetrical.

In Fig. 8.13, five arcs have $D_{\min} = 11.56$, whereas the remaining arc has $D_{\min} = 21.23$. The reason for this is that the first arc and the last arc must both be defined to be arc_1 (see Fig. 8.10). Thus, for this example, $D_{\min_1} = 11.56$.

Example 8.5

The requirements for this example are shown at the top of Fig. 8.14a. For this example, assume that NZ = 5 and NIN = 1. Use the pole-placer program to find the optimum locations for NA = 1 and NB = 3.

Solution

The initial pole locations are shown in Fig. 8.14a. These poles are iteratively moved as shown in Fig. 8.14b until the response is deemed satisfactory.

```
            F[1]=        .770016499
            F[2]=        .987631113
            F[3]=       1.61187851
            F[4]=       1.77667574

   FMIN[ 1]=5.374-01     AMIN= 56.23     DMIN= 21.23
   FMIN[ 2]=9.396-01     AMIN= 46.56     DMIN= 11.56
   FMIN[ 3]=1.000+00     AMIN= 46.56     DMIN= 11.56
   FMIN[ 4]=1.600+00     AMIN= 46.56     DMIN= 11.56
   FMIN[ 5]=1.656+00     AMIN= 46.56     DMIN= 11.56
   FMIN[ 6]=2.225+00     AMIN= 46.56     DMIN= 11.56
```

Fig. 8.13 Pole-placer results for a nongeometrically symmetric equiripple bandpass filter.

8.15 MODIFICATIONS OF THE POLE-PLACER PROGRAM

The pole-placer program just described can be modified so as to solve additional types of problems. For example, the next section shows how it can be modified so as to find attenuation poles for lowpass filters. A later chapter will modify it for "parametric" filters. It will also be modified so that it can be used for maximally flat filters instead of equiripple filters.

Further modifications are possible. For example, Smith and Temes describe how some of the poles can be fixed (i.e., not allowed to vary). Such modifications will not be discussed here, as it is not the purpose of this chapter to develop a general-purpose program for use by a neophyte designer. Instead, our aim is to describe the basic parts of a pole-placer program. Once the filter designer understands these fundamentals, he should be able to modify the program according to his specific problem.

```
DØ PART 1
      SA=1
      SB=2
   FS[0]=0
    A[0]=26
   FS[1]=990
    A[1]=0
   FS[2]=1055
    A[2]=52
   FS[3]=1100
    A[3]=42
    AMAX=.2
      FA=995
      FB=1052
      NZ=5
     NIN=1
      NA=1
      NB=3


    F[1]=989
    F[2]=1056
    F[3]=1057
    F[4]=1065

FMIN[ 1]=9.841+02    AMIN= 28.20    DMIN=  2.20
FMIN[ 2]=9.900+02    AMIN= 28.90    DMIN=  2.90
FMIN[ 3]=1.055+03    AMIN= 45.25    DMIN= -6.75
FMIN[ 4]=1.056+03    AMIN= 72.26    DMIN= 20.26
FMIN[ 5]=1.060+03    AMIN= 56.49    DMIN=  4.49
FMIN[ 6]=1.082+03    AMIN= 51.09    DMIN=  -.91
```

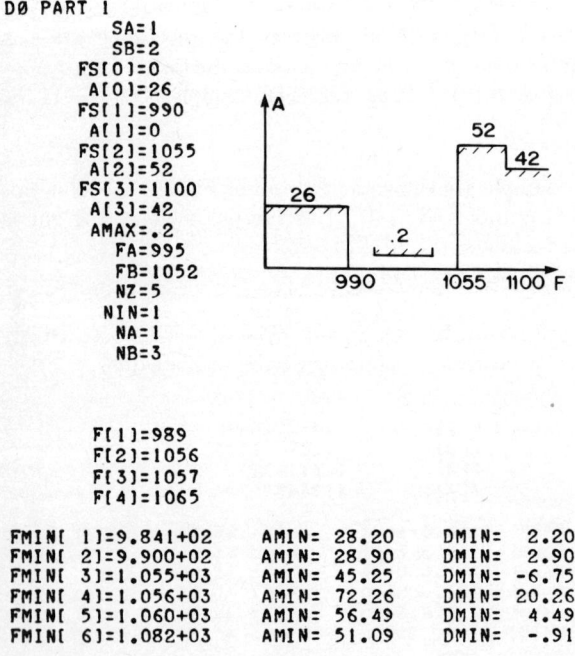

Fig. 8.14 Sample use of bandpass pole-placer program for an arbitrary stopband. (*Continued on next page.*)

D∅ PART 49

```
              F[1]=      988.551909
              F[2]=     1055.14377
              F[3]=     1056.99453
              F[4]=     1066.65385

FMIN[ 1]=9.834+02      AMIN= 29.25      DMIN=   3.25
FMIN[ 2]=9.900+02      AMIN= 26.02      DMIN=    .02
FMIN[ 3]=1.055+03      AMIN= 60.43      DMIN=   8.43
FMIN[ 4]=1.056+03      AMIN= 56.74      DMIN=   4.74
FMIN[ 5]=1.060+03      AMIN= 52.10      DMIN=    .10
FMIN[ 6]=1.084+03      AMIN= 51.62      DMIN=   -.38
```

>D∅ PART 49

```
              F[1]=      988.766249
              F[2]=     1055.22151
              F[3]=     1057.66187
              F[4]=     1067.97159

FMIN[ 1]=9.838+02      AMIN= 28.99      DMIN=   2.99
FMIN[ 2]=9.900+02      AMIN= 27.41      DMIN=   1.41
FMIN[ 3]=1.055+03      AMIN= 54.55      DMIN=   2.55
FMIN[ 4]=1.056+03      AMIN= 53.80      DMIN=   1.80
FMIN[ 5]=1.061+03      AMIN= 53.37      DMIN=   1.37
FMIN[ 6]=1.086+03      AMIN= 53.35      DMIN=   1.35
```

>D∅ PART 49

```
              F[1]=      988.788269
              F[2]=     1055.24055
              F[3]=     1057.753
              F[4]=     1068.13516

FMIN[ 1]=9.838+02      AMIN= 28.97      DMIN=   2.97
FMIN[ 2]=9.900+02      AMIN= 27.57      DMIN=   1.57
FMIN[ 3]=1.055+03      AMIN= 53.61      DMIN=   1.61
FMIN[ 4]=1.056+03      AMIN= 53.58      DMIN=   1.58
FMIN[ 5]=1.061+03      AMIN= 53.57      DMIN=   1.57
FMIN[ 6]=1.086+03      AMIN= 53.57      DMIN=   1.57
```

>

Fig. 8.14 (Continued) Sample use of bandpass pole-placer program for an arbitrary stopband.

8.16 LOWPASS POLE-PLACER PROGRAM

The program just described can be easily modified for lowpass filters. To do this, we consider a lowpass to be a special case of a bandpass: it is a bandpass with a lower passband edge of $\omega_A = 0$. Also, there are no attenuation poles at the origin; that is, $NZ = 0$.

To see that the above considerations allow us to modify the program

```
1; LØWPASS PØLE PLACER
1.1 DEMAND S
1.12 DEMAND FS[I],A[I] FØR I=1:1:S
1.2 DEMAND AMAX,FB,NIN,N
1.25 TYPE #,#,#
1.3 DEMAND F[I] FØR I=1:1:N
1.4 EE=10^(.1*AMAX)-1,FBB=FB*FB,FH=FS[1],A[0]=0
1.42 Z[I]=SQRT(1-FBB/F[I]/F[I]) FØR I=1:1:N
1.45 DØ PART 6 FØR F=FH
1.46 AH=A
1.6 TYPE #
1.7 TØ PART 3

3; CØNTRØL FØR UPPER STØPBAND
3.1 FM=FH,AM=AH,DM=AH-A[1]
3.2 FX=FH,FY=F[1],I=2
3.24 DØ PART 30
3.4 TØ PART 3.6 IF N=1
3.5 ZA=Z[I-1],ZB=Z[I]
3.52 DØ PART 8
3.54 FX=F[I-1],FY=F[I]
3.56 DØ PART 30 FØR I=I+1
3.58 TØ PART 3.5 IF I<N+1
3.6 ZA=Z[N],ZB=1
3.64 DØ PART 8 IF NIN>0
3.66 DØ PART 6.55 FØR Z=.9999999 IF NIN=0
3.68 FM=10^20,AM=A,DM=A-A[S] IF NIN=0
3.69 FX=F[N],FY=10*FS[S],I=N+2
3.7 DØ PART 30
3.82 TYPE #,#,#

6; DETERMINATIØN ØF LØSS
6.4 Z=SQRT(1-FBB/F/F)
6.55 L=((Z+1)/ Z-1 )^(NIN/2)
6.57 L=L*(Z+Z[I3])/(Z-Z[I3]) FØR I3=1:1:N
6.8 A=10*LØG(1+EE/4*(L+1/L)^2)

8; DETERMINATIØN ØF ARC MINIMUM   (FM=FREQ., AM=LØSS)
8.1 ZZO=ZA*ZB,I1=1,ZZA=ZA*ZA,ZZB=ZB*ZB
8.14 ZZO=(NIN/2*ZA+1)/(ZA+NIN/2)*ZZO IF ZB=1
8.16 DØ PART 8.9 FØR ZZ=ZZO
8.2 DO=D,ZZ1=ZZO-.00001
8.3 DØ PART 8.9 FØR ZZ=ZZ1
8.32 D1=D,ZZ2=(ZZO*D1-ZZ1*DO)/(D1-DO)
8.33 ZZ2=ZZA+.001 IF ZZ2<ZZA
8.34 ZZ2=ZZB-.001 IF ZZ2>ZZB
8.42 DO=D1,ZZO=ZZ1,ZZ1=ZZ2
8.5 TØ PART 8.3 IF  ZZ1-ZZO >.0000001
8.6 F=SQRT(FBB/(1-ZZ1))
8.7 DØ PART 6.55 FØR Z=SQRT(ZZ1)
8.82 AR=A[I1] IF FS[I1]<F
8.84 FM=F,AM=A,DM=A-AR,I1=I1+1
8.85 TØ PART 8.82 IF I1<S+1
8.9 D=NIN/2/(1-ZZ)
8.91 D=D+Z[I2]/(Z[I2]*Z[I2]-ZZ) FØR I2=1:1:N

>
```

Fig. 8.15 Pole-placer program for equiripple lowpass filters. (*Continued on next page.*)

Attenuation Poles for Equiripple Passband Filters

```
30; DETERMINATION OF MINIMUM DIFFERENCE (DMIN)
30.1 J=1
30.2 TO PART 30.6 IF FS[J]<FX OR FS[J]>FY
30.3 AR=MAX(A[J-1],A[J])
30.4 DO PART 6 FOR F=FS[J]
30.5 FM=F,AM=A,DM=A-AR IF DM>A-AR
30.6 J=J+1
30.7 TO PART 30.2 IF J<S+1
30.75 P=I-1
30.8 FMIN[P]=FM,AMIN[P]=AM,DMIN[P]=DM
30.9 TYPE P,FM,AM,DM IN FORM 1

49; DETERMINATION OF NEW POLES
49.1 N1=N+1,C1=LN(10)/40,I1=1
49.2 ZZ=(1-FBB/FMIN[I1]^2),Z=SQRT(ZZ)
49.21 A[I1,I2]=Z/(ZZ-Z[I2]^2) FOR I2=1:1:N
49.215 I1=I1+1
49.22 TO PART 49.2 IF I1<N1+1
49.3 A[I1,N1]=-C1,C[I1]=-C1*DMIN[I1] FOR I1=1:1:N1
49.4 TO PART 50

50; CROUT REDUCTION
50.1 R=1,J1=1,J2=N1
50.2 J=R+1
50.22 A[R,J]=A[R,J]-A[R,K]*A[K,J] FOR K=1:1:R-1
50.24 A[R,J]=A[R,J]/A[R,R],J=J+1
50.26 TO PART 50.22 IF J<N1+1
50.27 C=R+1
50.28 A[I,C]=A[I,C]-A[I,K]*A[K,C] FOR K=1:1:C-1 FOR I=C:1:N1
50.29 R=R+1
50.3 TO PART 50.2 IF R<N+1
50.4 C[J1]=C[J1]-A[J1,K]*C[K] FOR K=1:1:J1-1
50.42 C[J1]=C[J1]/A[J1,J1],J1=J1+1
50.43 TO PART 50.4 IF J1<N1+1
50.5 X[J2]=C[J2]
50.52 X[J2]=X[J2]-A[J2,K]*X[K] FOR K=J2+1:1:N1
50.525 J2=J2-1
50.53 TO PART 50.5 IF J2>0
50.55 TYPE #,#
50.6 DO PART 50.8 FOR I=1:1:N
50.7 TO PART 1.43
50.8 Z[I]=Z[I]+X[I],ZZ=Z[I]*Z[I]
50.82 F[I]=SQRT(FBB/(1-ZZ))
50.83 TYPE F[I]
```

Fig. 8.15 (Continued) Pole-placer program for equiripple lowpass filters.

in Fig. 8.11 so that it can be used for lowpass filters, consider the frequency transformation

$$Z^2 = \frac{f^2 - f_B^2}{f^2 - f_A^2}$$

For the lowpass case this becomes

$$Z^2 = 1 - \left(\frac{f_B}{f}\right)^2$$

150 Approximation Methods for Electronic Filter Design

This, and similar considerations, allow the bandpass program of Fig. 8.11 to be simplified for the lowpass case. The simpler lowpass pole-placer program is shown in Fig. 8.15.

An example of the use of the program is given in Fig. 8.16. The requirements shown are the same as for the elliptic lowpass filter of Example 5.3; thus the results should compare. That the answers do agree can be seen by comparing Figs. 5.13 and 8.16.

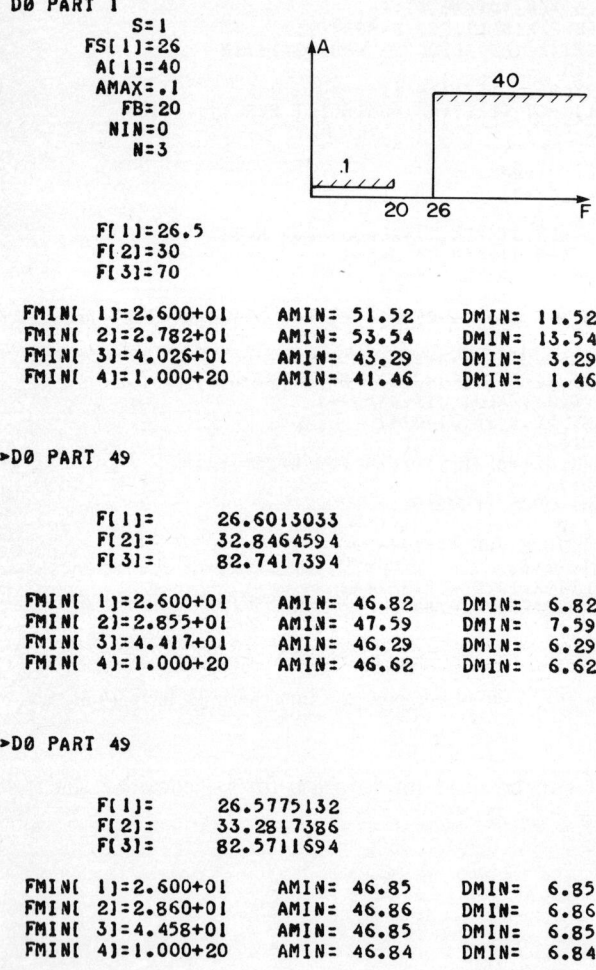

```
DØ PART 1
     S=1
     FS[1]=26
     A[1]=40
     AMAX=.1
     FB=20
     NIN=0
     N=3


     F[1]=26.5
     F[2]=30
     F[3]=70

 FMIN[ 1]=2.600+01    AMIN= 51.52    DMIN= 11.52
 FMIN[ 2]=2.782+01    AMIN= 53.54    DMIN= 13.54
 FMIN[ 3]=4.026+01    AMIN= 43.29    DMIN=  3.29
 FMIN[ 4]=1.000+20    AMIN= 41.46    DMIN=  1.46


►DØ PART 49


     F[1]=         26.6013033
     F[2]=         32.8464594
     F[3]=         82.7417394

 FMIN[ 1]=2.600+01    AMIN= 46.82    DMIN=  6.82
 FMIN[ 2]=2.855+01    AMIN= 47.59    DMIN=  7.59
 FMIN[ 3]=4.417+01    AMIN= 46.29    DMIN=  6.29
 FMIN[ 4]=1.000+20    AMIN= 46.62    DMIN=  6.62


►DØ PART 49


     F[1]=         26.5775132
     F[2]=         33.2817386
     F[3]=         82.5711694

 FMIN[ 1]=2.600+01    AMIN= 46.85    DMIN=  6.85
 FMIN[ 2]=2.860+01    AMIN= 46.86    DMIN=  6.86
 FMIN[ 3]=4.458+01    AMIN= 46.85    DMIN=  6.85
 FMIN[ 4]=1.000+20    AMIN= 46.84    DMIN=  6.84


►
```

Fig. 8.16 Sample use of lowpass pole-placer program for elliptic filter requirements.

Attenuation Poles for Equiripple Passband Filters

```
DØ PART 1
    S=2
    FS[1]=26
    A[1]=40
    FS[2]=40
    A[2]=10
    AMAX=.1
    FB=20
    NIN=0
    N=3
```

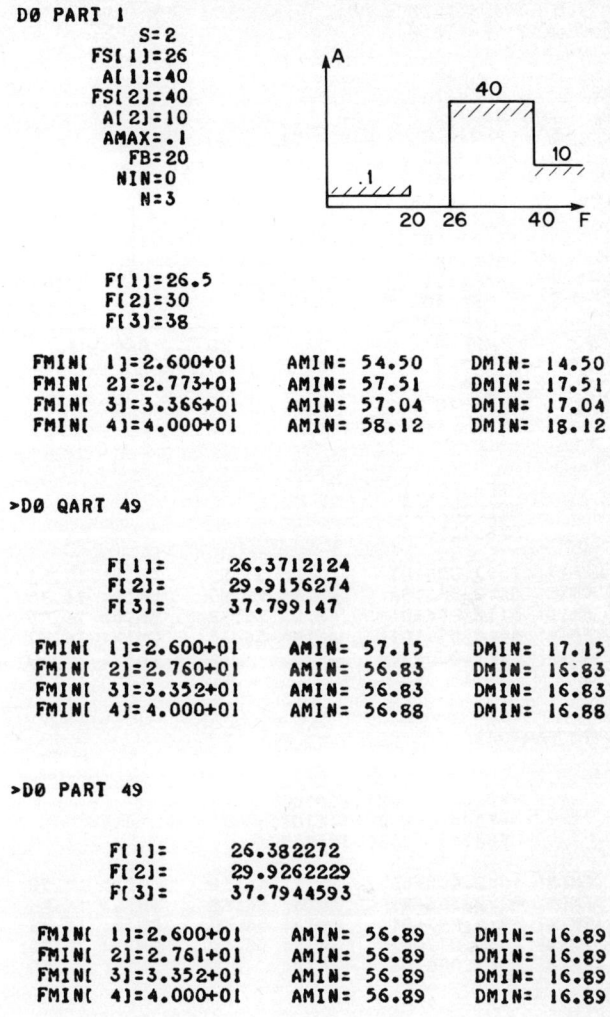

```
        F[1]=26.5
        F[2]=30
        F[3]=38

FMIN[ 1]=2.600+01    AMIN= 54.50    DMIN= 14.50
FMIN[ 2]=2.773+01    AMIN= 57.51    DMIN= 17.51
FMIN[ 3]=3.366+01    AMIN= 57.04    DMIN= 17.04
FMIN[ 4]=4.000+01    AMIN= 58.12    DMIN= 18.12


>DØ QART 49

        F[1]=      26.3712124
        F[2]=      29.9156274
        F[3]=      37.799147

FMIN[ 1]=2.600+01    AMIN= 57.15    DMIN= 17.15
FMIN[ 2]=2.760+01    AMIN= 56.83    DMIN= 16.83
FMIN[ 3]=3.352+01    AMIN= 56.83    DMIN= 16.83
FMIN[ 4]=4.000+01    AMIN= 56.88    DMIN= 16.88


>DØ PART 49

        F[1]=      26.382272
        F[2]=      29.9262229
        F[3]=      37.7944593

FMIN[ 1]=2.600+01    AMIN= 56.89    DMIN= 16.89
FMIN[ 2]=2.761+01    AMIN= 56.89    DMIN= 16.89
FMIN[ 3]=3.352+01    AMIN= 56.89    DMIN= 16.89
FMIN[ 4]=4.000+01    AMIN= 56.89    DMIN= 16.89

>
```

Fig. 8.17 Sample use of lowpass pole-placer program for an arbitrary stopband.

The program can also be used for lowpass filters that do not have equiripple stopbands, as is demonstrated in Fig. 8.17. In comparing this figure with Fig. 8.16, it should be noted that use of the pole-placer program allowed us to obtain about 10 dB more attenuation than if we had used an elliptic filter.

```
>8.2 DO=D,ZZ1=ZZ0-.001
>DELETE STEP 3.8
>DO PART 1
        SA=2
        SB=1
        FS(0)=0
        A(0)=10
        FS(1)=20
        A(1)=40
        FS(2)=800/26
        A(2)=0
        FS(3)=10^6
        A(3)=0
        AMAX=.1
        FA=40
        FB=10^5
        NZ=0
        NIN=0
        NA=3
        NB=0
```

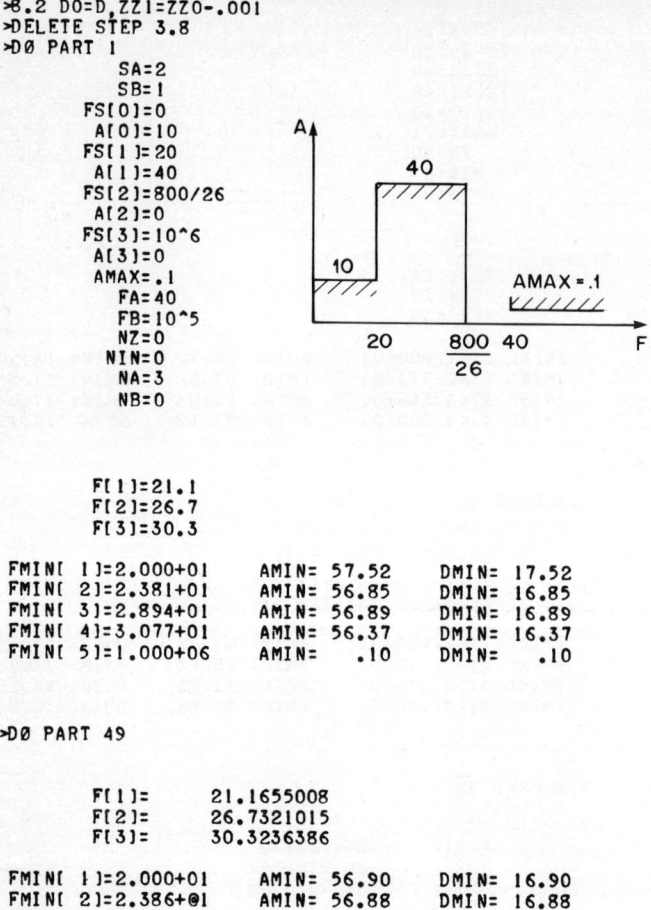

```
        F(1)=21.1
        F(2)=26.7
        F(3)=30.3

FMIN(1)=2.000+01    AMIN= 57.52    DMIN= 17.52
FMIN(2)=2.381+01    AMIN= 56.85    DMIN= 16.85
FMIN(3)=2.894+01    AMIN= 56.89    DMIN= 16.89
FMIN(4)=3.077+01    AMIN= 56.37    DMIN= 16.37
FMIN(5)=1.000+06    AMIN=   .10    DMIN=   .10


>DO PART 49


        F(1)=     21.1655008
        F(2)=     26.7321015
        F(3)=     30.3236386

FMIN(1)=2.000+01    AMIN= 56.90    DMIN= 16.90
FMIN(2)=2.386+01    AMIN= 56.88    DMIN= 16.88
FMIN(3)=2.897+01    AMIN= 56.88    DMIN= 16.88
FMIN(4)=3.077+01    AMIN= 56.90    DMIN= 16.90
FMIN(5)=1.000+06    AMIN=   .10    DMIN=   .10

>
```

Fig. 8.18 Sample use of bandpass pole-placer program for high-pass requirements.

8.17 HIGHPASS POLE-PLACER PROGRAM

Consider the highpass requirement shown in Fig. 8.18. The attenuation poles can be found in two different ways. One way is to make a highpass-to-lowpass transformation. If one makes the substitution

$$f \to \frac{800}{f}$$

```
DO PART 1
    SA=2
    SB=1
    FS[0]=0
    A[0]=10
    FS[1]=20
    A[1]=40
    FS[2]=800/26
    A[2]=0
    FS[3]=10^6
    A[3]=0
    AMAX=.1
    FA=40
    FB=100
    NZ=0
    NIN=0
    NA=3
    NB=0


    F[1]=21.1
    F[2]=26.7
    F[3]=30.3

    FMIN[ 1]=2.000+01   AMIN= 60.55   DMIN= 20.55
    FMIN[ 2]=2.384+01   AMIN= 59.68   DMIN= 19.68
    FMIN[ 3]=2.896+01   AMIN= 59.33   DMIN= 19.33
    FMIN[ 4]=3.077+01   AMIN= 58.64   DMIN= 18.64
    FMIN[ 5]=1.000+06   AMIN=  1.09   DMIN=  1.09

>DO PART 49

    F[1]=    21.1962197
    F[2]=    26.8048284
    F[3]=    30.3357408

    FMIN[ 1]=2.000+01   AMIN= 59.55   DMIN= 19.55
    FMIN[ 2]=2.394+01   AMIN= 59.51   DMIN= 19.51
    FMIN[ 3]=2.901+01   AMIN= 59.51   DMIN= 19.51
    FMIN[ 4]=3.077+01   AMIN= 59.54   DMIN= 19.54
    FMIN[ 5]=1.000+06   AMIN=  1.09   DMIN=  1.09

>
```

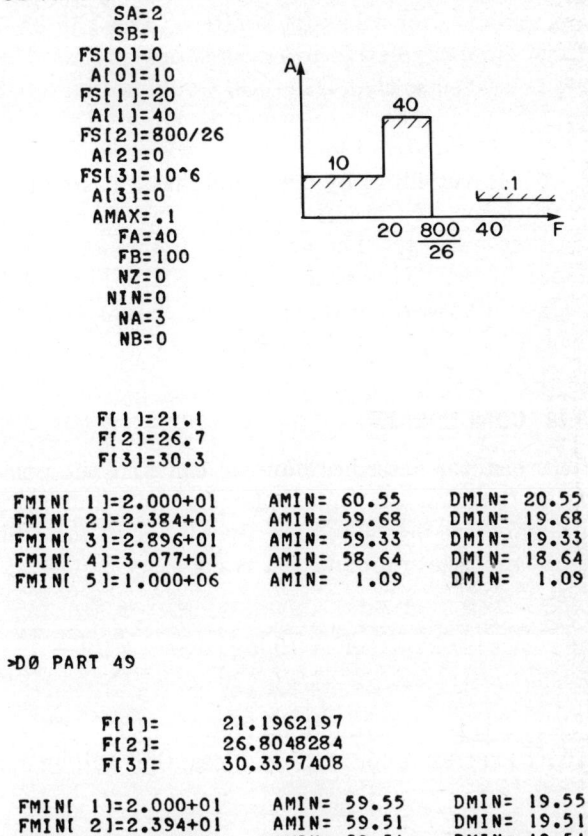

Fig. 8.19 Sample use of bandpass pole-placer program for "quasi" highpass filter.

the requirement in Fig. 8.18 is transformed to the requirement in Fig. 8.17. Thus, if we use NIN = 0, a set of attenuation poles is

$$f_1 = \frac{800}{37.794} = 21.167 \qquad f_2 = \frac{800}{29.926} = 26.732$$

$$f_3 = \frac{800}{26.382} = 30.324$$

Another way to find the attenuation poles is to proceed as if the requirement is for a bandpass filter. This can be done by letting f_B be a large number relative to f_A, as is demonstrated in Fig. 8.18. (Step 8.2 was modified so that D1 − D0 would not be zero in Step 8.32. Step 3.8 was deleted so that arc$_5$ would be ignored; that is, we do not care about the upper stopband loss.)

A highpass filter, by definition, has a passband that extends to infinite frequencies. In practice, one only need have the passband extend to a finite frequency. The lower this frequency is, the greater the stopband loss can be. This is demonstrated in Fig. 8.19, which indicates that $D_{\min} = 19.5$ for $F_B = 100$. This should be compared with Fig. 8.18, which had $D_{\min} = 16.9$ for $F_B = 10^5$.

8.18 CONCLUSION

This chapter describes how we can find the poles of attenuation for a filter that has an equiripple passband and arbitrary stopband. The programs we developed use the fact that, for a filter with an equiripple passband, the stopband loss is given by

$$A = 10 \log \left[1 + \frac{\epsilon^2}{4} \left(|L| + \frac{1}{|L|} \right)^2 \right]$$

where L is the loss function given by (8.4). We have not yet derived the above expression for stopband loss; that will be done in the next chapter.

REFERENCES

1. T. Laurent, *Frequency Filter Methods*, John Wiley & Sons, Inc., New York, 1963.
2. B. R. Smith and G. C. Temes, "An Iterative Approximation Procedure for Automatic Filter Synthesis," *IEEE Trans. on Circuit Theory*, March 1965, pp. 107–112.
3. F. B. Hildebrand, *Introduction to Numerical Analysis*, McGraw-Hill Book Company, New York, 1956, chap. 10.

PROBLEMS

8.1 This problem shows that the minimums of $|L| + 1/|L|$ occur where

$$D(Z) = \frac{NZ}{2} \left| \frac{f_B/f_A}{(f_B/f_A)^2 - Z^2} \right| + \frac{NIN}{2} \left| \frac{1}{1 - Z^2} \right| + \sum_{i=1}^{N} \frac{Z_i}{Z_i^2 - Z^2}$$

has its zeros.

    a. Show that if $F(z) = |L(z)| + 1/|L(Z)|$, then the minimums of $F(Z)$ occur where $|L(Z)|$ is minimum.

Attenuation Poles for Equiripple Passband Filters   155

    b. Show that for Z and Z_0 real numbers,

$$\frac{d}{dZ}\left|\frac{Z+Z_0}{Z-Z_0}\right|^M = M\left|\frac{Z+Z_0}{Z-Z_0}\right|^M \frac{2Z_0}{Z_0^2 - Z^2}$$

    c. Given that

$$L(Z) = \left(\frac{Z + \omega_B/\omega_A}{Z - \omega_B/\omega_A}\right)^{NZ/2} \left(\frac{Z+1}{Z-1}\right)^{NIN/2} \prod_{i=1}^{N} \frac{Z+Z_i}{Z-Z_i}$$

use part *b* to show that $d|L(z)|/dZ = 2|L|D(Z)$, where $D(Z)$ is as given above.

    d. Use the results in parts *a* and *b* to establish the statement at the beginning of the problem.

8.2 If
$$A = 10 \log\left|1 + \frac{\epsilon^2}{4}\left(|L| + \frac{1}{|L|}\right)^2\right| \quad (1)$$

where
$$L = \left(\frac{(f_A/f_B)Z + 1}{(f_A/f_B)Z - 1}\right)^{NZ/2}\left(\frac{Z+1}{Z-1}\right)^{NIN/2}\prod_{i=1}^{N}\frac{Z+Z_i}{Z-Z_i} \quad (2)$$

show that
$$\frac{\partial A}{\partial Z} = \frac{(40/\ln 10)(|L|^2 - 1/|L|^2)}{4/\epsilon^2 + (|L| + 1/|L|)^2} Z \frac{1}{Z^2 - Z_i^2} \quad (3)$$

8.3 In this problem we examine a lowpass filter that has its passband edge at $FB = 20$. The maximum loss in its passband is 0.1 dB. Also, it has three attenuation poles which are located at $F_1 = 26.58$, $F_2 = 33.29$, $F_3 = 82.61$.

    a. Find the loss at $F = 26$ by using the template method.
    b. Check the result of part *a* by using the exact formula for the loss.
    c. Repeat part *a* for a passband ripple of 0.5 dB.

8.4 Theorem 8.1 indicates that the stopband loss is given by

$$A = 10 \log\left|1 + \frac{\epsilon^2}{4}\left(|L| + \frac{1}{|L|}\right)^2\right| \quad (1)$$

For $|L|$ sufficiently large, this can be approximated as

$$A \approx 10 \log\left(\frac{\epsilon|L|}{2}\right)^2 \quad (2)$$

For a maximum passband ripple of 0.1 dB, what is the error due to the approximation if $|L| = 40$?

8.5 This problem uses one cycle of the Newton iteration procedure to help find the minimum of an arc. Assume an equiripple bandpass filter is described by $A_{max} = 0.1$, $FA = 0.9$, $FB = 1/0.9$, $NZ = 1$, $NIN = 1$, $F_1 = 0.5$, $F_2 = 0.7$, $F_3 = 1.3$, and $F_4 = 2$.

    a. Find Z_1, Z_2, Z_3, and Z_4.
    b. Find the approximate location of the minimum of the arc between F_3 and F_4 by using $Z_{m0}^2 = Z_3 Z_4$.
    c. Let $Z_{m1}^2 = Z_{m0}^2 - 0.01$. Find $D(Z_{m0})$ and $D(Z_{m1})$ by using (8.11).
    d. Find Z_{m2} by using (8.15), and then calculate the frequency that corresponds to Z_{m2}.

CHAPTER NINE

The Characteristic Function for Equiripple Passband Filters

9.1 INTRODUCTION

The characteristic function $K(s)$ is related to the input/output transfer function $H(s)$ by

$$H(s)H(-s) = 1 + K(s)K(-s) \qquad (9.1)$$

Thus, $H(s)$ and $K(s)$ have the same poles—the poles of attenuation. The previous chapter described a program that can be used to find the poles for an equiripple passband filter. This chapter derives the equation which is the basis for the pole-placer program. It also derives an expression for the zeros of the characteristic function.

9.2 THE CHARACTERISTIC FUNCTION AND THE TRANSFORMED VARIABLE

The characteristic function will be written as

$$K(s) = \frac{f(s)}{q(s)} \qquad (9.2)$$

The denominator $q(s)$ determines the poles of attenuation. These poles will be assumed to be purely imaginary, so that $q(s)$ can be written as

$$q(s) = s^{NZ} \prod_{i=1}^{N} (s^2 + \omega_i^2) \tag{9.3}$$

We assume that the pole locations ω_i have been determined by a pole-placer program, and that we want to find the zeros of attenuation [the roots of $f(s)$].

The solution will be found in terms of the transformed frequency, since $F(Z)$ is of much simpler form than $f(s)$. As a first step in the solution, rewrite (9.2) as

$$K(s)K(-s) = \frac{f(s)f(-s)}{q(s)q(-s)} \tag{9.4}$$

It can be seen from (9.3) that $q(s)$ is either an even or an odd function. Similarly, it will be shown that $f(s)$ is even. Thus (9.4) can be rewritten as*

$$K(s)K(-s) = \pm \frac{f^2(s)}{q^2(s)} \tag{9.5}$$

In terms of the transformed variable, $f(s)$ is

$$F(Z) = \frac{f(s)}{(s^2 + \omega_A^2)^{m/2}}$$

A similar expression can be written for $Q(Z)$. Thus, if we divide numerator and denominator of (9.5) by $(s^2 + \omega_A^2)^m$, it follows that†

$$K(s)K(-s) = \frac{F^2(Z)}{Q^2(Z)} \tag{9.6}$$

where $Z^2 = (s^2 + \omega_B^2)/(s^2 + \omega_A^2)$.

Equation (9.6) is the desired expression for the characteristic function in terms of the transformed variable. In the next two sections we find expressions for the terms $Q^2(Z)$ and $F^2(Z)$. It should be noted that we care only about the ratio of these terms.

The constant multiplier of $F^2(Z)/Q^2(Z)$ can be found by examining the behavior at $Z = 0$. From (9.6) it follows that

$$|K(j\omega_B)|^2 = \frac{F^2(0)}{Q^2(0)}$$

---

* The plus (minus) sign is for q even (odd).
† The plus-or-minus sign of (9.5) has been neglected, as the constant multiplier of $F^2(Z)/Q^2(Z)$ will be chosen to give the proper sign automatically.

where ω_B is the upper passband edge. By definition, the value of $K(s)$ will be ϵ at the upper passband edge.* Thus,

$$\epsilon^2 = \frac{F^2(0)}{Q^2(0)} \tag{9.7}$$

9.3 DETERMINATION OF $Q^2(z)$

The $q(s)$ we are investigating has a very special form. It is

$$q(s) = s^{NZ} \prod_{i=1}^{N} (s^2 + \omega_i^2)$$

From Theorem 7.2 we have

$$Q(Z) = \frac{q(s)}{(s^2 + \omega_A^2)^{m/2}}$$

$$= C_Q \left[1 - \left(\frac{\omega_A Z}{\omega_B}\right)^2 \right]^{NZ/2} (Z^2 - 1)^{NIN/2} \prod_{i=1}^{N} (Z^2 - Z_i^2) \tag{9.8}$$

where C_Q is real and

$$Z_i^2 = \frac{\omega_i^2 - \omega_B^2}{\omega_i^2 - \omega_A^2} \tag{9.9}$$

$$m = NZ + NIN + 2N \tag{9.10}$$

In the last section it was noted that we do not care about the coefficient of $Q(Z)$; we need only determine the ratio $F^2(Z)/Q^2(Z)$. Thus, ignoring the coefficient, we can rewrite (9.8) as

$$Q^2(Z) = \left[Z^2 - \left(\frac{\omega_B}{\omega_A}\right)^2 \right]^{NZ} (Z^2 - 1)^{NIN} \prod_{i=1}^{N} (Z^2 - Z_i^2)^2 \tag{9.11}$$

9.4 DETERMINATION OF $F^2(z)$

In Sec. 9.2 it was shown that the characteristic function can be written as

$$K(s)K(-s) = \frac{F^2(Z)}{Q^2(Z)}$$

Section 9.3 gave an expression for $Q^2(Z)$; this section derives an expression for $F^2(Z)$.

The solution is based on the fact that the passband is equiripple, as

* As $|H(j\omega_B)|^2 = 1 + |K(j\omega_B)|^2 = 1 + \epsilon^2$, which is in agreement with Eqs. (2.5) and (2.6).

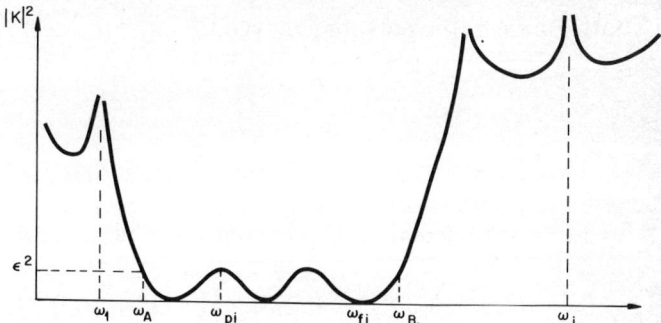

Fig. 9.1 Characteristic function for an equiripple bandpass filter.

indicated in Fig. 9.1. For a characteristic function such as this, we can write

$$|K|^2 = \frac{\epsilon^2}{1 - C_1^2 \dfrac{(\omega^2 - \omega_A^2)(\omega^2 - \omega_B^2) \prod_{i=1}^{m/2-1} (\omega^2 - \omega_{pi}^2)^2}{\prod_{i=1}^{m/2} (\omega^2 - \omega_{fi}^2)^2}} \quad (9.12)$$

To see that this is a valid representation of the behavior in Fig. 9.1, consider the following points:*

1. $|K|^2 = \epsilon^2$ at $\omega = \omega_A$ and $\omega = \omega_B$.
2. $|K|^2 = \epsilon^2$ and $d|K|^2/d\omega = 0$ at $\omega = \omega_{pi}$.
3. $|K|^2 = 0$ and $d|K|^2/d\omega = 0$ at $\omega = \omega_{fi}$.
4. There is one more minimum than maximum in the passband; thus the product with the ω_{fi} terms has one more component.

Equation (9.12) implies that

$$K(s)K(-s) = \frac{\epsilon^2 \prod_{i=1}^{m/2} (s^2 + \omega_{fi}^2)^2}{\prod_{i=1}^{m/2} (s^2 + \omega_{fi}^2)^2 - C_1^2(s^2 + \omega_A^2)(s^2 + \omega_B^2) \prod_{i=1}^{m/2-1} (s^2 + \omega_{pi}^2)^2} \quad (9.13)$$

But we also know that

$$K(s)K(-s) = \frac{f(s)f(-s)}{q(s)q(-s)} \quad (9.14)$$

---

* Problem 9.1 demonstrates that $m = NZ + NIN + 2N$. It should be noted in (9.12) that m must be divisible by 2; thus $NZ + NIN$ must be an even number.

Equating numerator polynomials yields

$$f(s) = C_f \prod_{i=1}^{m/2} (s^2 + \omega_{fi}^2) \qquad (9.15)$$

where C_f is an unimportant constant. Notice that $f(s)$ is an even function of s as promised in Sec. 9.2.

Equating the denominator polynomials of (9.13) and (9.14) yields

$$f^2(s) - C_p^2 (s^2 + \omega_A^2)(s^2 + \omega_B^2) \prod_{i=1}^{m/2-1} (s^2 + \omega_{pi}^2)^2 = C_q^2 q^2(s) \qquad (9.16)$$

where C_p and C_q are unimportant constants. Dividing both sides of this equation by $(s^2 + \omega_A^2)^m$ leads to

$$F^2(Z) - C_p^2 \frac{s^2 + \omega_B^2}{s^2 + \omega_A^2} \prod_{i=1}^{m/2-1} \left(\frac{s^2 + \omega_{pi}^2}{s^2 + \omega_A^2}\right)^2 = C^2 Q^2(Z) \qquad (9.17)$$

where C_p and C are unimportant constants. This can be rewritten as

$$F^2(Z) - Z^2 P^2(Z) = C^2 Q^2(Z) \qquad (9.18)$$

where
$$P(Z) = C_3 \prod_{i=1}^{m/2-1} (Z^2 - Z_{pi}^2) \qquad (9.19)$$

Using (9.11), we can rewrite (9.18) as

$$[F(Z) + ZP(Z)][F(Z) - ZP(Z)]$$

$$= C^2 \left[Z^2 - \left(\frac{\omega_B}{\omega_A}\right)^2\right]^{NZ} (Z^2 - 1)^{NIN} \prod_{i=1}^{N} (Z^2 - Z_i^2)^2$$

$$= C \left(Z + \frac{\omega_B}{\omega_A}\right)^{NZ} (Z + 1)^{NIN} \prod_{i=1}^{N} (Z + Z_i)^2$$

$$\times C \left(Z - \frac{\omega_B}{\omega_A}\right)^{NZ} (Z - 1)^{NIN} \prod_{i=1}^{N} (Z - Z_i)^2 \qquad (9.20)$$

It can be proven* that $F(Z) + ZP(Z)$ is Hurwitz (i.e., all the roots are in

---

* Let ω_{fi} represent a passband frequency of zero loss, and ω_{pi} represent a passband frequency of maximum loss, as indicated in Fig. 9.1. The functions $F(Z)$ and $P(Z)$ are then defined as

$$F(Z) = C_F \prod_{i=1}^{m/2} \left(Z^2 + \frac{\omega_B^2 - \omega_{fi}^2}{\omega_{fi}^2 - \omega_A^2}\right)$$

$$P(Z) = C_P \prod_{i=1}^{m/2-1} \left(Z^2 + \frac{\omega_B^2 - \omega_{pi}^2}{\omega_{pi}^2 - \omega_A^2}\right)$$

where the constants C_F and C_P are unimportant. The important fact is that the

the left half-plane); thus (9.20) implies that

$$F(Z) + ZP(Z) = C\left(Z + \frac{\omega_B}{\omega_A}\right)^{NZ} (Z + 1)^{NIN} \prod_{i=1}^{N} (Z + Z_i)^2 \quad (9.21)$$

From (7.11) and (9.19), $F(Z)$ and $P(Z)$ are both even functions of Z; thus (9.21) yields

$$F(Z) = \text{Ev}\left[C\left(Z + \frac{\omega_B}{\omega_A}\right)^{NZ} (Z + 1)^{NIN} \prod_{i=1}^{N} (Z + Z_i)^2\right] \quad (9.22)$$

where Ev means "even part of."

Equation (9.22) is the desired equation for $F(Z)$; the only unknown in it is C. This can be found by recalling that Eq. (9.7) states that $\epsilon^2 = F^2(0)/Q^2(0)$. It thus follows from (9.11) and (9.22) that

$$\frac{F^2(0)}{Q^2(0)} = \frac{\left[C(\omega_B/\omega_A)^{NZ} \prod_{i=1}^{N} Z_i^2\right]^2}{\left[(\omega_B/\omega_A)^{NZ} \prod_{i=1}^{N} Z_i^2\right]^2} = \epsilon^2 \quad (9.23)$$

Since this equation yields $C^2 = \epsilon^2$, we can summarize the results as

Theorem 9.1

The characteristic function $K(s)$ of an equiripple bandpass filter is given by

$$K(s)K(-s) = \frac{F^2(Z)}{Q^2(Z)}$$

where $F^2(Z) = \epsilon^2 \left\{ \text{Ev}\left[\left(Z + \frac{\omega_B}{\omega_A}\right)^{NZ} (Z + 1)^{NIN} \prod_{i=1}^{N} (Z + Z_i)^2\right] \right\}^2$

$$Q^2(Z) = \left[Z^2 - \left(\frac{\omega_B}{\omega_A}\right)^2\right]^{NZ} (Z^2 - 1)^{NIN} \prod_{i=1}^{N} (Z^2 - Z_i^2)^2 \quad (9.24)$$

$NZ + NIN =$ even number

9.5 DETERMINATION OF $Q^2(Z)$ AND $F^2(Z)$ FOR LOWPASS FILTERS

Theorem 9.1 described how we can find $Q^2(Z)$ and $F^2(Z)$ for bandpass filters that have equiripple passbands. Once these functions are known,

---

zeros of $F(Z)$ and $P(Z)$ are restricted to the imaginary axis where they alternate; this follows from Fig. 9.1. That is, first there is a zero of $F(Z)$, then a zero of $P(Z)$, then a zero of $F(Z)$, etc. Thus, the function $ZP(Z)/F(Z)$ is a reactance function, which implies that $F(Z) + ZP(Z)$ is Hurwitz. (E. A. Guillemin, *Synthesis of Passive Networks*, John Wiley & Sons, Inc., New York, 1957, p. 77.)

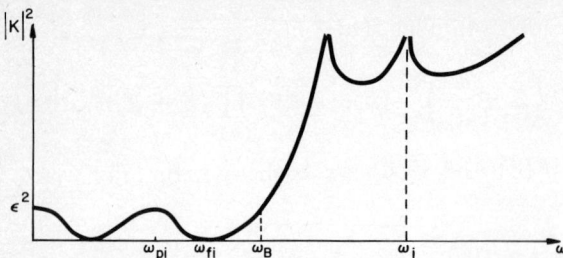

Fig. 9.2 Characteristic function for an even-order equiripple lowpass filter.

the characteristic function $K(s)$ can be evaluated at any frequency, since

$$K(s)K(-s) = \frac{F^2(Z)}{Q^2(Z)}$$

This section derives similar expressions for lowpass equiripple filters.

Consider the lowpass equiripple characteristic function shown in Fig. 9.2. The characteristic function for this case can be written as

$$|K|^2 = \frac{\epsilon^2}{1 - \dfrac{C_1{}^2\omega^2(\omega^2 - \omega_B{}^2)\prod_{i=1}^{m/2-1}(\omega^2 - \omega_{pi}{}^2)^2}{\prod_{i=1}^{m/2}(\omega^2 - \omega_{fi}{}^2)^2}} \quad (9.25)$$

Note that this is the same as (9.12) except that $\omega_A = 0$. Thus, if we set $\omega_A = 0$, the previous results still apply. Also, a lowpass filter has no attenuation poles at the origin, so that

$$m = \text{NZ} + \text{NIN} + 2N = \text{NIN} + 2N \quad (9.26)$$

If we are to apply Theorem 9.1, m must be even; thus, for lowpass filters NIN must be even. The lowpass version of Theorem 9.1 is

Theorem 9.2

A characteristic function $K(s)$ of an equiripple lowpass filter is given by

$$K(s)K(-s) = \frac{F^2(Z)}{Q^2(Z)}$$

The Characteristic Function for Equiripple Passband Filters

where
$$F^2(Z) = \epsilon^2 \left\{ \text{Ev} \left[(Z+1)^{\text{NIN}} \prod_{i=1}^{N} (Z+Z_i)^2 \right] \right\}^2$$

$$Q^2(Z) = (Z^2-1)^{\text{NIN}} \prod_{i=1}^{N} (Z^2 - Z_i^2)^2 \qquad (9.27)$$

NIN = even number

To see what happens if NIN is an odd number, consider Fig. 9.3. The characteristic function for this case can be written as

$$|K|^2 = \frac{\epsilon^2}{1 - \dfrac{C_1^2(\omega^2 - \omega_B^2) \prod_{i=1}^{(m-1)/2} (\omega^2 - \omega_{pi}^2)^2}{\omega^2 \prod_{i=1}^{(m-1)/2} (\omega^2 - \omega_{fi}^2)^2}} \qquad (9.28)$$

By analogy to Sec. 9.4, it follows that

$$f(s) = C_f \omega \prod_{i=1}^{(m-1)/2} (s^2 + \omega_{fi}^2)$$

$$F^2(Z) = \epsilon^2 \left\{ \text{Ev} \left[(Z+1)^{\text{NIN}} \prod_{i=1}^{N} (Z+Z_i)^2 \right] \right\}^2 \qquad (9.29)$$

$$Q^2(Z) = -(Z^2-1)^{\text{NIN}} \prod_{i=1}^{N} (Z^2 - Z_i^2)^2$$

These equations are the same as those in Theorem 9.2 except for the initial minus sign in the $Q^2(Z)$ expression.* Equations (9.27) and (9.29) can be combined as

---
* Since NIN is odd in (9.29), it follows that $F^2(0)/Q^2(0) = \epsilon^2$, which is a positive number as desired.

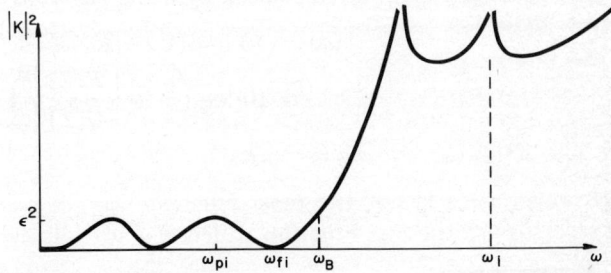

Fig. 9.3 Characteristic function for an odd-order equiripple lowpass filter.

Theorem 9.3

The characteristic function $K(s)$ of an equiripple lowpass filter is given by

$$K(s)K(-s) = \frac{F^2(Z)}{Q^2(Z)}$$

where

$$F^2(Z) = \epsilon^2 \left\{ \text{Ev} \left[(Z+1)^{\text{NIN}} \prod_{i=1}^{N} (Z+Z_i)^2 \right] \right\}^2 \quad (9.30)$$

$$Q^2(Z) = (-1)^{\text{NIN}} (Z^2-1)^{\text{NIN}} \prod_{i=1}^{N} (Z^2-Z_i^2)^2$$

9.6 THE LOSS FUNCTION $L(Z)$

The characteristic function is related to $F(Z)$ and $Q(Z)$ by

$$K(s)K(-s) = \frac{F^2(Z)}{Q^2(Z)} \quad (9.31)$$

where

$$F^2(Z) = \epsilon^2 \left\{ \text{Ev} \left[\left(Z + \frac{\omega_B}{\omega_A}\right)^{\text{NZ}} (Z+1)^{\text{NIN}} \prod_{i=1}^{N} (Z+Z_i)^2 \right] \right\}^2 \quad (9.32)$$

$$Q^2(Z) = (-1)^{\text{NZ+NIN}} \left[Z^2 - \left(\frac{\omega_B}{\omega_A}\right)^2 \right]^{\text{NZ}} (Z^2-1)^{\text{NIN}} \prod_{i=1}^{N} (Z^2-Z_i^2)^2$$

$$(9.33)$$

These formulas are valid for either lowpass or bandpass equiripple filters. For lowpass filters one sets NZ equal to zero, so that

$$\left(Z + \frac{\omega_B}{\omega_A}\right)^{\text{NZ}} = 1 = \left[Z^2 - \left(\frac{\omega_B}{\omega_A}\right)^2\right]^{\text{NZ}}$$

For bandpass filters NZ + NIN is even, so that $(-1)^{\text{NZ+NIN}}$ equals unity. The loss $A(\omega)$ can be calculated by using the above formulas. That is,

$$A(\omega) = 10 \log (1 + |K(j\omega)|^2)$$
$$= 10 \log \left(1 + \left|\frac{F(Z)}{Q(Z)}\right|^2\right)$$

where $Z^2 = (\omega^2 - \omega_B^2)/(\omega^2 - \omega_A^2)$.

However, this is not the most efficient way to calculate the loss; it is better to use the loss function which was introduced in Chap. 8:

$$L(Z) = \left(\frac{Z + \omega_B/\omega_A}{Z - \omega_B/\omega_A}\right)^{\text{NZ}/2} \left(\frac{Z+1}{Z-1}\right)^{\text{NIN}/2} \prod_{i=1}^{N} \frac{Z+Z_i}{Z-Z_i} \quad (9.34)$$

The Characteristic Function for Equiripple Passband Filters

The form of the loss function $L(Z)$ is very similar to the forms of $F^2(Z)$ and $Q^2(Z)$ in (9.32) and (9.33). Because of the similarity, the functions are related as shown in the following theorem.

Theorem 9.4

$$\frac{\epsilon}{2}\left|L(Z) + \frac{(-1)^{NZ+NIN}}{L(Z)}\right| = \left|\frac{F(Z)}{Q(Z)}\right| \qquad (9.35)$$

Proof

From (9.34) it follows that

$$L + \frac{(-1)^{NZ+NIN}}{L}$$

$$= \frac{(Z + \omega_B/\omega_A)^{NZ}(Z+1)^{NIN}\prod_{i=1}^{N}(Z+Z_i)^2 + (-1)^{NZ+NIN}(Z - \omega_B/\omega_A)^{NZ}(Z-1)^{NIN}\prod_{i=1}^{N}(Z-Z_i)^2}{[Z^2 - (\omega_B/\omega_A)^2]^{NZ/2}(Z^2-1)^{NIN/2}\prod_{i=1}^{N}(Z^2 - Z_i^2)}$$

$$= \frac{2\,\mathrm{Ev}\,[(Z + \omega_B/\omega_A)^{NZ}(Z+1)^{NIN}\prod_{i=1}^{N}(Z+Z_i)^2]}{[Z^2 - (\omega_B/\omega_A)^2]^{NZ/2}(Z^2-1)^{NIN/2}\prod_{i=1}^{N}(Z^2 - Z_i^2)}$$

The proof is completed by taking the magnitudes of both sides of the above equation and using (9.32) and (9.33).

Corollary 9.4

$$A(\omega) = 10\log\left(1 + \frac{\epsilon^2}{4}\left|L + \frac{(-1)^{NZ+NIN}}{L}\right|^2\right) \qquad (9.36)$$

The loss, as given by the above formula, is very easy to calculate. The next two sections demonstrate how it can be simplified even further by consideration of the stopband and passband separately.

9.7 STOPBAND PERFORMANCE

The loss function is given by

$$L(Z) = \left(\frac{Z + \omega_B/\omega_A}{Z - \omega_B/\omega_A}\right)^{NZ/2}\left(\frac{Z+1}{Z-1}\right)^{NIN/2}\prod_{i=1}^{N}\frac{Z+Z_i}{Z-Z_i} \qquad (9.37)$$

where, for $s = j\omega$,

$$Z = \left(\frac{\omega^2 - \omega_B{}^2}{\omega^2 - \omega_A{}^2}\right)^{1/2} \tag{9.38}$$

1. From (9.38) it follows that Z is a real number in the stopband; thus

$$\prod_{i=1}^{N} \frac{Z + Z_i}{Z - Z_i}$$

is real in the stopband.

2. In the lower stopband $Z > \omega_B/\omega_A > 1$; thus

$$\left(\frac{Z + \omega_B/\omega_A}{Z - \omega_B/\omega_A}\right)^{NZ/2} \left(\frac{Z + 1}{Z - 1}\right)^{NIN/2}$$

is real in the lower stopband (a lowpass has no lower stopband).

3. In the upper stopband $0 < Z < 1$, so

$$\left(\frac{Z + \omega_B/\omega_A}{Z - \omega_B/\omega_A}\right)^{NZ/2} \left(\frac{Z + 1}{Z - 1}\right)^{NIN/2}$$

is real (imaginary) if $NZ + NIN$ is even (odd).

4. Since a bandpass equiripple filter has $NZ + NIN$ even, (9.37) and 1 through 3 imply that *a bandpass filter has a real loss function $L(Z)$ in the stopband*.

5. Similarly, we can conclude that *a lowpass filter has a real (imaginary) loss function $L(Z)$ in the stopband if NIN is even (odd)*.

Since the loss is given by

$$A(\omega) = 10 \log \left(1 + \frac{\epsilon^2}{4} \left| L + \frac{(-1)^{NZ+NIN}}{L} \right|^2 \right)$$

application of 4 and 5 yields

Theorem 9.5

For a lowpass or bandpass equiripple filter, the stopband loss is given by

$$A(\omega) = 10 \log \left[1 + \frac{\epsilon^2}{4} \left(|L| + \frac{1}{|L|} \right)^2 \right] \tag{9.39}$$

9.8 PASSBAND PERFORMANCE

In this section, the loss function $L(Z)$ is examined in the passband. In general,

$$L(Z) = \left(\frac{Z + \omega_B/\omega_A}{Z - \omega_B/\omega_A}\right)^{NZ/2} \left(\frac{Z + 1}{Z - 1}\right)^{NIN/2} \prod_{i=1}^{N} \left(\frac{Z + Z_i}{Z - Z_i}\right) \tag{9.40}$$

The Characteristic Function for Equiripple Passband Filters

In the passband, we will choose to write this as

$$L(Z) = (e^{j\beta_Z})^{NZ/2}(e^{j\beta_{IN}})^{NIN/2} \sum_{i=1}^{N} e^{j\beta_i} \qquad (9.41)$$

A typical term in this expression is

$$e^{j\beta_i} = \frac{Z + Z_i}{Z - Z_i}$$

Since

$$Z = \left(\frac{\omega^2 - \omega_B^2}{\omega^2 - \omega_A^2}\right)^{1/2}$$

it follows that Z is pure imaginary in the passband; that is, $Z = j|Z|$. Thus,

$$e^{j\beta_i} = \frac{j|Z| + Z_i}{j|Z| - Z_i} = \frac{|Z|^2 - Z_i^2 - 2j|Z|Z_i}{|Z|^2 + Z_i^2}$$

From this it follows that:

Theorem 9.6

In the passband the loss function can be written as

$$L(Z) = e^{j\beta} \qquad (9.42)$$

where

$$\beta = \frac{NZ}{2}\beta_Z + \frac{NIN}{2}\beta_{IN} + \sum_{i=1}^{N}\beta_i \qquad (9.43)$$

$$\beta_Z = \tan^{-1}\frac{-2|Z|\omega_B/\omega_A}{|Z|^2 - (\omega_B/\omega_A)^2} \qquad (9.44a)$$

$$\beta_{IN} = \tan^{-1}\frac{-2|Z|}{|Z|^2 - 1} \qquad (9.44b)$$

$$\beta_i = \tan^{-1}\frac{-2|Z|Z_i}{|Z|^2 - Z_i^2} \qquad (9.44c)$$

To apply this theorem to find the loss in the passband, recall that

$$A(\omega) = 10 \log \left[1 + \frac{\epsilon^2}{4}\left|L + \frac{(-1)^{NZ+NIN}}{L}\right|^2\right] \qquad (9.45)$$

Substituting (9.42) into (9.45) yields

Theorem 9.7

For a lowpass (NIN even) or bandpass equiripple filter, the passband loss is given by

$$A(\omega) = 10 \log (1 + \epsilon^2 \cos^2 \beta)$$

```
1; LØSS ØF EQUIRIPPLE BPF
1.2 DEMAND AMAX,FA,FB,NZ,NIN,N
1.3 DEMAND F[I] FØR I=1:1:N
1.4 EE=10^(.1*AMAX)-1,FAA=FA*FA,FBB=FB*FB
1.42 Z[I]=SQRT((F[I]*F[I]-FBB)/(F[I]*F[I]-FAA)) FØR I=1:1:N
1.5 DEMAND F
1.6 TYPE #
1.7 TØ PART 6

6; DETERMINATIØN ØF STØPBAND LØSS
6.1 A=AMAX
6.2 TØ PART 6.9 IF F=FA
6.4 Z=SQRT( (F*F-FBB)/(F*F-FAA) )
6.5 TØ PART 10 IF F>FA AND F<FB
6.55 L=((Z+1)/ Z-1 )^(NIN/2)
6.56 L=L*((FA*Z/FB+1)/ FA*Z/FB-1 )^(NZ/2) IF NZ>0
6.57 L=L*(Z+Z[I])/(Z-Z[I]) FØR I=1:1:N
6.8 A=10*LØG(1+EE/4*(L+1/L)^2)
6.9 TYPE F,A IN FØRM 1

10; DETERMINATIØN ØF PASSBAND LØSS
10.55 B=NIN/2*ATN(-2*Z,Z*Z-1)
10.56 B=B+NZ/2*ATN(-2*Z*FB/FA,Z*Z-FBB/FAA)
10.57 B=B+ATN(-2*Z*Z[I],Z*Z-Z[I]*Z[I]) FØR I=1:1:N
10.8 A=10*LØG(1+EE*(CØS(B))^2)
10.9 TYPE F,A IN FØRM 1


FØRM        1
#.###^^^  ###.###
```

Fig. 9.4 Program to determine the loss of equiripple bandpass filters.

```
         AMAX=.1
           FA=.9
           FB=1/.9
           NZ=1
          NIN=1
            N=2
         F[1]=.7
         F[2]=1/.7
            F=.3

  3.000-01   40.488
>
DØ PART 6 FØR F=.901:.001:.91
   9.010-01   .081
   9.020-01   .065
   9.030-01   .051
   9.040-01   .039
   9.050-01   .029
   9.060-01   .021
   9.070-01   .014
   9.080-01   .009
   9.090-01   .005
   9.100-01   .002
>
```

Fig. 9.5 Sample use of the program in Fig. 9.4.

For a lowpass (NIN odd) equiripple filter, the passband loss is given by

$$A(\omega) = 10 \log (1 + \epsilon^2 \sin^2 \beta)$$

Figure 9.4 contains a program that can be used to calculate the loss of equiripple bandpass filters. This program is an extension of the one in

The Characteristic Function for Equiripple Passband Filters

```
1; LØSS ØF EQUIRIPPLE LPF
1.2 DEMAND AMAX,FB,NIN,N
1.3 DEMAND F[I] FØR I=1:1:N
1.4 EE=10^(.1*AMAX)-1,FBB=FB*FB
1.42 Z[I]=SQRT(1-FBB/F[I]/F[I]) FØR I=1:1:
1.5 DEMAND F
1.6 TYPE #
1.7 TØ PART 6

6; DETERMINATIØN ØF STØPBAND LØSS
6.4 Z=SQRT( 1-FBB/F/F )
6.5 TØ PART 10 IF F<FB
6.55 L=((Z+1)/ Z-1 )^(NIN/2)
6.57 L=L*(Z+Z[I])/(Z-Z[I]) FØR I=1:1:N
6.8 A=10*LØG(1+EE/4*(L+1/L)^2)
6.9 TYPE F,A IN FØRM 1

10; DETERMINATIØN ØF PASSBAND LØSS
10.55 B=NIN/2*ATN(-2*Z,Z*Z-1)
10.57 B=B+ATN(-2*Z*Z[I],Z*Z-Z[I]*Z[I]) FØR I=1:1:N
10.8 A=10*LØG(1+EE*(CØS(B))^2)
10.81 A=10*LØG(1+EE*(SIN(B))^2) IF NIN>2*IP(NIN/2)
10.9 TYPE F,A IN FØRM 1

FØRM     1
#.###^^^  ###.###
```

Fig. 9.6 Program to determine the loss of equiripple lowpass filters.

```
                    AMAX=.1
                      FB=1
                     NIN=1
                       N=3
                   F[1]=1.1
                   F[2]=1.5
                   F[3]=3
                       F=2

         2.000+00   48.154
>
>DØ PART 6 FØR F=.92
         9.200-01    .055
>
```

Fig. 9.7 Sample use of the program in Fig. 9.6 for odd-degree and even-degree equiripple lowpass filters.

```
         DØ PART 1
                    AMAX=.1
                      FB=1
                     NIN=2
                       N=3
                   F[1]=1.1
                   F[2]=1.5
                   F[3]=3
                       F=1.02

         1.020+00   2.329
>
>DØ PART 6 FØR F=.92
         9.200-01    .017
>
```

Fig. 8.7. The new program is based on Theorems 9.5 and 9.7; it can thus calculate the loss in either the stopband or passband. An example is given in Fig. 9.5.

Similarly, Fig. 9.6 contains a program that can be used to calculate the loss of equiripple lowpass filters. Examples are given in Fig. 9.7.

PROBLEMS

9.1 Equation (9.13) gives an expression for $K(s)K(-s)$, where K is the characteristic function.

    a. If the number of attenuation poles at infinity NIN is equal to 1, what is the constant C_1?

    b. Show that for NZ and NIN unequal to 0,

$$\omega_A \omega_B \prod_{i=1}^{m/2-1} \omega_{pi}^2 = \prod_{i=1}^{m/2} \omega_{fi}^2$$

    c. Recalling that the denominator of (9.13) is $q(s)q(-s)$, where

$$q(s) = s^{NZ} \prod_{i=1}^{N} s^2 + \omega_i^2$$

show that $m = NZ + NIN + 2N$.

9.2 *a.* Theorem 9.3 expresses the characteristic function for an equiripple lowpass filter. Simplify this for polynomial filters, that is, for Tschebycheff lowpass filters.

    b. Use the result in part a to show that the Tschebycheff polynomial can be written as

$$T_n(\omega) = \omega^n \, \text{Ev} \, (Z+1)^n$$

where

$$Z^2 = 1 + \frac{1}{s^2}\bigg|_{s=j\omega} = 1 - \frac{1}{\omega^2}$$

    c. Evaluate $T_3(\omega)$ by using the result in part b.

9.3 By analogy to the method used in Problem 9.1, show that the characteristic function for a Tschebycheff bandpass filter can be written as

$$|K(j\omega)| = \epsilon \left(\frac{\omega_A}{\omega} \frac{\omega^2 - \omega_A^2}{\omega_B^2 - \omega_A^2} \right)^{n/2} \text{Ev} \left[\left(Z + \frac{\omega_B}{\omega_A} \right)^{n/2} (Z+1)^{n/2} \right]$$

where n is the order of the filter and $Z^2 = (\omega^2 - \omega_B^2)/(\omega^2 - \omega_A^2)$.

9.4 In this problem we examine the bandpass filter described by:

$$A_{\max} = 0.1 \quad f_A = 0.9 \quad f_B = 1/0.9 \quad \text{NIN} = 1$$
$$f_1 = 0.7 \quad f_2 = 1/0.7 \quad \quad \quad \quad \quad \text{NZ} = 1$$

a. Find $F^2(Z)$ for the frequency $f = 0.3$.
b. Find $Q^2(Z)$ for the frequency $f = 0.3$.
c. Find the loss at $f = 0.3$ by using

$$A = 10 \log \left| 1 + \left[\frac{F(Z)}{Q(Z)} \right]^2 \right|$$

d. Find the loss function $L(Z)$ at $f = 0.3$.
e. Find the loss at $f = 0.3$ by using

$$A = 10 \log \left| 1 + \frac{\epsilon^2}{4} \left(|L| + \frac{1}{|L|} \right)^2 \right|$$

Hint: Use any results from the figures in the chapter.

9.5 For the filter in Example 9.4, find the loss at the passband frequency $f = 0.901$.

9.6 In this problem we examine the lowpass filter described by

$A_{max} = 0.1 \qquad f_B = 1 \qquad \text{NIN} = 1 \qquad f_1 = 1.1 \qquad f_2 = 1.5 \qquad f_3 = 3$

a. Find $L(Z)$ at $f = 2$ and $f = 0.92$.
b. Find the loss at $f = 2$ and $f = 0.92$.

CHAPTER TEN

Natural Modes for Equiripple Passband Filters

10.1 INTRODUCTION

The input/output transfer function $H(s)$ is related to the characteristic function $K(s)$ by

$$H(s)H(-s) = 1 + K(s)K(-s)$$

Because $H(s)$ and $K(s)$ have the same denominator, they can be written as

$$H(s) = \frac{e(s)}{q(s)} \qquad K(s) = \frac{f(s)}{q(s)}$$

Chapter 8 described a pole-placer program that could be used to find the denominator polynomial $q(s)$; that is, it presented a method that could be used to determine the attenuation poles.

If the passband is assumed to be equiripple, then the attenuation poles uniquely determine the attenuation zeros. Chapter 9 gave equations that could be used to find the zeros of the characteristic function; that is, the polynomial $f(s)$.

In order to construct a filter, we need to know not only the denominator $q(s)$ of the input/output transfer function, but also the numerator

$e(s)$. This chapter presents a practical* method of solving for the zeros of $H(s)$. The zeros of $H(s)$ will be termed *natural modes* because they are the natural frequencies of the transfer function.

10.2 $H(s)$ AND THE TRANSFORMED VARIABLE

$H(s) = e(s)/q(s)$ is related to the characteristic function by

$$H(s)H(-s) = \frac{e(s)e(-s)}{q(s)q(-s)} = 1 + K(s)K(-s) \qquad (10.1)$$

From (9.31) we can write $K(s)K(-s)$ as

$$K(s)K(-s) = \frac{F^2(Z)}{Q^2(Z)} \qquad (10.2)$$

where $F^2(Z)$ and $Q^2(Z)$ are given by (9.32) and (9.33). It can be written as a ratio of squared polynomials because $f(s)$ and $q(s)$ are either even or odd functions. However, $e(s)$ is neither even nor odd, so it is necessary to define two transformed functions $E(Z)$ and $E^*(Z)$. This definition is incorporated in the following theorem, which is merely a combination of (10.1), (10.2), (9.32), and (9.33).

Theorem 10.1

If $E(Z)$ and $E^*(Z)$ are defined as

$$E(Z) = \frac{e(s)}{(s^2 + \omega_A{}^2)^{m/2}} \qquad E^*(Z) = \frac{e(-s)}{(s^2 + \omega_A{}^2)^{m/2}} \qquad (10.3)$$

then
$$E(Z)E^*(Z) = F^2(Z) + Q^2(Z)$$
where

$$F^2(Z) = \epsilon^2 \left\{ \text{Ev}\left[\left(Z + \frac{\omega_B}{\omega_A}\right)^{NZ} (Z+1)^{NIN} \prod_{i=1}^{N} (Z + Z_i)^2 \right] \right\}^2 \qquad (10.5)$$

$Q^2(Z)$
$$= (-1)^{NZ+NIN}\left[Z^2 - \left(\frac{\omega_B}{\omega_A}\right)^2 \right]^{NZ} (Z^2-1)^{NIN} \prod_{i=1}^{N}(Z^2 - Z_i{}^2)^2 \qquad (10.6)$$

$m = NZ + NIN + 2N$

The remaining part of this chapter discusses how to find $e(s)$ (that is, the natural modes) once we have calculated $E(Z)E^*(Z)$.

---

* The method is practical because it uses the transformed variable to avoid numerical inaccuracies.

10.3 DETERMINATION OF $E(Z)E^*(Z)$

Figure 10.1 contains the bandpass zero-finder program which can be used to find the natural modes of equiripple bandpass filters. This section describes the parts that determine $E(Z)E^*(Z)$.

The initial part of the program (up to Step 1.46) is very similar to the initial part of the pole-placer program—it calculates some constants. Step 1.5 then directs the program to find $F^2(Z)$, and Step 1.54 directs the program to find $Q^2(Z)$. After this step, the computer has calculated the coefficients in the following equations:

$$F^2(Z) = C_0 + C_2 Z^2 + C_4 Z^4 + \cdots + C_{2M} Z^{2M}$$
$$Q^2(Z) = D_0 + D_2 Z^2 + D_4 Z^4 + \cdots + D_{2M} Z^{2M}$$

```
1; ZERO FINDER    (EQUIRIPPLE BANDPASS)
1.2 DEMAND AMAX,FA,FB,NZ,NIN,N
1.3 DEMAND F[I] FOR I=1:1:N
1.4 EE=10^(.1*AMAX)-1,E=SQRT(EE),FAA=FA*FA,FBB=FB*FB
1.41 M=NZ+NIN+2*N,TP=2*$PI,K=0,R=0,T=M
1.42 Z[I]=SQRT((F[I]*F[I]-FBB)/(F[I]*F[I]-FAA)) FOR I=1:1:N
1.43 W[I]=TP*F[I] FOR I=1:1:N
1.46 TYPE #,#
1.5 DO PART 50
1.54 DO PART 58
1.6 A[M-I/2]=C[I]+D[I] FOR I=0:2:2*M
1.64 DO PART 96
1.7 R=R+1,D=1+P[R]+Q[R]
1.72 B[R]=((P[R]/2+Q[R])*FAA+(1+P[R]/2)*FBB)/D
1.73 ZF[R]=((FBB*FBB+P[R]*FAA*FBB+Q[R]*FAA*FAA)/D)^.25
1.74 ZQ[R]=1/SQRT(2*(1-B[R]/ZF[R]^2)),ZW[R]=TP*ZF[R]
1.75 TYPE R,ZQ[R],R,ZF[R] IN FORM 1
1.77 TO PART 1.7 IF R<M/2
1.8 DO PART 6 FOR F=FB
1.82 K=AMAX-A,CH=10^(.05*K)
1.84 TYPE #,CH,#,#

6; CALCULATION OF LOSS
6.5 W=TP*F,WW=W*W,A=K-20*NZ*LOG(W)
6.6 A=A+10*LOG((WW-ZW[I]^2)^2+(W*ZW[I]/ZQ[I])^2) FOR I=1:1:M/2
6.7 A=A-10*LOG((WW-W[I]^2)^2) FOR I=1:1:N
6.8 TYPE F,A IN FORM 6 IF K<>0

50; DETERMINATION OF F(Z)=A[0]+A[2]*Z^2+A[4]*Z^4+...
50.2 S[I]=1 FOR I=1:1:NIN
50.3 S[I]=Z[I-NIN] FOR I=NIN+1:1:NIN+N
50.4 S[I]=Z[I-NIN-N] FOR I=NIN+N+1:1:NIN+2*N
50.5 S[I]=FB/FA FOR I=NIN+2*N+1:1:M
50.6 DO PART 90 FOR SN=I
50.8 A[I]=E*B[I] FOR I=0:2:M
50.93 DO PART 52 FOR I=0:2:2*M

52; DETERMINATION OF F(Z)^2=C[0]+C[2]*Z^2+C[4]*Z^4+...
52.1 JI=0,JF=I IF I<M+2
52.2 JI=I-M,JF=M IF I>M
52.3 C[I]=0
52.4 C[I]=C[I]+A[J]*A[I-J] FOR J=JI:2:JF
```

Fig. 10.1 Zero-finder program for equiripple bandpass filters. (*Continued on next page.*)

```
58; DETERMINATIØN ØF Q(Z)^2=D[0]+D[2]*Z^2+D[4]*Z^4+...
58.2 S[I]=-1 FØR I=1:1:NIN
58.3 S[I]=-Z[I-NIN]^2 FØR I=NIN+1:1:NIN+N
58.4 S[I]=S[I-N] FØR I=NIN+N+1:1:NIN+2*N
58.5 S[I]=-FBB/FAA FØR I=NIN+2*N+1:1:M
58.6 DØ PART 90 FØR SN=M
58.8 D[I]=B[I/2] FØR I=0:2:2*M


90; PRØD(X+S[I])=B[0]+B[1]*X+B[2]*X^2+...+B[SN]*X^SN
90.1 B[0]=S[1],B[1]=1,J=1
90.2 J=J+1
90.3 A[0]=S[J]*B[0]
90.4 A[I]=B[I-1]+S[J]*B[I] FØR I=1:1:J-1
90.5 B[I]=A[I] FØR I=0:1:J-1
90.6 B[J]=1
90.7 DØ PART 90.2 IF J<SN

96; FACTØR FINDER   (X^2+P*X+Q ØR X-A)
96.01; F(X)=A[0]*X^T+A[1]*X^(T-1)+...+A[T-1]*X+A[T]
96.1 A[I]=A[I]/A[0] FØR I=1:1:T
96.12 A[0]=1,B[0]=1,C[0]=1,I1=0
96.2 P=0,Q=0,I1=I1+1
96.3 B[1]=A[1]-P,C[1]=B[1]-P
96.32 B[I]=A[I]-P*B[I-1]-Q*B[I-2] FØR I=2:1:T
96.4 C[I]=B[I]-P*C[I-1]-Q*C[I-2] FØR I=2:1:T-1
96.5 X1=T-1,X2=T-2,X3=T-3,X4=C[X2]^2+C[X3]*(B[X1]-C[X1])
96.52 X4=.001 IF X4=0
96.6 DDP=(B[X1]*C[X2]-B[T]*C[X3])/X4,P=P+DDP
96.62 DQ=(B[T]*C[X2]-B[X1]*(C[X1]-B[X1]))/X4,Q=Q+DQ
96.64 TØ PART 96.3 IF  DDP + DQ >10^-6
96.7 P[I1]=P,Q[I1]=Q,A[1]=A[1]-P,T=T-2
96.72 A[I]=A[I]-P*A[I-1]-Q*A[I-2] FØR I=2:1:T
96.74 TØ PART 96.2 IF T>2
96.76 I1=I1+1,P[I1]=A[1],Q[I1]=A[2] IF T=2
96.78 A=-A[1] IF T=1


FØRM       1
ZQ[##]=###.####   ZF[##]=#.####^^^
FØRM       6
#.###^^^  ###.###
```

Fig. 10.1 (Continued)   Zero-finder program for equiripple bandpass filters.

Step 1.6 then yields the coefficients in

$$E(Z)E^*(Z) = F^2(Z) + Q^2(Z) = A_M + A_{M-1}Z^2 + \cdots + A_0 Z^{2M} \quad (10.7)$$

Since for bandpass filters M is an even number, we can write, as in Eq. (7.24a),

$$E(Z)E^*(Z) = C \prod_{i=1}^{M/2} Z^4 + p_i Z^2 + q_i \quad (10.8)$$

Given the coefficients in (10.7), the factors in (10.8) can be determined by using a root-finding method such as Bairstow's iterative procedure.[1] This method is used in Part 96.

10.4 DETERMINATION OF $(e)s$

From the previous section we have $E(Z)E^*(Z)$ in the form*

$$E(Z)E^*(Z) = C \prod_{i=1}^{M/2} Z^4 + p_i Z^2 + q_i \qquad (10.9)$$

Using Theorem 7.5 and Eq. (7.28), one obtains

$$e(s) = C_H \prod_{i=1}^{M/2} \left[s^2 + 2 \left(\frac{a_i - b_i}{2} \right)^{1/2} s + a_i \right] \qquad (10.10)$$

where
$$a_i^2 = \frac{\omega_B^4 + p_i \omega_A^2 \omega_B^2 + q_i \omega_A^4}{1 + p_i + q_i}$$
$$b_i = \frac{(p_i/2 + q_i)\omega_A^2 + (1 + p_i/2)\omega_B^2}{1 + p_i + q_i} \qquad (10.11)$$

Or, alternatively, we can write

$$e(s) = C_H \prod_{i=1}^{M/2} \left[s^2 + \frac{2\pi ZF_i}{ZQ_i} s + (2\pi ZF_i)^2 \right] \qquad (10.12)$$

where ZF_i is called the *undamped zero* of $e(s)$, and ZQ_i indicates the quality of the zero: the higher the quality, the closer the zero is to the imaginary axis.

It follows from (10.10) through (10.12) that

$$ZF_i = \left(\frac{f_B^4 + p_i f_A^2 f_B^2 + q_i f_A^4}{1 + p_i + q_i} \right)^{1/4} \qquad (10.13)$$

$$ZQ_i = \frac{1}{[2(1 - b_i/a_i)]^{1/2}} \qquad (10.14)$$

These calculations are performed in Steps 1.7 to 1.75.

To determine the constant C_H in (10.12), we can use the fact that $H(s)$ can be written as

$$H(s) = C_H \frac{\prod_{i=1}^{M/2} [s^2 + (ZW_i/ZQ_i)s + ZW_i^2]}{s^{NZ} \prod_{i=1}^{N} (s^2 + \omega_i^2)} \qquad (10.15)$$

---

* The constant C was not determined, as it will not be needed.

where $ZW_i = 2\pi ZF_i$. Thus the loss can be written as

$$A(\omega) = 20 \log C_H + 10 \log \left\{ \frac{\prod_{i=1}^{M/2} [(\omega^2 - ZW_i^2)^2 + (\omega ZW_i/ZQ_i)^2]}{\omega^{2NZ} \prod_{i=1}^{N} (\omega^2 - \omega_i^2)^2} \right\}$$

(10.16)

This equation can be written in simplified notation as

$$A(\omega) = k + A_1(\omega) \qquad (10.17)$$

where $k = 20 \log C_H$, and $A_1(\omega)$ is the remaining term in (10.16). The constant k can be determined from

$$k = A(\omega_B) - A_1(\omega_B) = A_{\max} - A_1(\omega_B) \qquad (10.18)$$

This calculation is performed by Steps 1.8 and 1.82. The constant C_H is then evaluated as

$$C_H = 10^{0.05k} \qquad (10.19)$$

Once C_H is determined, the loss can be found at any frequency F_0 by typing DO PART 6 FOR F = F_0.

Example 10.1

This example demonstrates the use of the program in Fig. 10.1. We shall find the natural modes of the following equiripple bandpass filter:

$A_{\max} = 0.1 \qquad f_A = 0.9 \qquad f_B = 1/0.9$
$NZ = 1 \qquad NIN = 1 \qquad f_1 = 0.7 \qquad f_2 = 1/0.7$

The example is a continuation of the problem that was started in Fig. 9.5.

```
DØ PART 1
       AMAX=.1
       FA=.9
       FB=1/.9
       NZ=1
       NIN=1
       N=2
       F[1]=.7
       F[2]=1/.7

ZQ[ 1]=  10.8205   ZF[ 1]=1.1364+00
ZQ[ 2]=   4.6153   ZF[ 2]=1.0000+00
ZQ[ 3]=  10.8205   ZF[ 3]=8.7998-01

CH=         5.2489 1408

>DØ PART 6 FØR F=.3,.905
3.000-01    40.488
9.050-01      .029
>
```

Fig. 10.2 Sample use of the equiripple bandpass zero finder.

Approximation Methods for Electronic Filter Design

Figure 10.2 shows how one can use the zero finder. In terms of the notation in Fig. 10.2, the input/output transfer function is

$$H(s) = C_H \frac{\prod_{i=1}^{3} s^2 + (2\pi ZF[i]/ZQ[i])s + (2\pi ZF[i])^2}{s[s^2 + (2\pi \times 0.7)^2][s^2 + (2\pi/0.7)^2]}$$

The loss is given by $A(f) = 20 \log |H(j2\pi f)|$. This can be calculated by using Part 6, as is illustrated at the bottom of Fig. 10.2.

Recall that we can also calculate the loss of an equiripple filter by using the loss function $L(Z)$. In fact, the loss for this filter was so found in Fig. 9.5. Those results agree with the results in Fig. 10.2.

```
1; ZERØ FINDER (EQUIRIPPLE LØWPASS)
1.2 DEMAND AMAX,FB,NIN,N
1.3 DEMAND F[I] FØR I=1:1:N
1.4 EE=10^(.1*AMAX)-1,E=SQRT(EE),FBB=FB*FB
1.41 M=NIN+2*N,EM=2*IP(M/2),TP=2*$PI,K=0,R=0,T=M
1.42 Z[I]=SQRT(1-FBB/F[I]/F[I]) FØR I=1:1:N
1.43 W[I]=TP*F[I] FØR I=1:1:N
1.46 TYPE #,#
1.5 DØ PART 50
1.54 DØ PART 58
1.59 C[2*M]=0 IF M>EM
1.6 A[M-I/2]=C[I]+D[I] FØR I=0:2:2*M
1.64 DØ PART 96
1.7 R=R+1,D=1+P[R]+Q[R]
1.72 B[R]=(1+P[R]/2)*FBB/D
1.73 ZF[R]=FB/D^.25
1.74 ZQ[R]=1/SQRT(2*(1-B[R]/ZF[R]^2)),ZW[R]=TP*ZF[R]
1.75 TYPE R,ZQ[R],R,ZF[R] IN FØRM 1
1.77 IØ PART 1.7 IF R<EM/2
1.78 TYPE SIGMA FØR SIGMA=SQRT(FBB/(A-1))*TP IF M>EM
1.8 DØ PART 6 FØR F=FB
1.82 K=AMAX-A,CH=10^(.05*K)
1.84 TYPE #,CH,#,#

6; CALCULATIØN ØF LØSS
6.5 W=TP*F,WW=W*W,A=K
6.6 A=A+10*LØG((WW-ZW[I]^2)^2+(W*ZW[I]/ZQ[I])^2) FØR I=1:1:M/2
6.7 A=A-10*LØG(WW-W[I]^2)^2) FØR I=1:1:N
6.75 A=A+10*LØG(WW+SIGMA^2) IF M>EM
6.8 TYPE F,A IN FØRM 6 IF K<>0

50; DETERMINATIØN ØF F(Z)=A[0]+A[2]*Z^2+A[4]*Z^4+...
50.2 S[I]=1 FØR I=1:1:NIN
50.3 S[I]=Z[I-NIN] FØR I=NIN+1:1:NIN+N
50.4 S[I]=Z[I-NIN-N] FØR I=NIN+N+1:1:NIN+2*N
50.6 DØ PART 90 FØR SN=I
50.8 A[I]=E*B[I] FØR I=0:2:EM
50.93 DØ PART 52 FØR I=0:2:2*EM

52; DETERMINATIØN ØF F(Z)^2=C[0]+C[2]*Z^2+C[4]*Z^4+...
52.1 JI=0,JF=I IF I<EM+2
52.2 JI=I-EM,JF=EM IF I>EM
52.3 C[I]=0
52.4 C[I]=C[I]+A[J]*A[I-J] FØR J=JI:2:JF
```

Fig. 10.3 Zero-finder program for equiripple lowpass filters. *(Continued on next page.)*

```
58; DETERMINATIØN ØF Q(Z)^2=D[0]+D[2]*Z^2+D[4]*Z^4+...
58.2 S[I]=-1 FØR I=1:1:NIN
58.3 S[I]=-Z[I-NIN]^2 FØR I=NIN+1:1:NIN+N
58.4 S[I]=S[I-N] FØR I=NIN+N+1:1:NIN+2*N
58.6 DØ PART 90 FØR SN=M
58.7 DD=(-1)^NIN
58.8 D[I]=DD*B[I/2] FØR I=0:2:2*M

90; PRØD(X+S[I])=B[0]+B[1]*X+B[2]*X^2+...+B[SN]*X^SN
90.1 B[0]=S[1],B[1]=1,J=1
90.2 J=J+1
90.3 A[0]=S[J]*B[0]
90.4 A[I]=B[I-1]+S[J]*B[I] FØR I=1:1:J-1
90.5 B[I]=A[I] FØR I=0:1:J-1
90.6 B[J]=1
90.7 DØ PART 90.2 IF J<SN

96; FACTØR FINDER  (X^2+P*X+Q ØR X-A)
96.01; F(X)=A[0]*X^T+A[1]*X^(T-1)+...+A[T-1]*X+A[T]
96.1 A[I]=A[I]/A[0] FØR I=1:1:T
96.12 A[0]=1,B[0]=1,C[0]=1,I1=0
96.2 P=0,Q=0,I1=I1+1
96.3 B[1]=A[1]-P,C[1]=B[1]-P
96.32 B[I]=A[I]-P*B[I-1]-Q*B[I-2] FØR I=2:1:T
96.4 C[I]=B[I]-P*C[I-1]-Q*C[I-2] FØR I=2:1:T-1
96.5 X1=T-1,X2=T-2,X3=T-3,X4=C[X2]^2+C[X3]*(B[X1]-C[X1])
96.52 X4=.001 IF X4=0
96.6 DDP=(B[X1]*C[X2]-B[T]*C[X3])/X4,P=P+DDP
96.62 DQ=(B[T]*C[X2]-B[X1]*(C[X1]-B[X1]))/X4,Q=Q+DQ
96.64 TØ PART 96.3 IF  DDP + DQ >10^-6
96.7 P[I1]=P,Q[I1]=Q,A[1]=A[1]-P,T=T-2
96.72 A[I]=A[I]-P*A[I-1]-Q*A[I-2] FØR I=2:1:T
96.74 TØ PART 96.2 IF T>2
96.76 I1=I1+1,P[I1]=A[1],Q[I1]=A[2] IF T=2
96.78 A=-A[1] IF T=1


FØRM     1
ZQ[##]=###.####   ZF[##]=#.####^^^
FØRM     6
#.###^^^  ###.###
```

Fig. 10.3 (Continued) Zero-finder program for equiripple lowpass filters.

10.5 DETERMINATION OF NATURAL MODES OF EQUIRIPPLE LOWPASS FILTERS

Figure 10.3 contains a lowpass zero-finder program that can be used to find the natural modes of an equiripple lowpass filter. This program is very similar to the bandpass program in Fig. 10.1. As was true in the bandpass zero finder, after Step 1.54 the computer has calculated the coefficients in the following equations:

$$F^2(Z) = C_0 + C_2 Z^2 + C_4 Z^4 + \cdots + C_{2M} Z^{2M}$$
$$Q^2(Z) = D_0 + D_2 Z^2 + D_4 Z^4 + \cdots + D_{2M} Z^{2M}$$

Step 1.6 then yields the coefficients in

$$E(Z)E^*(Z) = F^2(Z) + Q^2(Z) = A_M + A_{M-1} Z^2 + \cdots + A_0 Z^{2M} \quad (10.20)$$

As indicated in Eq. (7.24), this relation can be written in one of two ways:

M even:
$$E(Z)E^*(Z) = C \prod_{i=1}^{M/2} Z^4 + p_i Z^2 + q_i \quad (10.21a)$$

M odd:
$$E(Z)E^*(Z) = C(Z^2 - a) \prod_{i=1}^{(M-1)/2} Z^4 + p_i Z^2 + q_i \quad (10.21b)$$

Given the coefficients in (10.20), the factors in (10.21) can be determined by using a root-finding method such as the one mentioned in Sec. 10.3.

Once the roots have been determined, Theorem 7.5 and Eqs. (7.27) and (7.28) can be applied to yield*

M even:
$$e(s) = C_H \prod_{i=1}^{M/2} \left[s^2 + 2\left(\frac{a_i - b_i}{2}\right)^{1/2} s + a_i \right] \quad (10.22a)$$

M odd:
$$e(s) = C_H(s + \sigma_0) \prod_{i=1}^{(M-1)/2} \left[s^2 + 2\left(\frac{a_i - b_i}{2}\right)^{1/2} s + a_i \right] \quad (10.22b)$$

This can be rewritten as

M even:
$$e(s) = C_H \prod_{i=1}^{M/2} \left[s^2 + \frac{2\pi ZF_i}{ZQ_i} s + (2\pi ZF_i)^2 \right] \quad (10.23a)$$

M odd:
$$e(s) = C_H(s + \sigma_0) \prod_{i=1}^{(M-1)/2} \left[s^2 + \frac{2\pi ZF_i}{ZQ_i} s + (2\pi ZF_i)^2 \right] \quad (10.23b)$$

The program in Fig. 10.3 finds the natural modes of any equiripple lowpass filter by evaluating the preceding equations. Two examples of its application are shown in Fig. 10.4. The data at the top of the figure describe a fifth-order Tschebycheff lowpass filter.† It has a ripple of 0.5 dB and a passband edge of 1 Hz. In terms of the notation in Fig. 10.4, the input/output transfer function is

$$H(s) = C_H(s + \text{SIGMA}) \prod_{i=1}^{2} \left[s^2 + \frac{2\pi ZF_i}{ZQ_i} s + (2\pi ZF_i)^2 \right]$$

In particular, the qualities of the two zeros are

$$ZQ[1] = 4.5450 \quad ZQ[2] = 1.1778$$

These values agree with the qualities in Table 3.3.

---

* The constants a_i, b_i, and σ_0 can be found by applying (7.26).
† By definition, a Tschebycheff filter has no finite loss peaks; thus $N = 0$.

The filter just described is an odd-degree Tschebycheff lowpass filter; the bottom of Fig. 10.4 contains an even-degree Tschebycheff lowpass filter.

The equiripple zero finder can be used for elliptic filters as well as Tschebycheff filters. An elliptic filter, as described in Chap. 5, has an equiripple stopband and an equiminima stopband. That is, the attenuation poles are chosen such that the loss minimums in the stopband are all equal. Chapter 5 demonstrated that elliptic functions could be used to find the location of the attenuation peaks. Instead of using elliptic functions, one can use the pole-placer program of Chap. 8 to determine the attenuation poles of an elliptic filter; this was demonstrated in Fig. 8.12. Once the attenuation poles have been determined, the natural modes can be evaluated by using the zero-finder program. An example is given in Fig. 10.5, which uses the lowpass zero finder to analyze a lowpass elliptic filter. This is the filter that was first investigated in Fig. 5.13. The results in both figures are consistent, as demonstrated by the fact that both yield $A(26) = 46.854$ dB.

Of course, the lowpass zero finder can also be used for the more general arbitrary stopband case (as contrasted with an equiminima stopband). This is demonstrated in Fig. 10.6, where two filters are analyzed; they were previously encountered in Fig. 9.7. The validity of the results can

```
              DØ PART 1
                 AMAX=.5
                 FB= 1
                 NIN=5
                 N=0


        ZQ[ 1]=   4.5450    ZF[ 1]=1.0177+00
        ZQ[ 2]=   1.1778    ZF[ 2]=6.9048-01
            SIGMA=             2.27652134

                  CH=         .000570733547


        >


              DØ PART 1
                 AMAX=.5
                 FB= 1
                 NIN=6
                 N=0

        ZQ[ 1]=   6.5128    ZF[ 1]=1.0114+00
        ZQ[ 2]=   1.8104    ZF[ 2]=7.6812-01
        ZQ[ 3]=    .6836    ZF[ 3]=3.9623-01

                  CH=         .00018167013


        >
```

Fig. 10.4 Sample use of the equiripple lowpass zero finder for Tschebycheff filters.

182 Approximation Methods for Electronic Filter Design

```
DØ PART 1
        AMAX=.1
        FB= 20
        NIN=0
        N=3
        F[1]=82.6051
        F[2]=33.2858
        F[3]=26.5772


ZQ[ 1]=    7.8805    ZF[ 1]=2.0827+01
ZQ[ 2]=    1.7888    ZF[ 2]=1.8331+01
ZQ[ 3]=     .6250    ZF[ 3]=1.2975+01

        CH=      220.139445
```

Fig. 10.5 Sample use of the equiripple lowpass zero finder for an elliptic filter.

```
>DØ PART 6 FØR F=26
 2.600+01   46.854
>

 DØ PART 1
        AMAX=.1
        FB= 1
        NIN=1
        N=3
        F[1]=1.1
        F[2]=1.5
        F[3]=3


ZQ[ 1]=   15.9861    ZF[ 1]=1.0220+00
ZQ[ 2]=    3.1255    ZF[ 2]=9.5942-01
ZQ[ 3]=     .9810    ZF[ 3]=7.4670-01
        SIGMA=       3.32812512

        CH=       13.732973

>DØ PART 6 FØR F=.92,2
 9.200-01     .055
 2.000+00   48.154
>

 DØ PART 1
        AMAX=.1
        FB= 1
        NIN=2
        N=3
        F[1]=1.1
        F[2]=1.5
        F[3]=3


ZQ[ 1]=   18.4896    ZF[ 1]=1.0178+00
ZQ[ 2]=    4.0489    ZF[ 2]=9.5904-01
ZQ[ 3]=    1.4312    ZF[ 3]=7.7778-01
ZQ[ 4]=     .6062    ZF[ 4]=4.9918-01

        CH=       4.37134107
```

Fig. 10.6 Sample use of the equiripple lowpass zero finder for an arbitrary stopband.

```
>DØ PART 6 FØR F=.92,1.02
 9.200-01     .017
 1.020+00    2.329
>
```

be checked by comparing the losses that were calculated by two completely different methods.

10.6 CONCLUSIONS

We now know how to determine the input/output transfer function $H(s)$ for filters that have equiripple passbands. First, we use a pole-placer program to determine the denominator of $H(s)$. Next, we use a zero-finder program to determine the numerator of $H(s)$.

The pole-placer program was discussed in Chap. 8, and the zero-finder program in Chap. 10. These programs are not extremely efficient in their use of computer time. Also, they are written in the Telcomp language, which is not a universally available computer language. However, the purpose of presenting the programs was not efficiency or universality; it was to demonstrate that the computer can be used to find the transfer function $H(s)$. The Telcomp language was used because it is simple to understand.

The pole-placer program and zero finder were demonstrated for equiripple passbands and arbitrary stopbands. Since Tschebycheff filters and elliptic filters are special cases of the more general problem, the programs can also be used for these filters. However, they cannot be used for filters that have maximally flat passbands. These will be discussed in the next chapter.

REFERENCE

1. L. G. Kelly, *Handbook of Numerical Methods and Applications*, Addison-Wesley Publishing Company, Inc., Reading, Mass., 1967.

PROBLEMS

10.1 A certain bandpass filter has no attenuation poles at the origin or infinity. There is one attenuation pole located at the transformed frequency Z_1. Find $E(Z)E^*(Z)$ as a function of Z, Z_1, and ϵ.

10.2 Suppose that a bandpass filter has passband edges $\omega_A = 1$ and $\omega_B = 2$.
    a. Except for a constant multiplier, find $e(s)$ if $E(Z)E^*(Z) = 1.5Z^4 - 0.5Z^2 + 0.375$.
    b. From the answer to part a, find the quality of the natural mode.
    c. Find the quality of the natural mode by using Eq. (10.14).

10.3 Show that $ZQ_i = [2(1 - b_i/a_i)]^{-1/2}$, as stated in Eq. (10.14).

10.4 We want to find a transfer function that meets the following loss requirements.

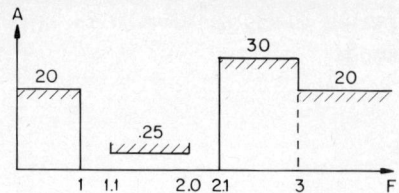

a. If we want one attenuation pole at infinity, one attenuation pole at the origin, and two poles at finite frequencies f_1, f_2:
   1. What is the form of $q(s)$?
   2. What is the form of $Q(z)$?
   3. What is the form of $F(z)$?
   4. What is the form of $f(s)$?

b. For the assumptions in part a, how would you answer the pole placer's responses to DO PART 1?

c. Assume that after the proper answers are given in part b, the computer indicates that the requirements are not met.
   1. How would the computer indicate that the requirements are not met?
   2. What would you do to meet the requirements?

d. Assume that the computer now indicates that the stopband requirements are satisfied with a 7.5-dB factor of safety. If we are worried that lossy elements will cause additional attenuation near the passband edges, what could we do?

e. Assume that the final design has one attenuation pole at infinity, one at the origin, and three poles at finite frequencies.
   1. How would you answer the zero finder's response to DO PART 1?
   2. Suppose, after a long wait, you decide that the factor-finder subroutine (Part 96) cannot find a factor. What could you do to help it?
   3. How can you determine whether or not the results of the zero finder agree with the results of the pole placer?

CHAPTER ELEVEN

Maximally Flat Passbands

11.1 INTRODUCTION

In Chap. 3 we discussed Tschebycheff filters. These filters have an equiripple passband, and all attenuations poles are located at infinity. We next discussed elliptic filters which, like the Tschebycheff filters, have an equiripple passband; however, the attenuation poles are located in such a manner as to yield a stopband with equal minimums. We then discussed filters that have an equiripple stopband but arbitrary passband. For these we used a pole-placer program to yield an "optimum" stopband.

We have also discussed filters that have maximally flat passbands instead of equiripple passbands. The first maximally flat filter discussed was the Butterworth, which is maximally flat at the origin. Like the Tschebycheff approximation, this has no finite attenuation poles. The inverse Tschebycheff filter is also maximally flat at the origin, but it has finite attenuation poles which were chosen so as to yield a stopband with equal minimums.

This chapter investigates filters that have maximally flat passbands and arbitrary stopbands. We shall first determine the form of the characteristic function that is necessary for a maximally flat passband. This will allow us to calculate the loss of a maximally flat filter—if we know the pole locations. In order to choose the pole locations, we shall again use a pole-placer program. Once we have the attenuation poles, we can calculate the natural modes by using a zero-finder program.

Thus, at the completion of this chapter we shall have two sets of tools at our disposal:

1. A pole-placer program and zero-finder program for equiripple filters.
2. A pole-placer program and zero-finder program for maximally flat filters.

Which set of tools the individual designer uses depends, to an extent, on his background. It has been this author's experience (as indicated in Sec. 11.7) that equiripple filters are usually preferable. However, there are special cases in which the maximally flat characteristic might be better. For example, an inverse Tschebycheff filter has approximate vestigial symmetry, which is useful in the transmission of data.[1]

The notation used throughout this chapter is the same as in the previous chapters which dealt with equiripple filters. In particular, the characteristic function $K(s)$ is written as $K(s) = f(s)/q(s)$. As usual, we shall assume that $q(s)$ is of the form

$$q(s) = s^{NZ} \prod_{i=1}^{N} (s^2 + \omega_i^2) \tag{11.1}$$

The problem we face in this chapter is, "Given $q(s)$, how does one find $f(s)$ so that the passband is maximally flat?"

Again, this problem will be solved in terms of the transformed variable. Thus our solution will be of the form

$$H(s)H(-s) = 1 + \frac{F^2(Z)}{Q^2(Z)} \tag{11.2}$$

Once we have found $F(Z)$ so that the passband is maximally flat, we can find $H(s)$ by the method used for equiripple filters.

11.2 DETERMINATION OF THE CHARACTERISTIC FUNCTION[2]

The input/output transfer function $H(s)$ is given by

$$H(s)H(-s) = 1 + \frac{F^2(Z)}{Q^2(Z)}$$

where for $q(s)$, as in (11.1),

$$Q^2(Z) = C_Q^2 \left[Z^2 - \left(\frac{\omega_B}{\omega_A}\right)^2 \right]^{NZ} (Z^2 - 1)^{NIN} \prod_{i=1}^{N} (Z^2 - Z_i^2)^2 \tag{11.3}$$

Section 9.4 showed how, given $Q^2(Z)$ as in (11.3), to find $F^2(Z)$ so that the passband is equiripple. This section shows how to find $F^2(Z)$ so that the passband is maximally flat.

The characteristic function for a maximally flat bandpass filter is shown in Fig. 11.1. By inspection of this figure* we have

$$|K|^2 = \frac{C_K^2(\omega^2 - \omega_0^2)^m}{\omega^{2NZ}\prod_{i=1}^{N}(\omega^2 - \omega_i^2)^2} \tag{11.4}$$

or, equivalently,

$$K(s)K(-s) = \frac{C_K^2(s^2 + \omega_0^2)^m}{s^{2NZ}\prod_{i=1}^{N}(s^2 + \omega_i^2)^2} \tag{11.5}$$

But we also know that

$$K(s)K(-s) = \frac{f(s)f(-s)}{q(s)q(-s)} \tag{11.6}$$

Equating numerator polynomials yields

$$f(s) = C_f(s^2 + \omega_0^2)^{m/2} \tag{11.7}$$

where C_f is an unimportant constant.

In terms of the transformed frequency, Eq. (11.7) is equivalent to†

$$\frac{f(s)}{(s^2 + \omega_A^2)^{m/2}} = F(Z) = C_F(Z^2 + Z_0^2)^{m/2} \tag{11.8}$$

where

$$Z_0^2 = \frac{\omega_B^2 - \omega_0^2}{\omega_0^2 - \omega_A^2} \tag{11.9}$$

---

* It is shown in Problem 11.1 that $m = NZ + NIN + 2N$, where $NZ + NIN$ must be even. Problem 11.2 shows why this is called a maximally flat characteristic.

† Note that, in (11.9), $-Z_0^2$ is the transformed variable that corresponds to ω_0^2. The minus sign has been introduced so that Z_0^2 is a positive number. The fact that it is positive is illustrated in (11.12).

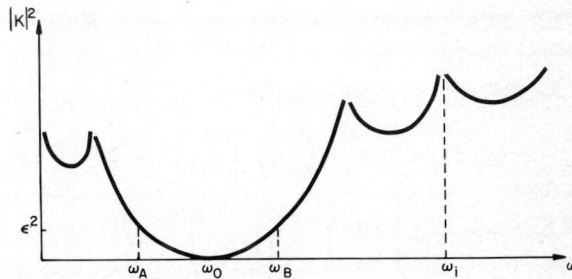

Fig. 11.1 Characteristic function for a maximally flat bandpass filter.

188 Approximation Methods for Electronic Filter Design

There are two unknowns in (11.8) and (11.9): C_F and ω_0. These unknowns can be found by recalling that

$$K(s)K(-s) = \frac{F^2(Z)}{Q^2(Z)}$$

where, from (9.11),

$$Q^2(Z) = \left[Z^2 - \left(\frac{\omega_B}{\omega_A}\right)^2\right]^{NZ} (Z^2 - 1)^{N_{IN}} \prod_{i=1}^{N} (Z^2 - Z_i^2)^2 \quad (11.10)$$

To find C_F and ω_0 we note from Fig. 11.1 that

$$|K(j\omega_A)|^2 = \epsilon^2 = \frac{F^2(\infty)}{Q^2(\infty)}$$

$$|K(j\omega_B)|^2 = \epsilon^2 = \frac{F^2(0)}{Q^2(0)}$$

But from (11.8) and (11.10)

$$\frac{F^2(\infty)}{Q^2(\infty)} = C_F^2 \qquad \frac{F^2(0)}{Q^2(0)} = \frac{C_F^2 Z_0^{2m}}{(\omega_B/\omega_A)^{2NZ} \prod_{i=1}^{N} Z_i^4}$$

These equations yield

$$C_F^2 = \epsilon^2 \quad (11.11)$$

$$Z_0^2 = \left[\left(\frac{\omega_B}{\omega_A}\right)^{2NZ} \prod_{i=1}^{N} Z_i^4\right]^{1/m} \quad (11.12)$$

These results are combined in the following theorem.

Theorem 11.1

The characteristic function $K(s)$ of a maximally flat bandpass filter is given by

$$K(s)K(-s) = \frac{F^2(Z)}{Q^2(Z)}$$

where $F^2(Z) = \epsilon^2(Z^2 + Z_0^2)^m$

$$Q^2(Z) = \left[Z^2 - \left(\frac{\omega_B}{\omega_A}\right)^2\right]^{NZ} (Z^2 - 1)^{N_{IN}} \prod_{i=1}^{N} (Z^2 - Z_i^2)^2$$

$$Z_0^2 = \left[\left(\frac{\omega_B}{\omega_A}\right)^{2NZ} \prod_{i=1}^{N} Z_i^4\right]^{1/m}$$

$$m = NZ + N_{IN} + 2N = \text{even number}$$

Corollary 11.1

The point of maximum flatness is given by

$$\omega_0 = \left(\frac{\omega_B{}^2 + Z_0{}^2\omega_A{}^2}{1 + Z_0{}^2}\right)^{1/2}$$

where $Z_0{}^2 = \left[\left(\dfrac{\omega_B}{\omega_A}\right)^{2NZ} \prod_{i=1}^{N} Z_i{}^4\right]^{1/m}$

11.3 DETERMINATION OF $Q^2(Z)$ AND $F^2(Z)$ FOR LOWPASS FILTERS

Theorem 11.1 describes how we can find $Q^2(Z)$ and $F^2(Z)$ for bandpass filters that have maximally flat passbands. This section, which is analogous to Sec. 9.5, derives similar expressions for maximally flat lowpass filters.

Consider the maximally flat lowpass characteristic function shown in Fig. 11.2. This characteristic function is the same as that in Fig. 11.1, except $\omega_0 = 0 = \omega_A$ and NZ = 0. Thus, by analogy to Eq. (11.5),

$$K(s)K(-s) = \frac{C_K{}^2 s^{2m}}{\prod_{i=1}^{N}(s^2 + \omega_i{}^2)^2} \tag{11.13}$$

where $m = \text{NIN} + 2N$.

Since the numerator of $K(s)$ is $f(s)$, it follows that

$$f(s) = C_f s^m \tag{11.14}$$

But for lowpass filters

$$F(Z) = \frac{f(s)}{s^m} \tag{11.15}$$

Combining (11.14) and (11.15), we can see that $F(Z)$ must simply be a

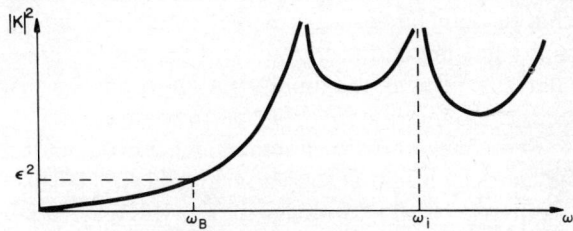

Fig. 11.2 Characteristic function for a lowpass filter maximally flat at the origin.

constant. This constant can be evaluated by recalling from (9.30) that

$$Q^2(Z) = (-1)^{NIN}(Z^2 - 1)^{NIN} \prod_{i=1}^{N} (Z^2 - Z_i^2)^2 \qquad (11.16)$$

Thus, since

$$|K(j\omega_B)|^2 = \epsilon^2 = \frac{F^2(0)}{Q^2(0)}$$

it follows that the constant $F(Z)$ must be given by

$$F^2(Z) = \epsilon^2 \prod_{i=1}^{N} Z_i^4 \qquad (11.17)$$

These results are summarized in the next theorem.

Theorem 11.2

The characteristic function of a lowpass filter that is maximally flat at the origin is given by

$$K(s)K(-s) = \frac{F^2(Z)}{Q^2(Z)}$$

where $F^2(Z) = \epsilon^2 \prod_{i=1}^{N} Z_i^4$

$$Q^2(Z) = (-1)^{NIN}(Z^2 - 1)^{NIN} \prod_{i=1}^{N} (Z^2 - Z_i^2)^2$$

Theorem 11.2 describes the characteristic function of a lowpass filter that is maximally flat at the origin as indicated in Fig. 11.2. Lowpass filters can also be maximally flat beyond the origin. P. Aronhime and A. Budak[3] discuss polynomial lowpass filters (i.e., no finite attenuation poles) that are maximally flat beyond the origin. The material in this chapter can be easily extended to include lowpass filters that are maximally flat beyond the origin and have finite attenuation poles. The fact that the filter is maximally flat beyond the origin can allow it to give better attenuation or delay performance.

The characteristic function of a lowpass filter that is maximally flat beyond the origin is shown in Fig. 11.3. If this lowpass characteristic is compared to the bandpass in Fig. 11.1, it can be seen that the lowpass may be considered to be a special case of the bandpass. That is, it is a bandpass with the lower passband edge at the origin ($\omega_A = 0$). Furthermore, there can be no attenuation poles at the origin (NZ = 0).

By analogy to (11.5), the characteristic function for the filter of Fig. 11.3 is of the form

$$K(s)K(-s) = \frac{C_K{}^2(s^2 + \omega_0{}^2)^m}{\prod_{i=1}^{N}(s^2 + \omega_i{}^2)^2} \qquad (11.18)$$

where $m = \text{NIN} + 2N$. Thus, since the numerator of $K(s)$ is $f(s)$, we must have

$$f(s) = C_f(s^2 + \omega_0{}^2)^{m/2} \qquad (11.19)$$

This $f(s)$ is the same as the $f(s)$ in (11.7); thus, the next theorem follows by analogy to Theorem 11.1.

Theorem 11.3

The characteristic function $K(s)$ of a lowpass filter that is maximally flat beyond the origin is given by

$$K(s)K(-s) = \frac{F^2(Z)}{Q^2(Z)}$$

where $F^2(Z) = \epsilon^2(Z^2 + Z_0{}^2)^m$

$$Q^2(Z) = (Z^2 - 1)^{\text{NIN}} \prod_{i=1}^{N} (Z^2 - Z_i{}^2)^2$$

$$Z_0{}^2 = \left(\prod_{i=1}^{N} Z_i{}^4\right)^{1/m}$$

$$m = \text{NIN} + 2N = \text{even number}$$

Corollary 11.3

The point of maximum flatness is given by

$$\omega_0 = \left(\frac{\omega_B{}^2}{1 + Z_0{}^2}\right)^{\frac{1}{2}} \qquad (11.20)$$

where $Z_0{}^2 = \left[\prod_{i=1}^{N} Z_i{}^4\right]^{1/m}$

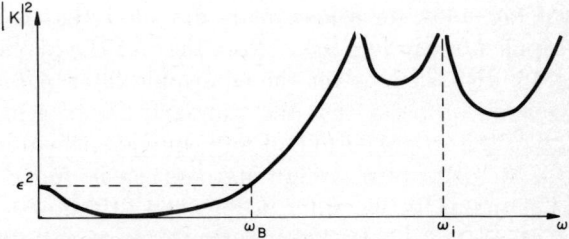

Fig. 11.3 Characteristic function for a lowpass filter maximally flat beyond the origin.

11.4 STOPBAND AND PASSBAND PERFORMANCE

The formulas given in the previous sections can be used to calculate the loss of maximally flat filters. This section describes programs that can be used for such calculations. The programs are based on the following formulas:

Bandpass (Figure 11.1)

$$F^2(Z) = \epsilon^2 (Z^2 + Z_0^2)^m$$

$$Q^2(Z) = \left[Z^2 - \left(\frac{\omega_B}{\omega_A}\right)^2\right]^{NZ} (Z^2 - 1)^{NIN} \prod_{i=1}^{N} (Z^2 - Z_i^2)^2$$

$$Z_0^2 = \left[\left(\frac{\omega_B}{\omega_A}\right)^{2NZ} \prod_{i=1}^{N} Z_i^4\right]^{1/m}$$

Lowpass (Figure 11.2)

$$F^2(Z) = \epsilon^2 \prod_{i=1}^{N} Z_i^4$$

$$Q^2(Z) = (-1)^{NIN}(Z^2 - 1)^{NIN} \prod_{i=1}^{N} (Z^2 - Z_i^2)^2$$

Lowpass (Figure 11.3)   NIN even

$$F^2(Z) = \epsilon^2 (Z^2 + Z_0^2)^m$$

$$Q^2(Z) = (Z^2 - 1)^{NIN} \prod_{i=1}^{N} (Z^2 - Z_i^2)^2$$

$$Z_0^2 = \left(\prod_{i=1}^{N} Z_i^4\right)^{1/m}$$

A program for the maximally flat bandpass is given at the top of Fig. 11.4. This program is similar to the program for equiripple bandpass filters in Fig. 9.4. A sample use of the program is given at the bottom of Fig. 11.4, for a maximally flat filter that is very similar to the equiripple filter in Fig. 9.5. Note that for the maximally flat filter $A(0.3) = 28.1$ dB, whereas for the equiripple filter $A(0.3) = 40.488$ dB. It is a general property that the stopband loss of a maximally flat filter is less than the stopband loss of an "equivalent" equiripple filter.

The point of maximum flatness can be found by using Corollary 11.1. For the particular filter in Fig. 11.4, it follows that $f_o = 1$.

A program for a lowpass that is maximally flat at the origin is given at the top of Fig. 11.5; one for a lowpass that is maximally flat beyond the origin is given at the top of Fig. 11.6. Both these programs are similar to the program for equiripple lowpass filters in Fig. 9.6.

Comparing the results in Figs. 11.5 and 11.6, one can see that the lowpass that is maximally flat beyond the origin (Fig. 11.6) has greater attenuation at $f = 2$ than does the lowpass that is maximally flat at the origin (Fig. 11.5). The next section presents another example that demonstrates that the stopband loss of a lowpass maximally flat beyond the origin is greater than the stopband loss of a lowpass maximally flat at the origin.

11.5 DETERMINATION OF ATTENUATION POLES FOR MAXIMALLY FLAT FILTERS

Chapter 8 described a pole-placer program that could be used to determine the attenuation poles for filters that have equiripple passbands.

```
1; LØSS ØF MAXIMALLY FLAT BPF
1.2 DEMAND AMAX,FA,FB,NZ,NIN,N
1.3 DEMAND F[I] FØR I=1:1:N
1.4 EE=10^(.1*AMAX)-1,FAA=FA*FA,FBB=FB*FB,M=NZ+NIN+2*N
1.42 ZZ[I]=(F[I]*F[I]-FBB)/(F[I]*F[I]-FAA) FØR I=1:1:N
1.43 ZZ0=(FBB/FAA)^NZ
1.44 ZZ0=ZZ0*ZZ[I]*ZZ[I] FØR I=1:1:N
1.45 ZZ0=ZZ0^(1/M)
1.5 DEMAND F
1.6 TYPE #
1.7 TØ PART 6

6.4 ZZ=(F*F-FBB)/(F*F-FAA)
6.55 QQZ=(ZZ-1)^NIN
6.56 QQZ=QQZ*(ZZ-FBB/FAA)^NZ
6.57 QQZ=QQZ*(ZZ-ZZ[I])^2 FØR I=1:1:N
6.58 FFZ=EE*(ZZ+ZZ0)^M
6.8 A=10*LØG(1+ FFZ/QQZ )
6.9 TYPE F,A IN FØRM 1
>

DØ PART 1
        AMAX=.1
         FA=.9
         FB=1/.9
         NZ=1
         NIN=1
         N=2
         F[1]=.7
         F[2]=1/.7
         F=.3

3.000-01   28.100
>

DØ PART 6 FØR F=SQRT((FBB+ZZ0*FAA)/(1+ZZ0))
1.000+00   0.000
>
```

Fig. 11.4 Program for the determination of the loss of a maximally flat bandpass filter and example of its use.

```
1; LØSS ØF MAXIMALLY FLAT LPF (FIGURE 11.2)
1.2 DEMAND AMAX,FB,NIN,N
1.3 DEMAND F[I] FØR I=1:1:N
1.4 EE=10^(.1*AMAX)-1,FBB=FB*FB
1.42 ZZ[I]=(1-FBB/F[I]/F[I]) FØR I=1:1:N
1.5 DEMAND F
1.6 TYPE #
1.7 TØ PART 6

6.4 ZZ=1-FBB/F/F
6.55 QQZ=(ZZ-1)^NIN
6.57 QQZ=QQZ*(ZZ-ZZ[I])^2 FØR I=1:1:N
6.58 FFZ=EE
6.59 FFZ=FFZ*ZZ[I]^2 FØR I=1:1:N
6.8 A=10*LØG(1+ FFZ/QQZ )
6.9 TYPE F,A IN FØRM 1

>

DØ PART 1
        AMAX=.1
        FB=1
        NIN=2
        N=3
        F[1]=1.1
        F[2]=1.5
        F[3]=3
        F=2

 2.000+00   10.898
>
```

Fig. 11.5 Program for the determination of the loss of a lowpass filter maximally flat at the origin and example of its use.

This section describes how the program can be modified if the passband is, instead, maximally flat. We first discuss the bandpass case, and then the lowpass case.

One of the major subroutines in the pole-placer program finds the relative minimums of arcs. That is, we want to find the minimums of

$$A = 10 \log \left[1 + \frac{F^2(Z)}{Q^2(Z)} \right] \tag{11.21}$$

where $\quad F^2(Z) = \epsilon^2 (Z^2 + Z_0^2)^m \tag{11.22}$

$$Q^2(Z) = \left[Z^2 - \left(\frac{\omega_B}{\omega_A}\right)^2 \right]^{NZ} (Z^2 - 1)^{NIN} \prod_{i=1}^{N} (Z^2 - Z_i^2)^2 \tag{11.23}$$

Problem 11.4 demonstrates that the minimums of this set of equations are given by the zeros of the function

$$D(Z) = \frac{NZ}{2} \frac{Z_0^2 + (f_B/f_A)^2}{(f_B/f_A)^2 - Z^2} + \frac{NIN}{2} \frac{Z_0^2 + 1}{1 - Z^2} + \sum_{i=1}^{N} \frac{Z_0^2 + Z_i^2}{Z_i^2 - Z^2} \tag{11.24}$$

This equation, similar to (8.11), is solved by using the Newton iteration procedure. The initial guesses for this procedure are

1. *Loss minimum between two finite (nonzero) loss peaks Z_j, Z_{j+1}*

$$Z_m^2 \approx \frac{Z_0^2(Z_j^2 + Z_{j+1}^2) + 2Z_j^2 Z_{j+1}^2}{Z_j^2 + Z_{j+1}^2 + 2Z_0^2} \quad (11.25)$$

2. *Loss minimum between the origin ($\omega = 0$) and the peak nearest the origin (Z_1).*

$$Z_m^2 \approx \frac{Z_0^2[(f_B/f_A)^2 + (NZ/2)Z_1^2] + (1 + NZ/2)(f_B/f_A)^2 Z_1^2}{(NZ/2)(f_B/f_A)^2 + Z_1^2 + (1 + NZ/2)Z_0^2} \quad (11.26)$$

3. *Loss minimum between the highest pole Z_N and infinity*

$$Z_m^2 \approx \frac{Z_0^2[1 + (NIN/2)Z_N^2] + (1 + NIN/2)Z_N^2}{NIN/2 + Z_N^2 + (1 + NIN/2)Z_0^2} \quad (11.27)$$

Equations (11.25) to (11.27) describe the initial guesses for the arc minimums. These guesses are necessary for the Newton iteration procedure in Part 8 of the pole placer (see Fig. 8.11). The only other major

```
1; LOSS OF MAXIMALLY FLAT LPF (FIGURE 11.3)
1.2 DEMAND AMAX,FB,NIN,N
1.3 DEMAND F[I] FOR I=1:1:N
1.4 EE=10^(.1*AMAX)-1,FBB=FB*FB,M=NIN+2*N
1.42 ZZ[I]=(1-FBB/F[I]/F[I]) FOR I=1:1:N
1.43 ZZ0=1
1.44 ZZ0=ZZ0*ZZ[I]*ZZ[I] FOR I=1:1:N
1.45 ZZ0=ZZ0^(1/M)
1.5 DEMAND F
1.6 TYPE #
1.7 TO PART 6

6.4 ZZ=1-FBB/F/F
6.55 QQZ=(ZZ-1)^NIN
6.57 QQZ=QQZ*(ZZ-ZZ[I])^2 FOR I=1:1:N
6.58 FFZ=EE*(ZZ+ZZ0)^M
6.8 A=10*LOG(1+ FFZ/QQZ )
6.9 TYPE F,A IN FORM 1

>
DO PART 1
        AMAX=.1
         FB=1
         NIN=2
         N=3
         F[1]=1.1
         F[2]=1.5
         F[3]=3
         F=2

2.000+00    40.745
>DO PART 6 FOR F=FB/SQRT(1+ZZ0)
 8.055-01    0.000
>
```

Fig. 11.6 Program for the determination of the loss of a lowpass filter maximally flat beyond the origin and example of its use.

modification required of the equiripple pole placer is the calculation of loss, which is done in Part 6. These modifications are shown in Fig. 11.7.*

To use the pole placer for maximally flat bandpass filters, one can

1. Load the pole placer of Fig. 8.11.
2. Load the modifications of Fig. 11.7.

This creates a pole placer for maximally flat bandpass filters. An example of the use of the program is given in Fig. 11.8.

The program in Fig. 11.7 is for a maximally flat bandpass filter. Suppose, instead, that we have a lowpass filter that is maximally flat at the origin. To find a pole-placer program for this, we must again find the relative minimums of the arcs. That is, we want to find the minimums of

$$A = 10 \log \left[1 + \frac{F^2(Z)}{Q^2(Z)} \right] \tag{11.28}$$

where
$$F^2(Z) = \epsilon^2 \prod_{i=1}^{N} Z_i^4 \tag{11.29}$$

$$Q^2(Z) = (-1)^{\text{NIN}}(Z^2 - 1)^{\text{NIN}} \prod_{i=1}^{N} (Z^2 - Z_i^2)^2 \tag{11.30}$$

The relative minimums of the arcs are thus the relative maximums of

* Problem 11.5 discusses the modification shown in Step 49.21.

```
1; PØLE PLACER  (MAXIMALLY FLAT BANDPASS)
1.01; THE FØLLØWING STEPS SHØULD BE ADDED TØ "BANDPASS PØLE PLACER"
1.42 ZZ[I]=(F[I]^2-FBB)/(F[I]^2-FAA),Z[I]=SQRT(ZZ[I]) FØR I=1:1:N
1.421 M=NZ+NIN+2*N,ZZ0=(FBB/FAA)^NZ
1.423 ZZ0=ZZ0*ZZ[I]^2 FØR I=1:1:N
1.425 ZZ0=ZZ0^(1/M)

6.55 ZZ=Z*Z
6.551 FFZ=EE*(ZZ+ZZ0)^M
6.56 QQZ=(ZZ-FBB/FAA)^NZ*(ZZ-1)^NIN
6.57 QQZ=QQZ*(ZZ-ZZ[I3])^2 FØR I3=1:1:N
6.8  A=10*LØG(1+ FFZ/QQZ )

8.1 Z1=ZA*ZA,Z2=ZB*ZB,ZZ0=(ZZ0*(Z1+Z2)+2*Z1*Z2)/(Z1+Z2+2*ZZ0)
8.11 I1=0,ZZA=Z1,ZZB=Z2
8.12 D=NZ/2*FBB/FAA+Z2+(1+NZ/2)*ZZ0 IF FB=FA*ZA
8.13 ZZ0=(ZZ0*(FBB/FAA+NZ/2*Z2)+(1+NZ/2)*FBB/FAA*Z2)/D IF FB=FA*ZA
8.14 D=NIN/2+Z1+(1+NIN/2)*ZZ0 IF ZB=1
8.15 ZZ0=(ZZ0*(1+NIN/2*Z1)+(1+NIN/2)*Z1)/D IF ZB=1
8.9 D=NIN/2*(ZZ0+1)/(1-ZZ)+NZ/2*(ZZ0+FBB/FAA)/(FBB/FAA-ZZ)
8.91 D=D+(ZZ0+ZZ[I2])/(ZZ[I2]-ZZ) FØR I2=1:1:N

49.21 A[I1,I2]=Z[I2]/(ZZ-Z[I2]^2) FØR I2=1:1:N

50.7 TØ PART 1.42
```

Fig. 11.7 Pole-placer program for maximally flat bandpass filters.

```
           SA= 1
           SB= 2
         FS[ 0]= 0
          A[ 0]= 30
         FS[ 1]= 6
          A[ 1]= 0
         FS[ 2]= 11
          A[ 2]= 40
         FS[ 3]= 12
          A[ 3]= 30
          AMAX= .1
            FA= 7
            FB= 10
            NZ= 1
           NIN= 1
            NA= 2
            NB= 3


          F[ 1]= 5
          F[ 2]= 5.9
          F[ 3]= 11.1
          F[ 4]= 11.6
          F[ 5]= 14.5

 FMIN[ 1]=3.414+00     AMIN= 30.79     DMIN=   .79
 FMIN[ 2]=5.600+00     AMIN= 33.50     DMIN=  3.50
 FMIN[ 3]=6.000+00     AMIN= 32.36     DMIN=  2.36
 FMIN[ 4]=1.100+01     AMIN= 38.89     DMIN= -1.11
 FMIN[ 5]=1.129+01     AMIN= 44.68     DMIN=  4.68
 FMIN[ 6]=1.200+01     AMIN= 40.98     DMIN=   .98
 FMIN[ 7]=2.254+01     AMIN= 34.90     DMIN=  4.90


>DO PART 49


          F[ 1]=     4.86135819
          F[ 2]=     5.90337915
          F[ 3]=    11.0614621
          F[ 4]=    11.6451252
          F[ 5]=    14.4423656

 FMIN[ 1]=3.295+00     AMIN= 31.99     DMIN=  1.99
 FMIN[ 2]=5.576+00     AMIN= 32.04     DMIN=  2.04
 FMIN[ 3]=6.000+00     AMIN= 32.01     DMIN=  2.01
 FMIN[ 4]=1.100+01     AMIN= 42.57     DMIN=  2.57
 FMIN[ 5]=1.127+01     AMIN= 41.96     DMIN=  1.96
 FMIN[ 6]=1.200+01     AMIN= 41.86     DMIN=  1.86
 FMIN[ 7]=2.240+01     AMIN= 34.98     DMIN=  4.98


>DO PART 49


          F[ 1]=     4.85921132
          F[ 2]=     5.90356801
          F[ 3]=    11.065057
          F[ 4]=    11.6497687
          F[ 5]=    14.4457144

 FMIN[ 1]=3.294+00     AMIN= 32.03     DMIN=  2.03
 FMIN[ 2]=5.576+00     AMIN= 32.03     DMIN=  2.03
 FMIN[ 3]=6.000+00     AMIN= 32.03     DMIN=  2.03
 FMIN[ 4]=1.100+01     AMIN= 42.06     DMIN=  2.06
 FMIN[ 5]=1.127+01     AMIN= 42.05     DMIN=  2.05
 FMIN[ 6]=1.200+01     AMIN= 42.05     DMIN=  2.05
 FMIN[ 7]=2.242+01     AMIN= 35.03     DMIN=  5.03
```

Fig. 11.8 Demonstration of the use of the maximally flat pole-placer program.

$Q^2(Z)$. Differentiation of (11.30) indicates that these maximums are given by the zeros of the function

$$D(Z) = \frac{\text{NIN}}{2} \frac{1}{1 - Z^2} \sum_{i=1}^{N} \frac{1}{Z_i^2 - Z^2} \qquad (11.31)$$

To solve this nonlinear set of equations by the Newton iterative procedure, we need the following initial guesses:

1. *Loss minimum between two finite loss peaks Z_j, Z_{j+1}*

$$Z_m^2 \approx \frac{Z_j^2 + Z_{j+1}^2}{2} \qquad (11.32)$$

2. *Loss minimum between the highest pole Z_N and infinity*

$$Z_m^2 \approx \frac{(\text{NIN}/2)Z_N^2 + 1}{(\text{NIN}/2) + 1} \qquad (11.33)$$

Equations (11.32) and (11.33) describe the initial guesses for the arc minimums. These guesses are necessary for the Newton iterative procedure and can be used to modify the equiripple lowpass pole placer of Fig. 8.15. These modifications are shown in Fig. 11.9 as Part 8. The other major modification required of the equiripple pole placer is the calculation of loss, which is done in Part 6.*

An example is shown in Fig. 11.10. In this example, the attenuation poles are chosen such that the stopband loss has equiminimums; that is, the maximally flat filter is of the inverse Tschebycheff type. In fact, this filter was previously studied in Example 4.1. In Fig. 11.10, the pole-placer program finds a set of poles that yield a minimum stopband loss of 30.22 dB, which is the same value that was found in Example 4.1.

The inverse Tschebycheff filters discussed in Chap. 4 had either none

* Problem 11.6 discusses the modification shown in Step 49.21.

```
1; POLE PLACER (MAXIMALLY FLAT LOWPASS   FIGURE 11.2)
1.01; THE FOLLOWING STEPS SHOULD BE ADDED TO "LOWPASS POLE PLACER"

6.55  ZZ=Z*Z,QQZ=(ZZ-1)^NIN
6.57  QQZ=QQZ*(ZZ-Z[I3]^2)^2 FOR I3=1:1:N
6.58  FFZ=EE
6.59  FFZ=FFZ*Z[I3]^4 FOR I3=1:1:N
6.8   A=10*LOG(1+ FFZ/QQZ )

8.1   ZZ0=(ZA*ZA+ZB*ZB)/2,I1=1,ZZA=ZA*ZA,ZZB=ZB*ZB
8.14  ZZ0=(NIN/2*ZA*ZA+1)/(NIN/2+1) IF ZB=1
8.91  D=D+1/(Z[I2]^2-ZZ) FOR I2=1:1:N

49.21 A[I1,I2]=ZZ/Z[I2]/(ZZ-Z[I2]^2) FOR I2=1:1:N

>
```

Fig. 11.9 Pole-placer program for lowpass filters maximally flat at the origin.

Maximally Flat Passbands

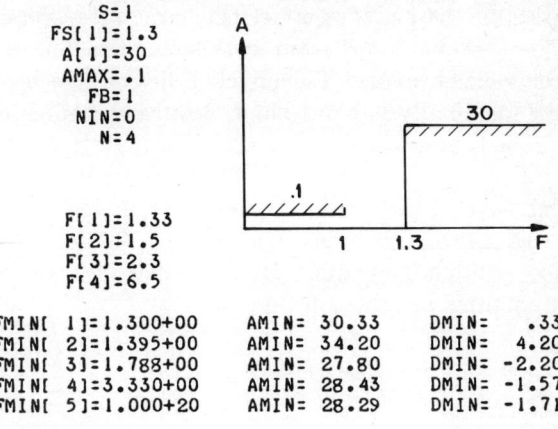

```
        S= 1
     FS[ 1]=1.3
      A[ 1]=30
     AMAX= .1
        FB= 1
       NIN=0
        N= 4


    F[ 1]=1.33
    F[ 2]=1.5
    F[ 3]=2.3
    F[ 4]=6.5

FMIN[ 1]=1.300+00      AMIN= 30.33      DMIN=   .33
FMIN[ 2]=1.395+00      AMIN= 34.20      DMIN=  4.20
FMIN[ 3]=1.788+00      AMIN= 27.80      DMIN= -2.20
FMIN[ 4]=3.330+00      AMIN= 28.43      DMIN= -1.57
FMIN[ 5]=1.000+20      AMIN= 28.29      DMIN= -1.71


>DØ PART 49


       F[ 1]=       1.32995509
       F[ 2]=       1.55848806
       F[ 3]=       2.32955235
       F[ 4]=       6.63816804

FMIN[ 1]=1.300+00      AMIN= 30.17      DMIN=   .17
FMIN[ 2]=1.406+00      AMIN= 30.51      DMIN=   .51
FMIN[ 3]=1.833+00      AMIN= 30.11      DMIN=   .11
FMIN[ 4]=3.384+00      AMIN= 29.98      DMIN=  -.02
FMIN[ 5]=1.000+20      AMIN= 29.98      DMIN=  -.02


>DØ PART 49


       F[ 1]=       1.32546903
       F[ 2]=       1.56346689
       F[ 3]=       2.33990642
       F[ 4]=       6.66342604

FMIN[ 1]=1.300+00      AMIN= 30.22      DMIN=   .22
FMIN[ 2]=1.407+00      AMIN= 30.22      DMIN=   .22
FMIN[ 3]=1.838+00      AMIN= 30.22      DMIN=   .22
FMIN[ 4]=3.397+00      AMIN= 30.22      DMIN=   .22
FMIN[ 5]=1.000+20      AMIN= 30.22      DMIN=   .22
```

Fig. 11.10 Demonstration of the use of the lowpass pole placer in Fig. 11.9.

or one attenuation pole at infinity (that is, NIN = 0 or NIN = 1). However, the material in this chapter now lets us find lowpass filters that

1. Are maximally flat at the origin.
2. Have an equiminimum stopband.
3. Have any number of attenuation poles at infinity.

Such filters have been termed *generalized inverse Tschebycheff filters*.[4,5] The pole-placer program in this chapter can, of course, be used to yield generalized inverse Tschebycheff filters, which comprise a special case of the maximally flat passband, arbitrary stopband filter.

The lowpass filters that were just investigated are maximally flat at the origin, as indicated in Fig. 11.2. Lowpass filters can also be maximally flat beyond the origin, as indicated in Fig. 11.3. To find a pole-placer program for this type of maximally flat lowpass, we can simply modify the bandpass results. In particular, by analogy to (11.24), the loss minimums are the solutions to

$$D(Z) = \frac{NIN}{2} \frac{Z_0^2 + 1}{1 - Z^2} + \sum_{i=1}^{N} \frac{Z_0^2 + Z_i^2}{Z_i^2 - Z^2} \qquad (11.34)$$

The initial guesses needed for the Newton iterative procedure are
1. *Loss minimum between two finite (nonzero) loss peaks Z_j, Z_{j+1}*

$$Z_m^2 \approx \frac{Z_0^2(Z_j^2 + Z_{j+1}^2) + 2Z_j^2 Z_{j+1}^2}{Z_j^2 + Z_{j+1}^2 + 2Z_0^2} \qquad (11.35)$$

2. *Loss minimum between the highest pole Z_N and infinity*

$$Z_m^2 \approx \frac{Z_0^2[1 + (NIN/2)Z_N^2] + (1 + NIN/2)Z_N^2}{NIN/2 + Z_N^2 + (1 + NIN/2)Z_0^2} \qquad (11.36)$$

Thus the pole placer for a lowpass that is maximally flat beyond the origin is very similar to the pole placer for a maximally flat bandpass.

```
1; PØLE PLACER (MAXIMALLY FLAT LØWPASS    FIGURE 11.3)
1.01; THE FOLLOWING STEPS SHØULD BE ADDED TØ "LØWPASS PØLE PLACER"
1.42 ZZ[I]=1-FBB/F[I]/F[I],Z[I]=SQRT(ZZ[I]) FØR I=1:1:N
1.421 M=NIN+2*N,ZZØ=1
1.423 ZZØ=ZZØ*ZZ[I]^2 FØR I=1:1:N
1.425 ZZØ=ZZØ^(1/M)

6.55  ZZ=Z*Z
6.551 FFZ=EE*(ZZ+ZZØ)^M
6.56  QQZ=(ZZ-1)^NIN
6.57  QQZ=QQZ*(ZZ-ZZ[I3])^2 FØR I3=1:1:N
6.8   A=10*LØG(1+ FFZ/QQZ )

8.1  Z1=ZA*ZA,Z2=ZB*ZB,ZZØ=(ZZØ*(Z1+Z2)+2*Z1*Z2)/(Z1+Z2+2*ZZØ)
8.11 I1=1,ZZA=Z1,ZZB=Z2
8.14 D=NIN/2+Z1+(1+NIN/2)*ZZØ IF ZB=1
8.15 ZZØ=(ZZØ*(1+NIN/2*Z1)+(1+NIN/2)*Z1)/D IF ZB=1
8.9  D=NIN/2*(ZZØ+1)/(1-ZZ)
8.91 D=D+(ZZØ+ZZ[I2])/(ZZ[I2]-ZZ) FØR I2=1:1:N

49.21 A[I1,I2]=Z[I2]/(ZZ-Z[I2]^2) FØR I2=1:1:N

50.7 TØ PART 1.42

>
```

Fig. 11.11 Pole-placer program for lowpass filters maximally flat beyond the origin.

Maximally Flat Passbands

In fact, the pole placer in Fig. 11.11 is almost identical to the pole placer in Fig. 11.7.

A sample use of the pole placer for maximally flat lowpasses ($\omega_0 \neq 0$) is given in Fig. 11.12. The requirements are the same as for the inverse Tschebycheff investigated in Fig. 11.10. It should be noted that the inverse Tschebycheff filter had a minimum stopband loss of 30.22 dB

```
            S=1
            FS[1]=1.3
            A[1]=30
            AMAX=.1
            FB=1
            NIN=0
            N=4


            F[1]=1.32
            F[2]=1.5
            F[3]=2
            F[4]=5.5

  FMIN[ 1]=1.300+00      AMIN= 51.45      DMIN= 21.45
  FMIN[ 2]=1.377+00      AMIN= 53.02      DMIN= 23.02
  FMIN[ 3]=1.678+00      AMIN= 56.07      DMIN= 26.07
  FMIN[ 4]=2.820+00      AMIN= 53.85      DMIN= 23.85
  FMIN[ 5]=1.000+20      AMIN= 54.46      DMIN= 24.46


>DØ PART 49


            F[1]=        1.31613026
            F[2]=        1.47385878
            F[3]=        2.0249153
            F[4]=        5.37947837

  FMIN[ 1]=1.300+00      AMIN= 53.98      DMIN= 23.98
  FMIN[ 2]=1.369+00      AMIN= 53.84      DMIN= 23.84
  FMIN[ 3]=1.664+00      AMIN= 53.90      DMIN= 23.90
  FMIN[ 4]=2.823+00      AMIN= 53.83      DMIN= 23.83
  FMIN[ 5]=1.000+20      AMIN= 53.83      DMIN= 23.83


>DØ PART 49


            F[1]=        1.31632861
            F[2]=        1.47379984
            F[3]=        2.02695629
            F[4]=        5.38337224

  FMIN[ 1]=1.300+00      AMIN= 53.87      DMIN= 23.87
  FMIN[ 2]=1.369+00      AMIN= 53.87      DMIN= 23.87
  FMIN[ 3]=1.664+00      AMIN= 53.87      DMIN= 23.87
  FMIN[ 4]=2.826+00      AMIN= 53.87      DMIN= 23.87
  FMIN[ 5]=1.000+20      AMIN= 53.87      DMIN= 23.87
```

Fig. 11.12 Demonstration of the use of the lowpass pole placer in Fig. 11.11.

whereas, if the point of maximum flatness is beyond the origin, the minimum stopband loss is 53.87 dB. That is, because $\omega_0 \neq 0$ the transition region is much steeper.

11.6 NATURAL MODES FOR MAXIMALLY FLAT PASSBAND FILTERS

The input/output transfer function $H(s)$ is a ratio of polynomials:

$$H(s) = \frac{e(s)}{q(s)}$$

The attenuation poles are determined by $q(s)$; these poles can be found as described in the previous section. This section discusses how $e(s)$ can be found. It is similar to the material in Chap. 10 for filters with equiripple passbands.

The function $e(s)$ is found by first finding the equivalent transformed function $E(Z)$. $E(Z)$ is given by

$$E(Z)E^*(Z) = F^2(Z) + Q^2(Z)$$

The formulas for $F^2(Z)$ and $Q^2(Z)$ for maximally flat filters were summarized in Sec. 11.4. It should be noted that the formulas for $Q^2(Z)$ for the maximally flat filter are the same as the formulas for equiripple filters; all that need be changed in the zero-finder programs are the formulas for $F^2(Z)$.

In the zero-finder programs, Parts 50 and 52 find $F^2(Z)$. The modifications of these parts for the various maximally flat filters are shown in Figs. 11.13 to 11.15. In particular, Fig. 11.13 shows the modifications necessary for a maximally flat bandpass characteristic. The modifications are applied to a specific example at the bottom of Fig. 11.13. This example finds the attenuation zeros for the filter that was first discussed in Fig. 11.4. The validity of the zero finder's results is demonstrated by the fact that both figures indicate that $A(0.3) = 28.1$ dB. Similarly, the results at the bottom of Fig. 11.14 check with those in Fig. 11.5, and the results in Fig. 11.15 check with those in Fig. 11.6.

11.7 COMPARISON OF MAXIMALLY FLAT AND EQUIRIPPLE FILTERS

Maximally flat filters concentrate all their "approximating power" at one frequency, whereas equiripple filters spread their approximating power across the entire passband. It is thus not surprising that equiripple filters can give better attenuation performance than maximally

```
1; ZERØ FINDER  (MAXIMALLY FLAT BANDPASS)
1.01; THE FØLLØWING STEPS SHØULD BE ADDED TØ "BANDPASS ZERØ FINDER"

50; DETERMINATIØN ØF F(Z)^2=C[0]+C[2]*Z^2+C[4]*Z^4+...
50.2 ZZØ=(FBB/FAA)^NZ
50.3 ZZØ=ZZØ*Z[I]^4 FØR I=1:1:N
50.4 ZZØ=ZZØ^(1/M)
50.5 S[I]=ZZØ FØR I=1:1:M
50.6 DØ PART 90 FØR SN=I
50.8 C[2*I]=EE*B[I] FØR I=0:1:M
50.93;

           AMAX=.1
             FA=.9
             FB=1/.9
             NZ=1
            NIN=1
              N=2
           F[1]=.7
           F[2]=1/.7

ZQ[ 1]=    6.2105    ZF[ 1]=1.1901+00
ZQ[ 2]=    2.2041    ZF[ 2]=1.0000+00
ZQ[ 3]=    6.2105    ZF[ 3]=8.4025-01

             CH=      1.25532942

>DØ PART 6 FØR F=.3
 3.000-01   28.100
>
```

Fig. 11.13 Zero-finder program for maximally flat bandpass filters and example of its use.

flat filters. As an example of the superiority of the equiripple attenuation performance, consider again the requirements at the top of Fig. 11.10. Recall that an eighth-order filter maximally flat at the origin exceeded the requirements by 0.22 dB; also, an eighth-order filter maximally flat beyond the origin exceeded the requirements by 23.87 dB. However, an eighth-order elliptic filter exceeds the requirements by even more—by 41.93 dB. This demonstrates that equiripple filters have better loss performance than do maximally flat filters.

Even though it is generally acknowledged that equiripple filters have better attenuation characteristics, maximally flat filters are often used because they are thought to have better delay performance.* We shall now discuss an example that demonstrates that (at least in some cases) equiripple filters can have better delay performance.

For this example, we again consider the requirements at the top of Fig. 11.10. However, suppose that not only do we want the loss to meet these requirements, but we also want to minimize the delay variation in the passband ($0 \leq f \leq 1$). It is shown in Chap. 14 that one

* See Chap. 14 for a discussion of delay.

way to improve delay performance in the region $0 \le f \le 1$ is to increase the passband edge f_B to f'_B. The frequency f'_B can be chosen such that the stopband loss just exceeds the requirements. As an example, we shall modify the filter described in Fig. 11.12, which was maximally flat beyond the origin. With the original passband edge $f_B = 1$, the stopband loss exceeded the requirements by 23.87 dB. If, as illustrated in Fig. 11.16, the passband edge is increased to $f'_B = 1.185$, then the stopband loss exceeds the requirements by only 0.48 dB. The transfer function that describes this new maximally flat filter is

$$H(s) = C_H \prod_{i=1}^{4} \frac{s^2 + (2\pi F_i/Q_i)s + (2\pi F_i)^2}{s^2 + (2\pi f_i)^2} \qquad (11.37)$$

where $C_H = 33.4627$ and

$$\begin{array}{lll} f_1 = 1.3105 & F_1 = 1.2403 & Q_1 = 14.7245 \\ f_2 = 1.4143 & F_2 = 1.2710 & Q_2 = 3.9267 \\ f_3 = 1.8109 & F_3 = 1.3159 & Q_3 = 1.5145 \\ f_4 = 4.3885 & F_4 = 1.2285 & Q_4 = 0.6203 \end{array}$$

```
1; ZERØ FINDER  (MAXIMALLY FLAT LØWPASS  FIGURE 11.2)
1.01; THE FØLLØWING STEPS SHØULD BE ADDED TØ "LØWPASS ZERØ FINDER"
50; DETERMINATIØN ØF F(Z)^2=C[0]+C[2]*Z^2+C[4]*Z^4+...
50.2 ZZØ=1
50.3 ZZØ=ZZØ*Z[I]^4 FØR I=1:1:N
50.4 C[0]=EE*ZZØ
50.6 C[I]=0 FØR I=2:2:2*M
50.8;
50.93;

        AMAX=.1
          FB=1
         NIN=2
           N=3
        F[1]=1.1
        F[2]=1.5
        F[3]=3

ZQ[ 1]=  16.3471    ZF[ 1]=1.0899+00
ZQ[ 2]=   3.1808    ZF[ 2]=1.3940+00
ZQ[ 3]=   1.0626    ZF[ 3]=2.0303+00
ZQ[ 4]=    .5483    ZF[ 4]=2.8345+00

           CH=      .00811843279

>DØ PART 6 FØR F=2
 2.000+00   10.898
>
```

Fig. 11.14 Zero-finder program for lowpass filters maximally flat at the origin and example of its use.

```
1; ZERØ FINDER  (MAXIMALLY FLAT LØWPASS  FIGURE 11.3)
1.01; THE FØLLØWING STEPS SHØULD BE ADDED TØ "LØWPASS ZERØ FINDER"

50; DETERMINATIØN ØF F(Z)^2=C[0]+C[2]*Z^2+C[4]*Z^4+...
50.2 ZZØ=1
50.3 ZZØ=ZZØ*Z[I]^4 FØR I=1:1:N
50.4 ZZØ=ZZØ^(1/M)
50.5 S[I]=ZZØ FØR I=1:1:M
50.6 DØ PART 90 FØR SN=I
50.8 C[2*I]=EE*B[I] FØR I=0:1:M
50.93;


        AMAX=.1
         FB=1
         NIN=2
          N=3
        F[1]=1.1
        F[2]=1.5
        F[3]=3


ZQ[ 1]=  13.0685    ZF[ 1]=1.0599+00
ZQ[ 2]=   2.7936    ZF[ 2]=1.1086+00
ZQ[ 3]=   1.1007    ZF[ 3]=1.0776+00
ZQ[ 4]=    .5799    ZF[ 4]=8.5615-01

          CH=        .534258006


>DØ PART 6 FØR F=2
 2.000+00   40.745
>
```

Fig. 11.15 Zero-finder program for lowpass filters maximally flat beyond the origin and example of its use.

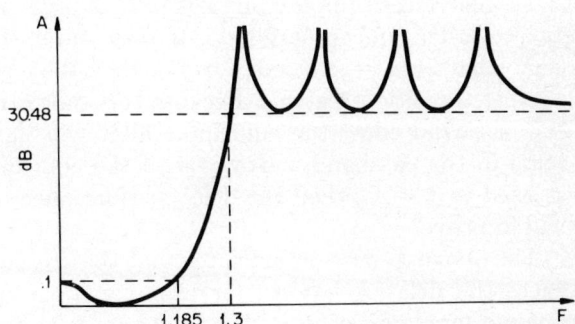

Fig. 11.16 Extension of passband to improve delay performance.

In order to compare the delay performance of the maximally flat filter with that of an equiripple filter, an elliptic filter was found that satisfies the loss requirements in Fig. 11.10. For an eighth-degree elliptic filter with a 0.1-dB ripple, it was possible to increase the passband edge to $f'_B = 1.27$. This resulted in a minimum stopband loss of 30.38 dB.

The elliptic transfer function is again of the form of (11.37), where now the coefficients are $C_H = 33.0547$ and

$$f_1 = 1.3042 \quad F_1 = 1.2779 \quad Q_1 = 57.7818$$
$$f_2 = 1.3545 \quad F_2 = 1.2605 \quad Q_2 = 10.4445$$
$$f_3 = 1.6095 \quad F_3 = 1.1850 \quad Q_3 = 2.4197$$
$$f_4 = 3.6773 \quad F_4 = 0.9582 \quad Q_4 = 0.6578$$

The delay performance of the maximally flat and equiripple filters was investigated by using the program discussed in Sec. 14.4. It was found that the delay of both filters increased monotonically in the passband; thus the passband delay variation can be found by obtaining the delay at the passband edges. This yields, for the lowpass maximally flat beyond the origin,

$$D(f = 1) - D(f = 0) = (0.875 - 0.329) \text{ s} = 0.546 \text{ s}$$

and for the elliptic lowpass filter,

$$D(f = 1) - D(f = 0) = (0.834 - 0.322) \text{ s} = 0.512 \text{ s}$$

In the above example, the eighth-degree elliptic filter had better delay performance than the eighth-degree maximally flat filter. This was because the elliptic filter had its "passband" extended to $f'_B = 1.27$, so that the transition between passband and stopband did not start at $f = 1$ (this was the highest frequency at which we were concerned with delay).

To generalize, suppose one has an nth-order maximally flat filter and an nth-order equiripple filter. If these filters have the same passband edge (that is, $A = A_{\max}$ at $f = f_B$), then the equiripple filter will have a sharper transition region. Because this sharp transition region starts at the passband edge, the equiripple filter will have a greater delay variation in the passband. However, if the equiripple passband edge is increased to $f = f'_B$, then the delay performance in the region $0 \leq f \leq f_B$ will improve.

The above discussion indicates that the attenuation performance influences the delay behavior; for example, a sharp transition region produces a large variation in delay. In fact, as discussed in Chap. 14, the attenuation performance of a minimum-phase transfer function uniquely determines its delay. Thus, if two filters have similar attenuation performance, they will have similar delay performance. Whether or not a filter is equiripple or maximally flat in its passband does not affect the delay performance as much as does the abruptness of the transition region and the amount of stopband loss. Thus it is not surprising that an equiripple filter with a sharper transition region will have poorer delay performance than a maximally flat filter. However, if the equiripple filter has a transition region and stopband loss comparable to those of a maximally flat filter, then their delay performance will be similar.

11.8 CONCLUSIONS

The simplest maximally flat filter is the Butterworth, which is a polynomial filter. Another commonly encountered maximally flat filter is the inverse Tschebycheff. These filters are special cases of the more general class of filters that has a maximally flat passband and arbitrary stopband.

Formulas are presented in this chapter that allow one to find the poles and zeros of a filter that has a maximally flat passband and arbitrary stopband. These formulas are for both bandpass and lowpass filters. In fact, we discussed two types of lowpass filters: one was maximally flat at the origin, and one was maximally flat beyond the origin.

The generalized inverse Tschebycheff filter was shown to be a special case of a lowpass that is maximally flat at the origin and has an arbitrary passband. A lowpass that is maximally flat beyond the origin can have better stopband performance than a lowpass that is maximally flat at the origin. It will be shown in Chap. 14 that it can also have better delay performance.

Finally, the maximally flat filters were compared with equiripple filters. An equiripple filter can have a sharper transition region and greater stopband loss than a maximally flat filter of the same degree. This type of behavior would result in the equiripple filter having poorer delay performance. However, if an equiripple filter (e.g., a sixth-order elliptic filter) has transition region and stopband performance similar to those of a maximally flat filter (e.g., an eighth-order inverse Tschebycheff filter), then their delay performance will be similar. In this case it would be advantageous to use the elliptic filter, because the lower degree would require fewer elements in its synthesis.

REFERENCES

1. W. R. Bennett and J. R. Davey, *Data Transmission*, McGraw-Hill Book Company, New York, 1965.
2. H. J. Orchard and G. C. Temes, "Filter Design Using Transformed Variables," *IEEE Trans. on Circuit Theory*, December 1968, pp. 385–408.
3. P. Aronhime and A. Budak, "Maximally Flat Magnitude Beyond the Origin," *IEEE Trans. on Circuit Theory*, May 1971, pp. 409–411.
4. C. Y. Chang, "Maximally Flat Amplitude Low-Pass Filters with an Arbitrary Number of Pairs of Real Frequency Transmission Zeros," *IEEE Trans. on Circuit Theory*, December 1968, pp. 465–467.
5. C. Y. Chang, "Attenuation Characteristics of Odd-Order Generalized Inverse Chebyshef Filters," *IEEE Trans. on Circuit Theory*, December 1968, pp. 467–471.

PROBLEMS

11.1 The characteristic function for maximally flat bandpass filters can be written as

$$K(s) = \frac{C_K(s^2 + \omega_0^2)^{m/2}}{s^{NZ} \prod_{i=1}^{N} (s^2 + \omega_i^2)}$$

a. What is NIN, the number of poles at infinity?
b. Show that NZ + NIN must be even.

11.2 The maximally flat characteristic function is given by

$$|K|^2 = \frac{C_K^2(\omega^2 - \omega_0^2)^m}{\omega^{2NZ} \prod_{i=1}^{N} (\omega^2 - \omega_i^2)^2}$$

Find, at $\omega = \omega_0$,

$$\frac{d^j|K|^2}{d\omega^j} \quad \text{for} \quad j = 1, 2, \ldots, m$$

11.3 This problem is a generalization of the example in Fig. 11.4. Assume that $f_B = 1/f_A$, that NZ = NIN, and that for every attenuation pole f_i below the passband there is one f_j above the passband such that $f_i = 1/f_j$. Show that the point of maximum flatness is $f_o = 1$.

11.4 This problem shows that the minimums of $F^2(Z)/Q^2(Z)$ occur where the following function is zero:

$$D(Z) = \frac{NZ}{2} \frac{Z_0^2 + (f_B/f_A)^2}{(f_B/f_A)^2 - Z^2} + \frac{NIN}{2} \frac{Z_0^2 + 1}{1 - Z^2} + \sum_{i=1}^{N} \frac{Z_0^2 + Z_i^2}{Z_i^2 - Z^2}$$

a. Show that the minimums of $F^2(Z)/Q^2(Z)$ occur where the following function $G(Z)$ has its minimums:

$$\frac{NZ}{2} \ln \left| \frac{Z^2 + Z_0^2}{Z^2 - f_B^2/f_A} \right| + \frac{NIN}{2} \ln \left| \frac{Z^2 + Z_0^2}{Z^2 - 1} \right| + \sum_{i=1}^{N} \ln \left| \frac{Z^2 + Z_0^2}{Z^2 - Z_i^2} \right|$$

b. Show that

$$\frac{d}{dZ} \ln \left| \frac{Z^2 + Z_0^2}{Z^2 - Z_i^2} \right| = \frac{2Z(Z_0^2 + Z_i^2)}{(Z^2 + Z_0^2)(Z_i^2 - Z^2)}$$

c. Use the results in *a* and *b* to establish the statement at the beginning of the problem.

11.5 Section 8.12 discussed the calculation of new pole locations for the equiripple pole placer. This required the calculation of $\partial A/\partial Z_i$ (the variation of A

with respect to the ith pole). For the equiripple case we used

$$\frac{\partial A}{\partial Z_i} \approx \frac{40}{\ln 10} \frac{Z}{Z^2 - Z_i^2} \tag{8.21}$$

This problem demonstrates that a similar approximation can be used for maximally flat filters.

a. Show that for a maximally flat bandpass filter, the stopband loss can be approximated by

$$A \approx \frac{10}{\ln 10} \ln \frac{2(Z^2 + Z_0^2)^m}{[Z^2 - (\omega_B/\omega_A)^2]^{NZ}(Z^2 - 1)^{NIN} \prod_{i=1}^{N} (Z^2 - Z_i^2)^2}$$

b. Show that

$$\frac{\partial A}{\partial Z_i} \approx \frac{40}{\ln 10} \frac{Z_i}{Z^2 - Z_i^2}$$

11.6 This problem is the same as Problem 11.5 except it is concerned with maximally flat lowpass filters.

a. Show that for the type of filter in Fig. 11.2 the stopband loss can be approximated by

$$A \approx \frac{10}{\ln 10} \ln \frac{\epsilon^2 \prod_{i=1}^{N} Z_i^4}{(-1)^{NIN}(Z^2 - 1)^{NIN} \prod_{i=1}^{N} (Z^2 - Z_i^2)^2}$$

b. Show that

$$\frac{\partial A}{\partial Z_i} \approx \frac{40}{\ln 10} \frac{Z^2/Z_i}{Z^2 - Z_i^2}$$

CHAPTER TWELVE

Parametric Filters

12.1 INTRODUCTION

The filters we have discussed so far have either had equiripple or maximally flat passbands. This put certain restrictions on the transfer functions, some of which are

1. Equiripple or maximally flat bandpass filters must have NZ + NIN equal to an even number (i.e., the number of attenuation poles at the origin plus the number at infinity must be an even number).
2. If NIN is an even number for an equiripple lowpass filter, then the zero-frequency loss is $A(0) = A_{max}$.
3. If NIN is an odd number for an equiripple lowpass filter, then $A(0) = 0$.
4. If a lowpass filter is maximally flat beyond the origin, then NIN must be an even number.

This chapter discusses *parametric* approximations, which can be used to eliminate the above restrictions. These approximations are not equiripple (or maximally flat), but can be arbitrarily close to such an approximation.

Parametric approximations are useful not only because they allow

NZ + NIN to be an odd number; they also are useful because passive filters require that the characteristic function have a special form if the filter is to be realizable in certain configurations. For example, passive filters are normally designed as lossless ladder networks terminated in resistors. If it is desired that the ladder behave as a short circuit at zero frequency and an open circuit at infinite frequency, then the characteristic function must have a zero of attenuation on the real axis. The equiripple filters we have discussed do not have a zero on the real axis,* but the parametric filters do.

12.2 THE BASIC TRICK

Let $H_1(s)$ denote the input/output transfer function of an equiripple bandpass filter. This can be written in terms of the transformed variable Z as

$$H_1(s)H_1(-s) = 1 + \frac{F_1^2(Z)}{Q_1^2(Z)} \tag{12.1}$$

where $\quad Q_1^2(Z) = \left[Z^2 - \left(\frac{\omega_B}{\omega_A}\right)^2\right]^{NZ} (Z^2 - 1)^{NIN} \prod_{i=1}^{N} (Z^2 - Z_i^2)^2 \tag{12.2}$

We shall assume that this is an equiripple approximation; thus, the following two facts must be true:

NZ + NIN = even number (12.3)

$$F_1^2(Z) = \epsilon_1^2 \left\{ \mathrm{Ev}\left[\left(Z + \frac{\omega_B}{\omega_A}\right)^{NZ} (Z + 1)^{NIN} \prod_{i=1}^{N} (Z + Z_i)^2\right]\right\}^2 \tag{12.4}$$

We would like to add a simple pole at the origin ($Z = \omega_B/\omega_A$) or a simple pole at infinity ($Z = 1$); however, (12.3) will not allow this. We shall get around this constraint by settling for an approximation that is not equiripple but that can be made to be arbitrarily close to an equiripple approximation.

12.3 ADDITION OF A POLE AT INFINITY

Consider the transfer function

$$H(s)H(-s) = 1 + \frac{Z^2 - \alpha^2}{Z^2 - 1} \frac{F_1^2(Z)}{Q_1^2(Z)} \tag{12.5}$$

---

* For a more detailed discussion of passive parametric filters, consult Chap. 14 of *Synthesis of Filters*[1] or Chap. 6 of *Computer Oriented Circuit Design*.[2]

where $F_1^2(Z)$ and $Q_1^2(Z)$ are as given by (12.4) and (12.2). If the factor $(Z^2 - \alpha^2)/(Z^2 - 1)$ is approximately unity in the passband, then the passband will be approximately equiripple. Also, the denominator $Z^2 - 1$ has added a simple pole at infinity, as desired.

The factor $(Z^2 - \alpha^2)/(Z^2 - 1)$ produces what is termed a *parametric* pole at infinity. It is termed parametric because of the parameter α, which may be varied by the designer to produce different effects. For example, the parameter can be varied so that passive realizations have reasonable element values.

The frequency $Z^2 = \alpha^2$ corresponds to zero loss, as this is where $H(s)$ is unity. From Fig. 7.1c, if $\alpha^2 < 1$ then this zero loss occurs in the upper stopband. On the other hand, if $\alpha^2 > (\omega_B/\omega_A)^2$ then the zero loss occurs in the lower stopband. Since we certainly do not want such "holes" in the stopbands, we require

$$1 < \alpha^2 < \left(\frac{\omega_B}{\omega_A}\right)^2 \tag{12.6}$$

Referring to Fig. 7.1a, we see that for α in this range the characteristic function will have a zero on the σ axis (a real attenuation zero).

We have said that for the passband to be equiripple the factor $(Z^2 - \alpha^2)/(Z^2 - 1)$ must be approximately unity. We shall now determine just how "equiripple" the passband is. From (12.5) it follows that

$$|H(j\omega_A)|^2 = \left[1 + \left(\frac{Z^2 - \alpha^2}{Z^2 - 1}\right)\frac{F_1^2(Z)}{Q_1^2(Z)}\right]_{Z \to \infty} = 1 + \frac{F_1^2(\infty)}{Q_1^2(\infty)} \tag{12.7}$$

Thus the ripple amplitude of the parametric filter at $\omega = \omega_A$ is the same as the ripple amplitude of the equiripple filter; that is,

$$A(\omega_A) = 10 \log (1 + \epsilon_1^2) = A_{\max_1} \tag{12.8}$$

However, the ripple at $\omega = \omega_B$ has changed. This can be demonstrated by evaluating (12.5) at $Z = 0$, which leads to

$$A(\omega_B) = 10 \log (1 + \alpha^2 \epsilon_1^2) \tag{12.9}$$

The relative change in ripple is a measure of how equiripple the passband is. This quantitative measure is defined next.

Definition 12.1

The relative change in ripple R_C of a parametric equiripple bandpass filter is defined as

$$R_C = \frac{|A(\omega_B) - A(\omega_A)|}{A_{\max_1}} \tag{12.10}$$

where $A_{\max_1}$ is the ripple amplitude of the prototype equiripple filter.

Theorem 12.1

For an equiripple filter with a parametric pole at infinity as described by

$$H(s)H(-s) = 1 + \frac{Z^2 - \alpha^2}{Z^2 - 1} \frac{F_1^2(Z)}{Q_1^2(Z)}$$

the relative change in ripple is given by

$$R_C = \frac{\log(1 + \alpha^2 \epsilon_1^2)}{\log(1 + \epsilon_1^2)} - 1 \tag{12.11}$$

The losses at the passband edges are

$$A(\omega_A) = A_{\max_1} \qquad A(\omega_B) = A_{\max_1}(1 + R_C) \tag{12.12}$$

Proof

This theorem is simply a combination of (12.8), (12.9), and (12.10).

The parameter ϵ_1 is usually a small number, so we can use the following approximation, which is based on the Taylor-series expansion of (12.11):

$$R_C \approx \alpha^2 - 1 \tag{12.13}$$

Example 12.1

If the prototype equiripple filter has a ripple amplitude $A_{\max_1} = 0.5$ dB and $\alpha^2 = 1.2$, then (12.11) yields $R_C = 0.187$. The approximation in (12.13) yields $R_C \approx 0.2$. The approximation in (12.13) is even better if the ripple amplitude is smaller. For example, if $A_{\max_1} = 0.001$ dB and $\alpha^2 = 1.2$, then (12.11) and (12.13) agree to three significant figures.

In Example 12.1, the original ripple amplitude was $A_{\max_1} = 0.5$ dB. From (12.12) the parametric ripple amplitude at $\omega = \omega_B$ is 0.59 dB. This exceeds the original ripple; if this is objectionable we have two options:

1. We can make α closer to unity, for as α approaches unity the passband becomes more and more equiripple. This is indicated by the fact that $R_C \approx \alpha^2 - 1$, so that as α approaches unity the relative change in ripple approaches zero. However, reducing α will reduce the stopband attenuation.

2. If the parametric multiplier is made to be

$$P = \frac{1}{\alpha^2} \frac{Z^2 - \alpha^2}{Z^2 - 1}$$

then, since $P(Z = 0) = 1$, the loss at $s = j\omega_B$ will be

$$A(\omega_B) = A_{\max_1}$$

The loss at the lower stopband ($s = j\omega_A$) will be less than $A_{\max_1}$. This is summarized in the following theorem.

Theorem 12.2

For an equiripple filter with a parametric pole at infinity as described by

$$H(s)H(-s) = 1 + \frac{1}{\alpha^2} \frac{Z^2 - \alpha^2}{Z^2 - 1} \frac{F_1^2(Z)}{Q_1^2(Z)} \qquad (12.14)$$

the relative change in ripple is given by

$$R_C = 1 - \frac{\log(1 + \epsilon_1^2/\alpha^2)}{\log(1 + \epsilon_1^2)} \approx 1 - \frac{1}{\alpha^2} \qquad (12.15)$$

The losses at the passband edges are

$$A(\omega_A) = A_{\max_1}(1 - R_C) \qquad A(\omega_B) = A_{\max_1} \qquad (12.16)$$

12.4 ADDITION OF A POLE AT THE ORIGIN

If a parametric pole is added to a transfer function that is originally equiripple or maximally flat, we can write the parametric transfer function $H(s)$ as

$$H(s)H(-s) = 1 + P(Z) \frac{F_1^2(Z)}{Q_1^2(Z)} \qquad (12.17)$$

where the parametric multiplier $P(Z)$ determines the location of the parametric pole.

If $P(Z)$ is to produce a pole at the origin, this implies that $P(Z = \omega_B/\omega_A)$ should be infinite. The $P(Z)$ that we shall choose in order to produce a parametric pole at the origin is

$$P(Z) = \frac{Z^2 - \alpha^2}{Z^2 - (\omega_B/\omega_A)^2} \qquad (12.18)$$

As for the parametric pole at infinity, we require $1 < \alpha^2 < (\omega_B/\omega_A)^2$ so that there will not be a zero of attenuation in the stopband. The closer α is to ω_B/ω_A, the closer $H(s)$ approximates the original prototype filter. For example, if $H_1(s)$ is an equiripple filter, then as α approaches ω_B/ω_A the transfer function $H(s)$ becomes more and more equiripple. Again R_C, the relative change in ripple, can be defined as

$$R_C = \frac{|A(\omega_B) - A(\omega_A)|}{A_{\max_1}}$$

Using the facts that

$$A(\omega_A) = 10 \log |H(j\omega)|^2_{Z=\infty}$$

and

$$A(\omega_B) = 10 \log |H(j\omega)|^2_{Z=0}$$

we can easily establish the following theorem.

Theorem 12.3

For an equiripple filter with a parametric pole at the origin as described by

$$H(s)H(-s) = 1 + \frac{Z^2 - \alpha^2}{Z^2 - (\omega_B/\omega_A)^2} \frac{F_1^2(Z)}{Q_1^2(Z)}$$

the relative change in ripple is given by

$$R_C = 1 - \frac{\log\left[1 + \alpha^2 \epsilon_1^2/(\omega_B^2/\omega_A^2)\right]}{\log(1 + \epsilon_1^2)} \approx 1 - \frac{\alpha^2}{\omega_B^2/\omega_A^2}$$

The losses at the passband edges are

$$A(\omega_A) = A_{\max_1} \qquad A(\omega_B) = A_{\max_1}(1 - R_C)$$

12.5 STOPBAND AND PASSBAND PERFORMANCE

This section describes how the stopband or passband loss can be calculated for parametric equiripple bandpass filters. Similar calculations could be performed for equiripple lowpass filters or for maximally flat filters.

For a parametric filter we can write

$$H(s)H(-s) = 1 + P(Z)\frac{F_1^2(Z)}{Q_1^2(Z)} \qquad (12.19)$$

The parametric factor $P(Z)$ determines the location of the parametric peak. For a parametric peak at the origin,

$$P(Z) = \frac{Z^2 - \alpha^2}{Z^2 - (\omega_B/\omega_A)^2} \qquad (12.20)$$

whereas for a parametric peak at infinity,

$$P(Z) = \frac{1}{\alpha^2} \frac{Z^2 - \alpha^2}{Z^2 - 1} \qquad (12.21)$$

The program in Fig. 9.4 was used to calculate the passband or stopband loss of equiripple bandpass filters. It did this by evaluating

$$A_1(\omega) = 10 \log\left[1 + \frac{F_1^2(Z)}{Q_1^2(Z)}\right] \qquad (12.22)$$

at the transformed frequency $Z^2 = (\omega^2 - \omega_B^2)/(\omega^2 - \omega_A^2)$. Actually, neither the quantity $F_1^2(Z)$ nor the quantity $Q_1^2(Z)$ was evaluated separately; instead the entire ratio $F_1^2(Z)/Q_1^2(Z)$ was determined by using the loss function $L(Z)$.

To modify the program in Fig. 9.4 so that it calculates the loss of parametric equiripple bandpass filters, we can modify (12.22) to

$$A(\omega) = 10 \log \left[1 + P(Z) \frac{F_1^2(Z)}{Q_1^2(Z)} \right]$$

where the parametric factor $P(Z)$ is as given in (12.20) or (12.21). Thus, for a parametric filter we can proceed to calculate the loss as if it were an equiripple filter, and then modify the result by inserting the parametric factor $P(Z)$.

Figure 12.1 contains a program that can be used to calculate the loss of a parametric equiripple filter that has a parametric pole at infinity. The major modification of the prototype program is Step 6.41, which calculates the parametric factor given by (12.21).

The programs in Figs. 9.4 and 12.1 were used to obtain the results in Fig. 12.2. At the top of Fig. 12.2, an equiripple bandpass filter is analyzed; at the bottom, two parametric filters (parametric pole at infinity). The parametric multipliers are

$$P(Z) = \frac{1}{1.1} \frac{Z^2 - 1.1}{Z^2 - 1} \quad \text{and} \quad P(Z) = \frac{1}{1.2} \frac{Z^2 - 1.2}{Z^2 - 1}$$

```
1; LØSS ØF EQUIRIPPLE BPF, PARAMETRIC PØLE AT INFINITY
1.2 DEMAND AMAX,FA,FB,NZ,NIN,N
1.3 DEMAND F[I] FØR I=1:1:N
1.31 DEMAND AA
1.32 NIN=NIN-1
1.4 EE=10^(.1*AMAX)-1,FAA=FA*FA,FBB=FB*FB
1.42 Z[I]=SQRT((F[I]*F[I]-FBB)/(F[I]*F[I]-FAA)) FØR I=1:1:N
1.5 DEMAND F
1.6 TYPE #
1.7 TØ PART 6

6; DETERMINATIØN ØF STØPBAND LØSS
6.1 A=10*LØG(1+EE/AA)
6.2 TØ PART 6.9 IF F=FA
6.4 Z=SQRT( '(F*F-FBB)/(F*F-FAA)´)
6.41 P=1/AA*(Z*Z-AA)/(Z*Z-1)
6.5 TØ PART 10 IF F>FA AND F<FB
6.55 L=((Z+1)/´Z-1´)^(NIN/2)
6.56 L=L*((FA*Z/FB+1)/´FA*Z/FB-1´)^(NZ/2) IF NZ>0
6.57 L=L*(Z+Z[I])/(Z-Z[I]) FØR I=1:1:N
6.8 A=10*LØG(1+P*EE/4*(L+1/L)^2)
6.9 TYPE F,A IN FØRM 1

10; DETERMINATIØN ØF PASSBAND LØSS
10.55 B=NIN/2*ATN(-2*Z,Z*Z-1)
10.56 B=B+NZ/2*ATN(-2*Z*FB/FA,Z*Z-FBB/FAA)
10.57 B=B+ATN(-2*Z*Z[I],Z*Z-Z[I]*Z[I]) FØR I=1:1:N
10.8 A=10*LØG(1+P*EE*(CØS(B))^2)
10.9 TYPE F,A IN FØRM 1
```

Fig. 12.1 Program for the determination of the loss of an equiripple bandpass filter that has a parametric pole at infinity.

```
            DØ PART 1
                AMAX=.1
                FA=.9
                FB=1/.9
                NZ=1
                NIN=1
                N=2
                F[1]=.7
                F[2]=1/.7
                F=.3

           3.000-01    40.488
>
>DØ PART 6 FØR F=.001,.8,1.3,100
           1.000-03    90.365
           8.000-01    17.714
           1.300+00    24.832
           1.000+02    70.365
>
```

```
DØ PART 1                           DØ PART 1
    AMAX=.1                             AMAX=.1
    FA=.9                               FA=.9
    FB=1/.9                             FB=1/.9
    NZ=1                                NZ=1
    NIN=2                               NIN=2
    N=2                                 N=2
    F[1]=.7                             F[1]=.7
    F[2]=1/.7                           F[2]=1/.7
    AA=1.1                              AA=1.2
    F=.3                                F=.3

3.000-01    39.267                  3.000-01    37.897
>                                   >
>DØ PART 6 FØR F=.001,.8,1.3,100    >DØ PART 6 FØR F=.001,.8,1.3,100
 1.000-03    89.032                  1.000-03    87.486
 8.000-01    17.134                  8.000-01    16.582
 1.300+00    25.235                  1.300+00    25.544
 1.000+02   103.673                  1.000+02   106.304
```

Fig. 12.2 Demonstration of the use of the loss program in Fig. 12.1.

By examining Fig. 12.2, we can see that the addition of a parametric pole at infinity has increased the upper stopband loss (e.g., for $\alpha^2 = 1.1$, at $f = 100$ the loss increased from 70.365 dB to 103.673 dB); however, the loss in the lower stopband is less.

Figure 12.3 contains a program that can be used to find the loss of a parametric equiripple filter that has a parametric pole at the origin. This program was used to obtain the results at the bottom of Fig. 12.4. The parametric multipliers for these filters are

$$P(Z) = \frac{Z^2 - 1.4}{Z^2 - (1/0.81)^2} \quad \text{and} \quad P(Z) = \frac{Z^2 - 1.35}{Z^2 - (1/0.81)^2}$$

By examining Fig. 12.4, we can see that the addition of a parametric pole at the origin has increased the lower stopband loss (e.g., for $\alpha^2 = 1.4$,

```
1; LØSS ØF EQUIRIPPLE BPF, PARAMETRIC PØLE AT THE ØRIGIN
1.2 DEMAND AMAX,FA,FB,NZ,NIN,N
1.3 DEMAND F[I] FØR I=1:1:N
1.31 DEMAND AA
1.32 NZ=NZ-1
1.4 EE=10^(.1*AMAX)-1,FAA=FA*FA,FBB=FB*FB
1.42 Z[I]=SQRT((F[I]*F[I]-FBB)/(F[I]*F[I]-FAA)) FØR I=1:1:N
1.5 DEMAND F
1.6 TYPE #
1.7 TØ PART 6

6; DETERMINATIØN ØF STØPBAND LØSS
6.1 A=AMAX
6.2 TØ PART 6.9 IF F=FA
6.4 Z=SQRT( (F*F-FBB)/(F*F-FAA) )
6.41 P=(Z*Z-AA)/(Z*Z-FBB/FAA)
6.5 TØ PART 10 IF F>FA AND F<FB
6.55 L=((Z+1)/ Z-1 )^(NIN/2)
6.56 L=L*((FA*Z/FB+1)/ FA*Z/FB-1 )^(NZ/2) IF NZ>0
6.57 L=L*(Z+Z[I])/(Z-Z[I]) FØR I=1:1:N
6.8 A=10*LØG(1+P*EE/4*(L+1/L)^2)
6.9 TYPE F,A IN FØRM 1

10; DETERMINATIØN ØF PASSBAND LØSS
10.55 B=NIN/2*ATN(-2*Z,Z*Z-1)
10.56 B=B+NZ/2*ATN(-2*Z*FB/FA,Z*Z-FBB/FAA)
10.57 B=B+ATN(-2*Z*Z[I],Z*Z-Z[I]*Z[I]) FØR I=1:1:N
10.8 A=10*LØG(1+P*EE*(CØS(B))^2)
10.9 TYPE F,A IN FØRM 1

FØRM 1
  #.###^^^ ###.###
```

Fig. 12.3 Program for the determination of the loss of an equiripple bandpass filter that has a parametric pole at the origin.

at $f = 0.001$ the loss increased from 90.365 dB to 143.195 dB); however, the loss in the upper stopband is less.

The programs in Figs. 12.1 and 12.3 are for parametric equiripple bandpass filters. These programs were obtained by modifying the equiripple bandpass filter program in Fig. 9.4. Similarly, the equiripple lowpass filter program of Fig. 9.6 could be modified for parametric equiripple lowpass filters. Also, the maximally flat programs in Figs. 11.4 through 11.6 can be easily modified for parametric maximally flat filters.

12.6 DETERMINATION OF ATTENUATION POLES FOR PARAMETRIC FILTERS

Chapter 8 described a pole-placer program that could be used to determine the attenuation poles for filters that have equiripple passbands. This section describes how the program can be modified for parametric equiripple filters.

The pole-placer program must be modified so as to find the minimum of an arc for a parametric filter. That is, we want to find the minimum of

$$A = 10 \log \left[1 + P(Z) \frac{F_1{}^2(Z)}{Q_1{}^2(Z)} \right] \qquad (12.23)$$

where, for a parametric peak at infinity,

$$P(Z) = \frac{1}{\alpha^2} \frac{Z^2 - \alpha^2}{Z^2 - 1} \qquad (12.24)$$

and, for a parametric peak at the origin,

$$P(Z) = \frac{Z^2 - \alpha^2}{Z^2 - (\omega_B/\omega_A)^2} \qquad (12.25)$$

```
      DØ PART 1
        AMAX=.1
        FA=.9
        FB=1/.9
        NZ=1
        NIN=1
        N=2
        F[1]=.7
        F[2]=1/.7
        F=.3

      3.000-01   40.488
>
>DØ PART 6 FØR F=.001,.8,1.3,100
      1.000-03   90.365
      8.000-01   17.714
      1.300+00   24.832
      1.000+02   70.365
>
```

```
DØ PART 1                       DØ PART 1
  AMAX=.1                         AMAX=.1
  FA=.9                           FA=.9
  FB=1/.9                         FB=1/.9
  NZ=2                            NZ=2
  NIN=1                           NIN=1
  N=2                             N=2
  F[1]=.7                         F[1]=.7
  F[2]=1/.7                       F[2]=1/.7
  AA=1.4                          AA=1.35
  F=.3                            F=.3

3.000-01   45.104               3.000-01   46.120
>                               >
>DØ PART 6 FØR F=.001,.8,1.3,100   >DØ PART 6 FØR F=.001,.8,1.3,100
  1.000-03  143.195                1.000-03  144.665
  8.000-01   17.975                8.000-01   18.076
  1.300+00   24.262                1.300+00   24.010
  1.000+02   69.191                1.000+02   68.611
>                                 >
```

Fig. 12.4 Demonstration of the use of the loss program in Fig. 12.3.

Since $F_1^2(Z)/Q_1^2(Z)$ is for the stopband of an equiripple bandpass filter, we can write

$$\frac{F_1^2(Z)}{Q_1^2(Z)} = \frac{\epsilon^2}{4}\left(|L| + \frac{1}{|L|}\right)^2 \qquad (12.26)$$

where*
$$L(Z) = \left(\frac{Z + \omega_B/\omega_A}{Z - \omega_B/\omega_A}\right)^{NZ/2} \left(\frac{Z+1}{Z-1}\right)^{NIN/2} \prod_{i=1}^{N} \frac{Z + Z_i}{Z - Z_i} \qquad (12.27)$$

From Eqs. (12.23) to (12.27) it follows that the minimum of an arc can be found by obtaining the minimum of

$$G(Z) = \left|\frac{Z^2 - \alpha^2}{Z^2 - \beta^2}\right|^{1/2} \left(|L| + \frac{1}{|L|}\right) \qquad (12.28)$$

where $\beta = 1$ for a parametric peak at infinity, and $\beta = \omega_B/\omega_A$ for a parametric peak at the origin. Problem 12.1 demonstrates that the minimum of $G(Z)$ is given by the zeros of the function

$$D(Z) = \frac{\beta^2 - \alpha^2}{(\beta^2 - Z^2)(Z^2 - \alpha^2)} ZG(Z) + 2\left|\frac{Z^2 - \alpha^2}{Z^2 - \beta^2}\right|^{1/2}$$

$$\times \left(|L| - \frac{1}{|L|}\right)\left[\frac{NZ}{2} \frac{f_B/f_A}{(f_B/f_A)^2 - Z^2} + \frac{NIN}{2} \frac{1}{1 - Z^2} + \sum_{i=1}^{N} \frac{Z_i}{Z_i^2 - Z^2}\right]$$

$$(12.29)$$

This equation is solved via the Newton iteration procedure which was

* In this expression, NZ is the number of poles at the origin for the equiripple filter. If a parametric peak is added to the origin, then the number of peaks would become $NZ + 1$.

```
1: PØLE PLACER (PARAMETRIC EQUIRIPPLE BANDPASS)
1.01: THE FØLLØWING STEPS SHØULD BE ADDED TØ "BANDPASS PØLE PLACER"
1.31 DEMAND AA,B; AA=ALPHA^2, B=0 (B=1) FØR PAR. PØLE AT 0 (INF.)
1.32 NINP=NIN, NZP=NZ-1, BB=(FB/FA)^2 IF B=0
1.33 NINP=NIN-1, NZP=NZ, BB=1 IF B=1

6.55 L=((Z+1)/`Z-1`)^(NINP/2)
6.56 L=L*((FA*Z/FB+1)/`FA*Z/FB-1`)^(NZP/2) IF NZP>0
6.58 P=1/AA*(Z*Z-AA)/(Z*Z-1) IF B=1
6.59 P=(Z*Z-AA)/(Z*FBB/FAA) IF B=0
6.8 A=10*LØG(1+P*EE/4*(L+1/L)^2)

8.9 D=NINP/2/(1-ZZ)+NZP/2*FB/FA/(FBB/FAA-ZZ)
8.92 Z=SQRT(ZZ)
8.93 L=((Z+FB/FA)/`Z-FB/FA`)^(NZP/2)
8.94 L=L*((Z+1)/`Z-1`)^(NINP/2)
8.95 L=L*(Z+Z[I2])/(Z-Z[I2]), FØR I2=1:1:N
8.96 L=`L`,P1=SQRT(`ZZ-AA`/`ZZ-BB`),G=P1*(L+1/L)
8.97 D=(BB-AA)/(BB-ZZ)*Z/(ZZ-AA)*G+2*P1*(L-1/L)*D

>
```

Fig. 12.5 Pole-placer program for parametric equiripple bandpass filters.

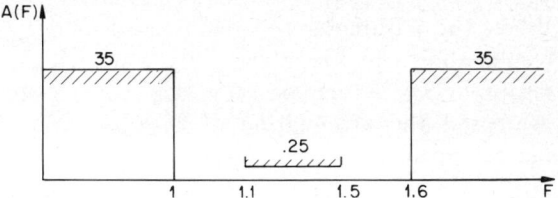

Fig. 12.6 Requirements for a bandpass filter example.

used in Part 8 of the original pole placer (Fig. 8.11). For the parametric pole placer, Part 8 can be modified as shown in Fig. 12.5. This Part 8, when added to the original Part 8, calculates $D(Z)$ as given by (12.29).

Another modification of the original pole placer is that the loss must now be calculated for the parametric case; thus Part 6 must be modified as in Fig. 12.5.

Adding the program shown in Fig. 12.5 to the equiripple pole placer creates a parametric pole placer which is used exactly as the equiripple pole placer was. (However, it should be noted that, for the parametric bandpass filter, NZ + NIN must be odd instead of even.)

12.7 RESULTS FROM THE POLE-PLACER PROGRAM

This section uses the pole placer to find parametric characteristics that meet the requirements in Fig. 12.6. These are the same as the requirements that were given in Example 8.4. Example 8.4 met the requirements with an equiripple filter which had NZ = 1 = NIN and four finite attenuation poles as given in Fig. 8.13. The results in Fig. 8.13 show that the requirements were exceeded everywhere by at least 11.56 dB. The first arc exceeded the requirements by 21.23 dB.

Now consider what happens if a parametric pole is added at the origin. In that case,

$$NZ = 2 \qquad NIN = 1$$

When the parameter α^2 is chosen to be 1.8, the pole placer gives the results in Fig. 12.7. If these results are compared with those of Example 8.4, two observations can be made:

1. The minimum attenuation has decreased from 11.56 dB to 11.42 dB.
2. The minimum of arc_1 has increased from 21.23 dB to 22.62 dB.

Thus the parametric pole at the origin has increased the attenuation of the first arc at the expense of the overall characteristic.

The overall characteristic can be improved by instead adding a parametric pole at infinity. In that case,

$$NZ = 1 \qquad NIN = 2$$

When the parameter α^2 is chosen to be 1.1, the pole placer gives the results shown in Fig. 12.8a. By comparing these results with those in Example 8.4, it can be seen that the parametric pole at infinity has decreased the attenuation of the first arc, but improved the overall performance.

The above results are for $\alpha^2 = 1.1$, which is very near the minimum allowable value of unity. If α^2 is increased to 1.4, the attenuation of the first arc decreases even more, but the overall performance improves, as demonstrated by the results shown in Fig. 12.8b.

Additional pole-placer runs were made for different values of α^2. Increasing α^2 beyond 1.4 did improve the overall performance, but only slightly. In fact, if α^2 was increased much beyond 1.4, the overall performance was actually degraded.

In Example 8.4 the requirement was exceeded by 11.56 dB. When a parametric pole was added at infinity (with $\alpha^2 = 1.4$), the requirement was exceeded by 12.26 dB. Thus a parametric pole can improve the stopband performance, but not by very much.

If parametric poles do not offer significant improvement of stopband performance, why are they used? The main reason is to provide a transfer function that is realizable by a specific configuration. For example, Watanabe[3] showed that if the characteristic function has two real zeros (two zeros on the $s = \sigma$ axis), then it may be realized by a symmetrical bandpass filter that has all its inductors being used to produce infinite loss at finite frequencies. The two zeros can be equal in magnitude and opposite in phase; thus, the parametric multiplier can be[4]

$$P(Z) = \frac{1}{\alpha^2} \frac{(Z^2 - \alpha^2)^2}{(Z^2 - 1)[Z^2 - (\omega_B/\omega_A)^2]} \tag{12.30}$$

```
D0 PART 49

          F[1]=        .76103591
          F[2]=        .987550156
          F[3]=       1.61188592
          F[4]=       1.77694809

FMIN[ 1]=5.492-01    AMIN= 57.62    DMIN= 22.62
FMIN[ 2]=9.390-01    AMIN= 46.42    DMIN= 11.42
FMIN[ 3]=1.000+00    AMIN= 46.42    DMIN= 11.42
FMIN[ 4]=1.600+00    AMIN= 46.42    DMIN= 11.42
FMIN[ 5]=1.656+00    AMIN= 46.42    DMIN= 11.42
FMIN[ 6]=2.227+00    AMIN= 46.42    DMIN= 11.42
```

>

Fig. 12.7 Demonstration of the use of the pole placer with a parametric pole at the origin.

```
           F[1]=      .799016459
           F[2]=      .988072855
           F[3]=     1.611671
           F[4]=     1.77016783

FMIN[ 1]=5.624-01    AMIN=  53.61    DMIN=  18.61
FMIN[ 2]=9.426-01    AMIN=  46.85    DMIN=  11.85
FMIN[ 3]=1.000+00    AMIN=  46.85    DMIN=  11.85
FMIN[ 4]=1.600+00    AMIN=  46.85    DMIN=  11.85
FMIN[ 5]=1.655+00    AMIN=  46.85    DMIN=  11.85
FMIN[ 6]=2.167+00    AMIN=  46.85    DMIN=  11.85
```

(a)

```
           F[1]=      .845064531
           F[2]=      .988983508
           F[3]=     1.61137614
           F[4]=     1.76198417

FMIN[ 1]=5.915-01    AMIN=  48.29    DMIN=  13.29
FMIN[ 2]=9.487-01    AMIN=  47.26    DMIN=  12.26
FMIN[ 3]=1.000+00    AMIN=  47.26    DMIN=  12.26
FMIN[ 4]=1.600+00    AMIN=  47.26    DMIN=  12.26
FMIN[ 5]=1.653+00    AMIN=  47.26    DMIN=  12.26
FMIN[ 6]=2.114+00    AMIN=  47.26    DMIN=  12.26
```

(b)

Fig. 12.8 Demonstration of the use of the pole placer with a parametric pole at infinity: (a) $\alpha^2 = 1.1$; (b) $\alpha^2 = 1.4$.

If this factor is used to modify an equiripple characteristic function as indicated by

$$H(s)H(-s) = 1 + P(Z)\frac{F_1^2(Z)}{Q_1^2(Z)} \tag{12.31}$$

then a parametric pole is added at the origin ($Z = \omega_B/\omega_A$), and another is added at infinity ($Z = 1$). In fact, the parametric multiplier in (12.30) can be considered to be a product of two parametric multipliers:

$$P_1 = \frac{Z^2 - \alpha^2}{Z^2 - (\omega_B/\omega_A)^2} \qquad P_2 = \frac{1}{\alpha^2}\frac{Z^2 - \alpha^2}{Z^2 - 1}$$

The multiplier P_1 produces a parametric pole at the origin, while P_2 produces a parametric pole at infinity.

A pole-placer program could be written for the parametric filters that are described by (12.30), but that will not be done here.

12.8 PARAMETRIC LOWPASS FILTERS

The previous section demonstrated that the addition of a parametric peak can increase stopband attenuation; however, this increase is usually slight. The main purpose of adding a parametric peak is to alter the characteristic function in some desired manner. For example, if we are designing a lowpass filter, we may want the dc loss to be zero; that is, $A(0) = 0$. In the design of passive filters, having zero dc loss allows the source and load impedances to be equal; but this implies that an equiripple lowpass filter should be of odd degree. Chapter 6 discussed one way to obtain an even-degree lowpass filter that has $A(0) = 0$: Transform the frequency scale so that the lowest attenuation zero is shifted to the origin. Another way is to add a parametric pole (at infinity) to an odd-degree equiripple lowpass filter. This would produce an even-degree parametric lowpass filter that has $A(0) = 0$.

The previous sections demonstrated that the equiripple bandpass filter theory could be easily modified to treat parametric equiripple bandpass filters. Similarly, the equiripple lowpass filter theory can be easily modified to treat parametric equiripple lowpass filters; however, this will not be done here.

12.9 NATURAL MODES FOR PARAMETRIC FILTERS

The treatment of parametric filters has paralleled the treatment of equiripple filters. We first found how to calculate the passband or stopband loss of parametric filters. We next created a pole-placer program so as to be able to determine the attenuation poles. That is, the input/output transfer function $H(s)$ is a ratio of polynomials:

$$H(s) = \frac{e(s)}{q(s)} \tag{12.32}$$

The pole-placer program can be used to determine $q(s)$; this section discusses a zero-finder program that can be used to determine $e(s)$.

The function $e(s)$ will be found by first obtaining the equivalent transformed function $E(Z)$ that is described by

$$E(Z)E^*(Z) = F^2(Z) + Q^2(Z) \tag{12.33}$$

The parametric functions $F^2(Z)$ and $Q^2(Z)$ are related to the prototype equiripple functions $F_1^2(Z)$ and $Q_1^2(Z)$. To find the relations, consider

$$H(s)H(-s) = 1 + \frac{F^2(Z)}{Q^2(Z)} = 1 + P(Z)\frac{F_1^2(Z)}{Q_1^2(Z)} \tag{12.34}$$

where, for a parametric peak at infinity,

$$P(Z) = \frac{1}{\alpha^2}\frac{Z^2 - \alpha^2}{Z^2 - 1} \tag{12.35}$$

and, for a parametric peak at the origin,

$$P(Z) = \frac{Z^2 - \alpha^2}{Z^2 - (\omega_B/\omega_A)^2} \tag{12.36}$$

Comparing Eqs. (12.34) through (12.36), we see that we can write

$$F^2(Z) = C(Z^2 - \alpha^2)F_1^2(Z) \tag{12.37}$$

where C is $1/\alpha^2$ or unity, depending on whether the parametric pole is at infinity or the origin.

We have just seen that $F^2(Z)$ can be easily found by first calculating $F_1^2(Z)$ for the prototype equiripple filter and then modifying it as in (12.37). It is thus an easy task to modify an equiripple zero-finder program into a parametric equiripple zero finder. The necessary modifications are contained in Fig. 12.9. Adding this program to the one in Fig. 10.1 creates a parametric zero finder for bandpass filters.

In Fig. 12.9, Part 53 performs the calculations described in (12.37). The only other major modification in Fig. 12.9 is to allow for the fact that $e(s)$ has a real root. From Theorem 7.5, this root is given by

$$\sigma_0^2 = \frac{\omega_B^2 - a\omega_A^2}{a - 1} \tag{12.38}$$

This is calculated in Step 1.78.

```
1: ZERO FINDER (PARAMETRIC EQUIRIPPLE BANDPASS)
1.01; THE FOLLOWING STEPS SHOULD BE ADDED TO "BANDPASS ZERO FINDER"
1.31 DEMAND AA,B; AA=ALPHA^2, B=0 (B=1) FOR PAR. POLE AT 0 (INF.)
1.32 NIN=NIN-1 IF B=1
1.33 NZ=NZ-1 IF B=0
1.415 T=T+1
1.51 DO PART 53
1.77 TO PART 1.7 IF R<IP(M/2)
1.78 TYPE SIGMA FOR SIGMA=SQRT((FBB-A*FAA)/(A-1))*TP
1.8 DO PART 6 FOR F=FB IF B=1
1.81 DO PART 6 FOR F=FA IF B=0

6.75 A=A+10*LOG(WW+SIGMA^2)

53; DETERMINATION OF C*(Z*Z-AA)*F(Z)^2=C[0]+C[2]*Z^2+C[4]*Z^4+...
53.1 A[0]=-AA*C[0],A[2*M+2]=C[2*M],P1=1
53.2 A[I]=C[I-2]-AA*C[I] FOR I=2:2:2*M
53.3 P1=1/AA IF B=1
53.4 C[I]=P1*A[I] FOR I=0:2:2*M+2

58.1 NIN=NIN+1,M=M+1 IF B=1
58.11 NZ=NZ+1,M=M+1 IF B=0

>
```

Fig. 12.9 Zero-finder program for parametric equiripple bandpass filters.

226 Approximation Methods for Electronic Filter Design

Figure 12.10 contains two examples of the use of the parametric zero finder. The top part of the figure is for a parametric peak at infinity; the lower part is for a parametric peak at the origin. These filters were previously encountered in Figs. 12.2 and 12.4. The validity of the results can be checked by noting that the loss at $f = 0.3$ was the same when calculated by the zero-finder program (Fig. 12.9) as when calculated by the loss program (Fig. 12.3).

```
DØ PART 1
        AMAX=.1
        FA=.9
        FB=1/.9
        NZ=1
        NIN=2
        N=2
        F[1]=.7
        F[2]=1/.7
        AA=1.2
        B=1


ZQ[ 1]=  11.0239    ZF[ 1]=1.1365+00
ZQ[ 2]=   4.5022    ZF[ 2]=1.0027+00
ZQ[ 3]=  10.2663    ZF[ 3]=8.7764-01
        SIGMA=         7.19724324

          CH=          .523408043

>DØ PART 6 FØR F=.3
 3.000-01   37.897
>
```

(a)

```
DØ PART 1
        AMAX=.1
        FA=.9
        FB=1/.9
        NZ=2
        NIN=1
        N=2
        F[1]=.7
        F[2]=1/.7
        AA=1.4
        B=0


ZQ[ 1]=  10.5631    ZF[ 1]=1.1378+00
ZQ[ 2]=   4.5589    ZF[ 2]=9.9879-01
ZQ[ 3]=  10.9201    ZF[ 3]=8.7992-01
        SIGMA=         3.15085566

          CH=          4.58530671

>DØ PART 6 FØR F=.3
 3.000-01   45.104
>
```

(b)

Fig. 12.10 Demonstration of the use of the zero finder: (a) parametric pole at infinity; (b) parametric pole at the origin.

12.10 CONCLUSIONS

For a parametric filter, the input/output transfer function $H(s)$ can be written as

$$H(s)H(-s) = 1 + P(Z)\frac{F_1^2(Z)}{Q_1^2(Z)}$$

where $F_1(Z)/Q_1(Z)$ is the characteristic function of a prototype filter that is either maximally flat or equiripple. The parametric multiplier can be chosen so that the parametric filter is essentially maximally flat or equiripple.

The addition of a parametric pole can improve the stopband performance slightly, but the main reason for introducing a parametric peak is to provide a transfer function that is realizable by a specific configuration.

REFERENCES

1. J. L. Herrero and G. Willoner, *Synthesis of Filters*, Prentice-Hall, Inc., Englewood Cliffs, N.J., 1966.
2. F. F. Kuo and W. G. Magnuson, *Computer Oriented Circuit Design*, Prentice-Hall, Inc., Englewood Cliffs, N.J., 1968.
3. H. Wantanabe, "On the Circuit with a Minimum Number of Coils," *IRE Trans. on Circuit Theory*, March 1960, pp. 77–78.
4. J. Bingham, "The Approximation Problem for Both Conventional and Parametric Band-Pass Filters," *IEEE Trans. on Circuit Theory*, September 1964, pp. 408–410.

PROBLEMS

12.1 This problem shows that the minimums of

$$G(z) = \left|\frac{Z^2 - \alpha^2}{Z^2 - \beta^2}\right|^{1/2}\left(|L| + \frac{1}{|L|}\right)$$

occur where the following function is zero:

$$D(Z) = \frac{\beta^2 - \alpha^2}{(\beta^2 - Z^2)(Z^2 - \alpha^2)} ZG(Z) + 2\left|\frac{Z^2 - \alpha^2}{Z^2 - B^2}\right|^{1/2}\left(|L| - \frac{1}{|L|}\right)$$

$$\times \left[\frac{NZ}{2}\frac{f_B/f_A}{(f_B/f_A)^2 - Z^2} + \frac{NIN}{2}\frac{1}{1 - Z^2} + \sum_{i=1}^{N}\frac{Z_i}{Z_i^2 - Z^2}\right]$$

a. Show that

$$\frac{d}{dZ}\left|\frac{Z^2 - \alpha^2}{Z^2 - \beta^2}\right|^{1/2} = \left|\frac{Z^2 - \alpha^2}{Z^2 - \beta^2}\right|^{1/2}\frac{\beta^2 - \alpha^2}{(\beta^2 - Z^2)(Z^2 - \alpha^2)}Z$$

b. Show that for

$$L(Z)\left(\frac{Z+\omega_B/\omega_A}{Z-\omega_B/\omega_A}\right)^{NZ/2}\left(\frac{Z+1}{Z-1}\right)^{NIN/2}\prod_{i=1}^{N}\frac{Z+Z_i}{Z-Z_i}$$

$$\frac{d|L|}{dZ} = 2|L|\left[\frac{NZ}{2}\frac{f_B/f_A}{(f_B/f_A)^2-Z^2} + \frac{NIN}{2}\frac{1}{1-Z^2} + \sum_{i=1}^{N}\frac{Z_i}{Z_i^2-Z^2}\right]$$

c. Use the results in Parts a and b to establish the result at the beginning of the problem.

12.2 Assume that a prototype equiripple filter is described by $f_A = 1$, $f_B = 1.1$, and $A_{max} = 0.1$ dB. If a parametric pole described by $\alpha^2 = 1.05$ is added at infinity [such that $A(f = 1.1) = 0.1$ dB],

a. What is the loss at $f = 1$?
b. What is the loss at $f = 1.5$ if the prototype filter had a loss of 20 dB there?
c. What is the loss at the origin if the prototype filter had a loss of 30 dB there?

12.3 As in Problem 12.2, assume that the prototype equiripple filter is described by $f_A = 1$, $f_B = 1.1$, and $A_{max} = 0.1$ dB. This time the parametric filter will be formed by adding a parametric pole at the origin. Let this pole be described by $\alpha^2 = 1.2$.

a. What is the relative change in ripple R_C?
b. What is the loss at $f = f_B$?
c. What is the loss at infinity if the prototype filter had a loss of 30 dB there?

12.4 Equation (12.30) describes a parametric multiplier that introduces a parametric pole at the origin and another at infinity. Assume that the prototype filter is described by $f_A = 1$, $f_B = 1.1$, and $A_{max} = 0.1$ dB. If $\alpha^2 = 1.1$,

a. What is the loss at $f = 1$?
b. What is the loss at $f = 1.1$?

CHAPTER THIRTEEN

Optimization Techniques for Approximation Theory

13.1 INTRODUCTION

Chapter 8 described a pole-placer program that can be used to find the attenuation poles for an equiripple passband filter. The "optimum" poles for an arbitrary stopband requirement can be found by using the program iteratively. The pole-placer program was relatively simple, because it used the fact that the attenuation poles uniquely determine the loss of an equiripple passband filter. That is, given that the passband is equiripple, one can calculate the loss without having to determine the zeros of $H(s)$.

Chapter 11 described how the original pole-placer program could be modified to yield results for maximally flat passband filters. Chapter 12 then gave further modifications for parametric filters. The important feature about all these programs was that the zeros of $H(s)$ did not have to be calculated; because the passband had a special shape, the loss was uniquely determined by the attenuation poles.

If the passband does not have a special shape, but is instead arbitrary, then the zeros of $H(s)$ are not uniquely determined by the poles. To choose a transfer function to meet arbitrary passband and arbitrary stopband requirements optimally, we must use more sophisticated optimization techniques than those contained in the pole-placer program.

This chapter briefly discusses various optimization techniques that can be used for solving approximation problems. These techniques can be used to find transfer functions that meet arbitrary loss requirements. However, the techniques have a much broader range of application than this—for example, they can be applied to satisfy delay requirements. Of course, optimization techniques need not be applied solely to transfer functions; they can also be applied to networks. One can specify a specific network configuration and then use optimization techniques to choose the element values so as to meet some performance criterion.

We shall not discuss any one optimization method in great detail. The reader who is interested in a specific technique should consult the references.* The purpose of this chapter is to indicate how optimization techniques can be applied to help solve the approximation problem.

13.2 SYSTEM RESPONSE AND ERROR CRITERIA

The purpose of this book is to help a filter designer obtain a transfer function of the form

$$H(s) = \frac{b_0 + b_1 s + b_2 s^2 + \cdots + b_n s^n}{a_0 + a_1 s + a_2 s^2 + \cdots + a_m s^m} \qquad (13.1)$$

Most of the transfer functions we have discussed can be written as

$$H(s) = C_H \frac{\prod_{i}^{M} s^2 + (\omega_{zi}/Q_{zi})s + \omega_{zi}^2}{s^{NZ} \prod_{i=1}^{N} s^2 + \omega_i^2} \qquad (13.2)$$

Optimization techniques can be applied so as to enable the designer to choose the available parameters optimally. In (13.1) the available parameters would be a_i and b_i; in (13.2) they would be ω_i, ω_{zi}, Q_{zi}, and C_H.

If one wants to apply approximation techniques to networks instead of transfer functions, then the available parameters might be resistors. In filter design it is often better to work directly with the network, instead of with the transfer function. One reason for this is that constraints can be placed on the elements to guarantee that the circuit is realizable; for example, all resistors can be constrained to have positive element

---

* There are several references at the end of this chapter. Of these, the first three are survey articles that contain many additional references.

values. Or, the parameters can be constrained to be within a certain range, so that reasonable element values are generated by the optimization procedure. Also, the network can be modeled in such a way as to compensate for parasitic elements. For example, a parasitic capacitance could be included in the network model. If this is constrained to have a fixed value, the remaining elements can be optimized so as to produce the desired response.

In any optimization problem there will be some available parameters that should be chosen, perhaps subject to certain constraints, so as to produce a desired response. This response might be, for example, loss as a function of frequency. Thus, the optimization program must contain means for evaluating the system response.

An error function is generated by comparing the system response with the requirement. Because the optimization procedure will usually be performed on a digital computer, we shall assume that the error function is computed only at discrete points (which will be termed *frequencies* for this discussion). The number of frequencies M depends on the number of available parameters. Certainly, the number of frequencies should be greater than the number of available parameters. A good rule of thumb (which can be arrived at by analogy to the sampling theorem[4]) is that the number of frequencies should be twice the number of available parameters.

Chapter 1 discussed various error functions that could be used. One of these was the mean-square error,

$$E = \int_{\omega_A}^{\omega_B} [f(\omega) - h(\omega)]^2 \, d\omega \qquad (13.3)$$

If we are only considering discrete points, we could instead use

$$E = \sum_{i=1}^{M} (f_i - h_i)^2 \qquad (13.4)$$

where f_i represents the requirement, and h_i represents the system response. The square-error criterion in (13.4) can be generalized to the following pth-error criterion:

$$E = \sum_{i=1}^{M} \omega_i (f_i - h_i)^p \qquad (13.5)$$

where ω_i is a weighting function that can be used if certain frequencies are more important than others.

For large p, the pth-error criterion approaches the Tschebycheff error criterion. That is, for large p (in practice, p of the order of 10) we are minimizing the maximum error.

13.3 INITIAL PARAMETERS

The steps outlined above are the same steps that had to be performed for the pole-placer program in Chap. 8. In Chap. 8 we assumed that the transfer function was of the form

$$H(s) = C_H \frac{\prod_{i=1}^{M/2} s^2 + (\omega_{zi}/Q_{zi})s + \omega_{zi}^2}{s^{NZ} \prod_{i=1}^{N} s^2 + \omega_i^2}$$

The available parameters were the pole frequencies ω_i. We chose the available parameters so as to optimize the system response (the loss). The error criterion was of the Tschebycheff type, as we minimized the maximum error (the error was how much the requirement exceeded the loss).

To use the pole-placer program, we had to make an initial guess for the attenuation poles. This initial guess affected the rate of convergence of the iterative procedure. The same is true for any optimization method—one must make an initial guess for the available parameters. The better the initial guess, the faster will be the rate of convergence.

If we want to design a filter that has an arbitrary passband and an arbitrary stopband, how do we initially choose the available parameters? One way would be to simplify the problem by first assuming that the stopband is still arbitrary but the passband is equiripple. From previous chapters of this book, we know how to solve that problem. We can then use this solution as an initial guess for the more difficult problem: the case of the arbitrary passband and arbitrary stopband.

13.4 MINIMIZATION TECHNIQUES

Once we have obtained a measure of error for our initial parameters, we want to know how to change the parameters so as to minimize the error. In the pole-placer program we used a Taylor-series approximation to linearize a set of equations. The linear set of equations was then solved to yield a new set of pole frequencies. There are many other methods that can be used to minimize the error; in this section we divide the methods into three classes and give some examples for each class.

1. *Simple Methods* (*Nonderivative*)

The sophisticated optimization techniques evaluate the first and/or higher derivatives; however, there are some procedures that do not require derivatives. Perhaps one of the simplest ways to minimize the error function is to vary one available parameter at a time.[5] For example, first minimize the error for parameter p_i. Then minimize the error for the parameter p_{i+1}; after doing this for all parameters, one would start over again at p_1. Since each step in this minimization procedure is a single parameter search, the golden-section method[6] may be used to find the minimum.

The simplex method[7] can be used for multiparameter searches; that is, it does not seek a minimum by examining the behavior of the error function for one parameter at a time. In n-dimensional space, a set of $n + 1$ points forms a simplex In the simplex method we start with an arbitrary simplex and then let it tumble and shrink toward the region where the error is a minimum. The direction in which we allow the simplex to move is determined by evaluating the error function at each of the $n + 1$ points of the simplex.

2. Slope-following Methods

Slope-following methods evaluate the first derivatives of the error function ($\partial e/\partial p_i$) and use this information to indicate how the parameters should be changed in order to minimize the error. The first derivatives determine the gradient of the error function. The gradient points in the direction of the greatest change in error; thus, to minimize the error one proceeds in the direction opposite to the gradient. This is the basis of the steepest-descent[8] method, which uses the gradient to predict parameter changes for error minimization. Steepest descent can be considered to be an attempt to change parameter values so as to proceed down an error slope most rapidly.

3. Second-Order Methods

The slope-following methods tend to reduce the error rapidly in the initial stages of an optimization procedure; however, their convergence is rather slow as the minimum is approached. To improve the rate of convergence, one can use not only the first derivatives of the error function ($\partial e/\partial p_i$), but also the second derivatives ($\partial^2 e/\partial p_i\, \partial p_j$). Just as the first derivatives determined the gradient of the error function, the second derivatives determine the Hessian.

The various second-order methods differ mainly in the way they try to approximate the second derivatives. The second derivatives are not usually found by using a perturbation scheme based on varying elements; they are instead approximated by using knowledge of the error function and its gradient at previous iterations. The Fletcher-Powell[8] minimization procedure is one of the best known second-order methods. A commercially available optimization program, Match,[9] uses a version of Fletcher-Powell to optimize circuits for loss or delay performance.

13.5 PRACTICAL OPTIMIZATION PROGRAMS FOR THE APPROXIMATION PROBLEM

The previous section indicated that many different optimization algorithms exist. No one procedure can be shown to be best; which one converges most rapidly depends on the specific problem to be solved. Thus, an optimization program that helps to solve the approximation problem should contain different optimization algorithms.

A successive-approximation program, Suprox, has been used for a period of years at Bell Telephone Laboratories. It uses a steepest-descent algorithm to help improve the parameter values for the initial stages of optimization. As the minimum is approached and the rate of convergence slows, Suprox switches to a second-order optimization method, least squares.

Recently, a more powerful general-purpose optimization program, Gpop, has been developed at Bell Telephone Laboratories. This contains more optimization procedures than Suprox and is written in such a way that new algorithms can be incorporated in the future without requiring extensive programming effort.

Besides containing a variety of optimization algorithms, an optimization program for the approximation problem must contain subroutines that can be used to calculate the error function. In one problem the error might be determined by the loss of a network; in another it might be determined by the delay. Thus subroutines for loss, delay, etc., should be included in the optimization program. The "etc." implies that the program should be flexible enough so that additional subroutines can be added easily.

The optimization programs described in this section require a large computer because they are very general. We need such a general program if we want to find a filter that has an "arbitrary" passband and stopband. We were able to use a simpler pole-placer program for filters that had equiripple passbands and arbitrary stopbands. Similarly, one could use a simpler program for filters that have arbitrary passbands and equiminimum stopbands. A brief discussion of such filters is contained in the next section.

13.6 ARBITRARY PASSBAND, EQUIMINIMUM STOPBAND

A bandpass characteristic with an equiminimum stopband is shown in Fig. 13.1. If the signal to be suppressed has a flat energy spectrum, we would probably want a filter that has an equiminimum stopband; thus such a requirement is frequently encountered.

Temes and Gyi[10] discuss how to design filters with an arbitrary passband and equiminimum stopband. Szentirmai[11] extended their work so that the levels of the upper and lower stopbands could differ. The work presented here will use Szentirmai's approach, but will include only the case in which both stopbands have the same level. This is the problem most frequently encountered; for the more general problem, the reader is referred to the original article.

In order to design a filter that approximates a zero-loss passband, we usually work with the characteristic function $K(s)$ instead of the transfer function $H(s)$. These functions are related by

$$H(s)H(-s) = 1 + K(s)K(-s) \tag{13.6}$$

The constant of unity value in (13.6) was introduced because the filter had zero passband loss. For $H(s)$ as in (13.6), the loss is given by

$$A = 10 \log (1 + |K(j\omega)|^2) \tag{13.7}$$

Since Temes and Gyi were interested in approximating a constant stopband (instead of a constant passband), they defined a function $x(\omega)$ which allowed (13.7) to be written as

$$A = A_{\min} + 10 \log [1 + x(\omega)] \tag{13.8}$$

This equation focuses attention on the stopband. In the stopband, the minimum attenuation is $A_{\min}$; thus $x(\omega)$ must be greater than zero in the stopband. In fact, from Fig. 13.1 we know that $x(\omega)$ must be zero

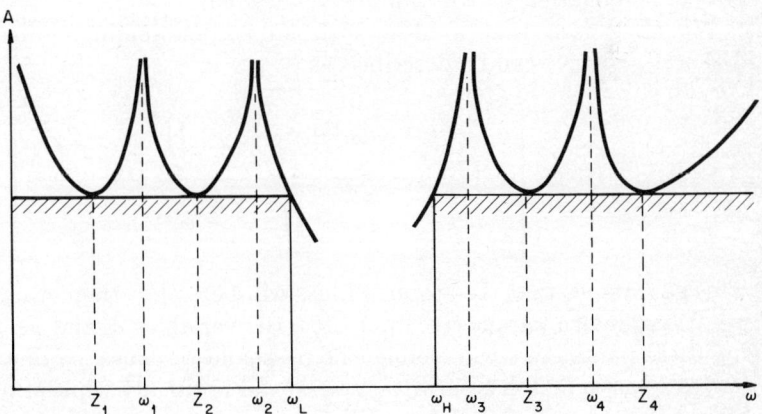

Fig. 13.1 Bandpass filter with an arbitrary passband and equiminimum stopband.

at ω_L, ω_H, and Z_i ($i = 1, 2, \ldots, N$). Thus (13.8) can be rewritten as

$$A = A_{\min} + 10 \log \left[1 + C \frac{(\omega^2 - \omega_L^2)(\omega^2 - \omega_H^2) \prod_{i=1}^{N} (\omega^2 - Z_i^2)^2}{\omega^2 \prod_{i=1}^{N} (\omega^2 - \omega_i^2)^2} \right] \quad (13.9)$$

Some observations* can be made about this equation:

1. In the passband ($\omega_L < |\omega| < \omega_H$) the loss is less than $A_{\min}$.
2. At the stopband frequencies ω_L, ω_H, Z_i the attenuation is $A_{\min}$. At all other stopband frequencies the attenuation is greater than $A_{\min}$.
3. The attenuation is infinite at the pole frequencies ω_i.
4. The constant C can be chosen such that the passband attenuation has a specific value at a certain frequency, for example,

$$A(\omega_B) = A_{\max}$$

5. There is a simple attenuation pole at the origin (NZ = 1).
6. There is a simple attenuation pole at infinity (NIN = 1).

If one does not want NZ equal to unity, then the exponent of ω in (13.9) must be changed. Similarly, the number of poles at infinity can be changed by replacing the constant C with a function $f(\omega^2)$, for example,

$$f(\omega^2) = C(\omega^2 + \alpha^2)$$

This function has no positive real zeros in ω^2 and thus does not change the locations where $A = A_{\min}$. However, introducing the function does add an attenuation pole at infinity, as desired.

We have seen that a filter with an equiminimum stopband (and NZ = 1 = NIN) can be described as

$$A = A_{\min} + 10 \log \left[1 + \frac{C(\omega^2 - \omega_L^2)(\omega^2 - \omega_H^2) \prod_{i=1}^{N} (\omega^2 - Z_i^2)^2}{\omega^2 \prod_{i=1}^{N} (\omega^2 - \omega_i^2)^2} \right] \quad (13.10)$$

This guarantees that the stopband loss will be no less than $A_{\min}$. There are $2N$ unknown parameters in (13.10), the variables Z_i and ω_i. Changing these parameters will change the passband response, so they may be varied so as to match an arbitrary passband optimally while still guaranteeing that the stopband loss will be no less than $A_{\min}$. If we apply

* These observations assume that C is a positive number.

optimization techniques to an equation such as (13.10) instead of to a more general equation such as

$$A = \frac{b_0 + b_1\omega^2 + b_2\omega^4 + \cdots}{a_0 + a_1\omega^2 + a_2\omega^4 + \cdots}$$

the results will converge faster, as we have helped the computer by giving it a special form for the solution.

REFERENCES

1. P. E. Fleischer, "Optimization Techniques in System Design," pp. 175–217 of *System Analysis by Digital Computer*, edited by F. F. Kuo and J. F. Kaiser, John Wiley & Sons, Inc., New York, 1966.
2. G. C. Temes and D. A. Calahan, "Computer-Aided Network Optimization: The State of the Art," *Proc. IEEE*, November 1967, pp. 1832–1863.
3. T. J. Aprille, "An Application of Computer Optimization Techniques to Network Synthesis," thesis at University of Illinois, Urbana, 1968.
4. A. Papoulis, *The Fourier Integral and its Applications*, McGraw-Hill Book Company, New York, 1962.
5. M. Friedman and L. S. Savage, *Selected Techniques of Statistical Analysis*, McGraw-Hill Book Company, New York, 1947.
6. D. Pierre, *Optimization Theory with Applications*, John Wiley & Sons, Inc., New York, 1969.
7. J. A. Nelder and R. Mead, "A Simplex Method for Function Minimization," *Comput. J.*, January 1965, pp. 308–313.
8. R. Fletcher and M. J. D. Powell, "A Rapidly Convergent Descent Method for Minimization," *Comput. J.*, July 1963, pp. 163–168.
9. H. B. Lee, et al., "Program Refines Circuit from Rough Design Data," *Electronics*, November 23, 1970, pp. 58–65.
10. G. C. Temes and M. Gyi, "Design of Filters with Arbitrary Passband and Chebyshev Stopband Attenuation," *1967 IEEE Int. Conf. Rec.*, pp. 2–12.
11. G. Szentirmai, "On a Filter Approximation Problem," *IEEE Trans. on Circuit Theory*, May 1970, pp. 280–282.

CHAPTER FOURTEEN

Delay and Related Subjects

14.1 INTRODUCTION

The input/output transfer function $H(j\omega)$ is a complex quantity; that is, it has not only an amplitude $|H(j\omega)|$, but also a phase arg $H(j\omega)$. Previous chapters have considered only the amplitude response, which may be adequate for many applications, but sometimes one must consider the phase characteristic too.

The phase characteristic affects the transient response, which is of particular interest in pulse systems such as radar and digital communications. In these cases one is interested in quantities such as rise time and amount of overshoot, because they indicate how much the pulse is distorted. Rise time and overshoot are time-domain quantities; however, it is often easier to approach the approximation problem in the frequency domain. For this reason one often prefers to work with a quantity termed *delay*.

14.2 DEFINITION OF DELAY

We would like to define a frequency-domain quantity that indicates the quality of a filter's transient response. If a filter is such that an output pulse is an exact replica of the input pulse, except for a time delay, then the transient response is ideal. Such a situation is portrayed in Fig. 14.1.

The transfer function for the "ideal" filter in Fig. 14.1 can be determined by observing that

$$v_2(t) = v_1(t - T) \tag{14.1}$$

If the Laplace transformation of $v(t)$ is denoted as $V(s)$, it follows that

$$V_2(s) = e^{-sT} V_1(s) \tag{14.2}$$

This implies that

$$\frac{V_1(s)}{V_2(s)} = H(s) = e^{sT} \tag{14.3}$$

The above result indicates that if (as in Fig. 14.1) the output pulse is a delayed replica of the input pulse, then $\arg H(j\omega) = \omega T$, where T indicates how much the pulse is delayed. This motivates the following definition.

Definition 14.1

The delay $D(\omega)$ of a filter that has an input/output transfer function $H(j\omega)$ is defined to be*

$$D(\omega) = \frac{d}{d\omega} \arg H(j\omega) \tag{14.4}$$

---

* Some authors prefer to call the function in (14.4) *envelope delay* or *group delay* to differentiate it from *phase delay*, which is equal to [arg $H(\omega)$]/ω. Guillemin[22] also considers *signal delay*. We shall only use the expression in (14.4); thus there will be no confusion if it is simply referred to as *delay*.

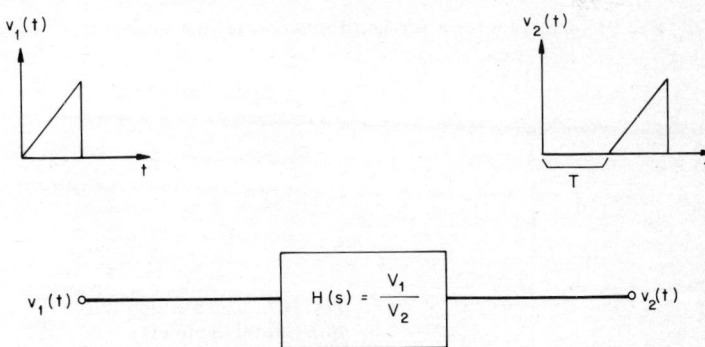

Fig. 14.1 An ideal transient response is a delayed replica of the input.

If an output pulse is a delayed replica of the input, then the delay $D(\omega)$ is a constant. On the other hand, if the output pulse is distorted (e.g., if it has a finite rise time), then the delay will vary with frequency, indicating that different frequencies are delayed by different amounts of time as they pass through the network. Since a pulse can be considered to be composed of many different frequencies, it is easy to understand why a delay that is a function of frequency leads to distortion.*

14.3 CALCULATION OF DELAY

This section gives some theorems that are useful in calculating the delay of a transfer function. Later sections will make use of this material to find the delay for different types of approximations. The theorems are relatively simple to establish and will thus not be proven here.

The input/output transfer function is a ratio of polynomials and can thus be written as

$$H(s) = \frac{e(s)}{q(s)}$$

In previous chapters the zeros of $q(s)$ were restricted to the $j\omega$ axis; however, a useful theorem can be stated for an even more general $q(s)$. The theorem that follows assumes that the zeros of $q(s)$ have quadrantal symmetry. Quadrantal symmetry implies that if s_0 is a zero of $q(s)$, then $-s_0$ is also a zero. Figure 14.2 shows the zeros of a typical $q(s)$ that has quadrantal symmetry.

Theorem 14.1

If the zeros of $q(s)$ have quadrantal symmetry or are only on the $j\omega$ axis, then $q(s)$ does not affect the delay.

Theorem 14.1 can be used to help establish the next theorem.

* See Problem 14.3 for a further discussion of this subject.

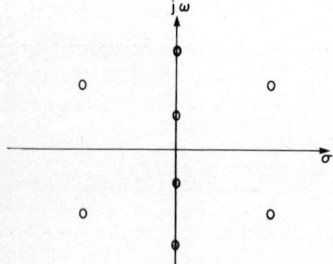

Fig. 14.2 Zeros which illustrate quadrantal symmetry.

Theorem 14.2

If
$$H(s) = \frac{\prod_{i=1}^{K}(s + \sigma_i) \prod_{i=1}^{L}(s^2 + A_i s + B_i)}{q(s)} \quad (14.5)$$

where the zeros of $q(s)$ have quadrantal symmetry or are only on the $j\omega$ axis, then the delay is given by

$$D(\omega) = \sum_{i=1}^{K} \frac{\sigma_i}{\sigma_i^2 + \omega^2} + \sum_{i=1}^{L} \frac{A_i(B_i + \omega^2)}{(B_i - \omega^2)^2 + (A_i \omega)^2} \quad (14.6)$$

Theorem 14.2 is useful in calculating the delay when the numerator of $H(s)$ is expressed as a product of factors. If the numerator is instead expressed as a polynomial, the next theorem should be used. This theorem is established in Problem 14.1.

Theorem 14.3

If $H(s) = \left(\sum_{i=0}^{n} h_i s^i \right)/q(s)$, where the zeros of $q(s)$ have quadrantal symmetry or are only on the $j\omega$ axis, then the delay is given by

$$\frac{\left[\sum_{k=0}^{k_1} (-1)^k h_{2k} \omega^{2k} \right] \left[\sum_{k=0}^{k_2} (-1)^k (2k+1) h_{2k+1} \omega^{2k} \right] - \left[\sum_{k=1}^{k_1} (-1)^k (2k) h_{2k} \omega^{2k-1} \right] \left[\sum_{k=0}^{k_2} (-1)^k h_{2k+1} \omega^{2k+1} \right]}{\left[\sum_{k=0}^{k_1} (-1)^k h_{2k} \omega^{2k} \right]^2 + \left[\sum_{k=0}^{k_2} (-1)^k h_{2k+1} \omega^{2k+1} \right]^2} \quad (14.7)$$

where k_1 = integer part of $n/2$
k_2 = integer part of $(n-1)/2$

14.4 A PROGRAM FOR THE CALCULATION OF DELAY

The first part of this book was devoted to approximations for amplitude characteristics. The passbands of these filters are either maximally flat or equiripple. When one designs a filter which is either maximally flat or equiripple, one may want to know what the delay performance is; if not adequate, it might have to be equalized with an allpass network. This section discusses a program that can be used to investigate the delay of any of the filters described earlier in this book.

The attenuation poles of the filters we have discussed are located on

```
1; CALCULATIØN ØF DELAY
1.1 DEMAND K,L
1.12 DEMAND SIGMA[I] FØR I=1:1:K
1.13 DEMAND ZF[I],ZQ[I] FØR I=1:1:L
1.14 DEMAND F
1.15 TYPE #
1.2 TP=2*$PI
1.3 ZW[I]=TP*ZF[I],A[I]=ZW[I]/ZQ[I],B[I]=ZW[I]^2 FØR I=1:1:L
1.4 DØ PART 6

6.1 W=TP*F,WW=W*W,D=0
6.2 D=D+SIGMA[I]/(SIGMA[I]^2+WW) FØR I=1:1:K
6.3 D=D+A[I]*(B[I]+WW)/((B[I]-WW)^2+(A[I]*W)^2) FØR I=1:1:L
6.4 TYPE F,D IN FØRM 1

 FØRM        1
#.###^^^  #.###^^^
```

Fig. 14.3 Program for the calculation of delay.

the $j\omega$ axis; thus, Theorem 14.1 indicates that only the numerator of $H(s)$ affects the delay. The numerator of $H(s)$ can be written as

$$e(s) = \prod_{i=1}^{K} (s + \sigma_i) \prod_{i=1}^{L} \left(s^2 + \frac{Z\omega_i}{ZQ_i} s + Z\omega_i^2 \right) \quad (14.8)$$

The notation in (14.8) is the same as that used in the programs that found $e(s)$ for equiripple filters (Figs. 10.1 and 10.3) or for maximally flat filters (Figs. 11.13 to 11.15). That is, $Z\omega_i$ is the undamped natural frequency of a zero, and ZQ_i is the quality of that zero. Comparing (14.5) and (14.8) yields

$$A_i = \frac{Z\omega_i}{ZQ_i} \qquad B_i = Z\omega_i^2$$

Thus from (14.6) the delay is given by

$$D(\omega) = \sum_{i=1}^{K} \frac{\sigma_i}{\sigma_i^2 + \omega^2} + \sum_{i=1}^{L} \frac{(Z\omega_i/ZQ_i)(Z\omega_i^2 + \omega^2)}{(Z\omega_i^2 - \omega^2)^2 + (Z\omega_i\omega/ZQ_i)^2} \quad (14.9)$$

The program in Fig. 14.3, which is based on this equation, can be used to find delay.

We now have programs available that allow us to do a detailed amplitude-and-delay investigation of any equiripple, maximally flat, or parametric filter. As an example, we shall examine the normalized Butterworth lowpass filter; later, we shall compare it with a "Bessel filter."

By definition, the normalized Butterworth lowpass filter is described by

$$F_B = \frac{1}{2\pi} \qquad A_{\max} = 3 \text{ dB}$$

Also, there are no peaks on the finite $j\omega$ axis ($N = 0$), and the number of peaks at infinity is NIN = n, where n is the order of the Butterworth filter. With these parameters as input, the program in Fig. 11.5 can be used to investigate the amplitude response. The result is shown in Fig. 14.4.

To find the delay performance we need $Z\omega_i$ and ZQ_i, which can be determined by using the program in Fig. 11.14. However, it is more direct to use Eq. (2.18):

n even:

$$H(s) = \prod_{i=1}^{n/2} [s^2 + 2(\cos \theta_i)s + 1]$$

where $\theta_i = (\pi/2n)(2i - 1)$

n odd:

$$H(s) = (s + 1) \prod_{i=1}^{(n-1)/2} [s^2 + (2 \cos \theta_i)s + 1]$$

where $\theta_i = \pi i/n$.

Thus, for example, for $n = 4$ the inputs for the delay program in Fig. 14.3 are

$$K = 0 \qquad ZF_1 = \frac{1}{2\pi} \qquad ZF_2 = \frac{1}{2\pi}$$

$$L = 2 \qquad ZQ_1 = \frac{1}{2 \cos (\pi/8)} \qquad ZQ_2 = \frac{1}{2 \cos (3\pi/8)}$$

The program in Fig. 14.3 was used to generate the data for the curves

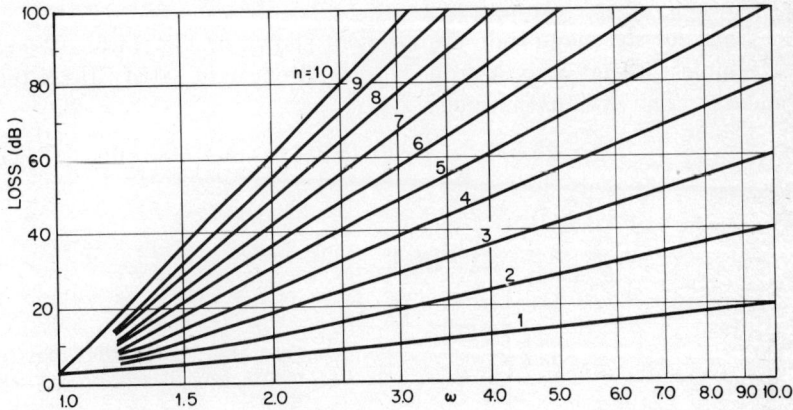

Fig. 14.4 Amplitude response of normalized Butterworth lowpass filters.

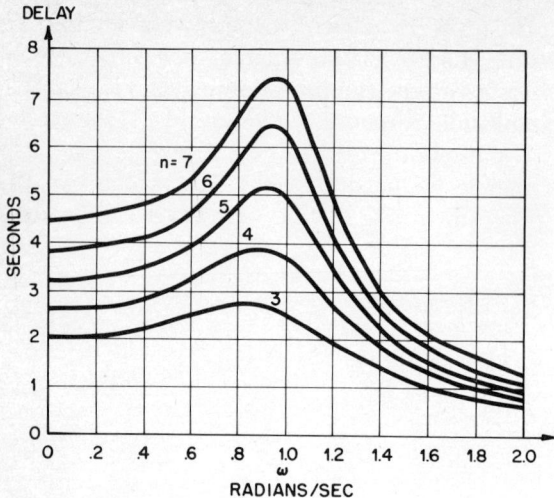

Fig. 14.5 Delay performance of normalized Butterworth lowpass filters.

shown in Fig. 14.5. These curves indicate that the delay peaks very near the passband edge. They also demonstrate that the sharper the transition region, the larger the delay peak.

14.5 RELATIONS BETWEEN MAGNITUDE AND DELAY

The previous section investigated the delay for the normalized Butterworth approximation. Figure 14.5 shows that the sharper the transition region (i.e., the larger n), the greater the variation in delay. This section investigates other relations between magnitude and delay. This is done by means of some examples.

Consider the magnitude requirement shown in Fig. 14.6. We saw in Example 5.3 that a sixth-order elliptic filter could satisfy these requirements. The attenuation poles are

$$F_1 = 26.577 \qquad F_2 = 33.286 \qquad F_3 = 82.605 \qquad (14.10)$$

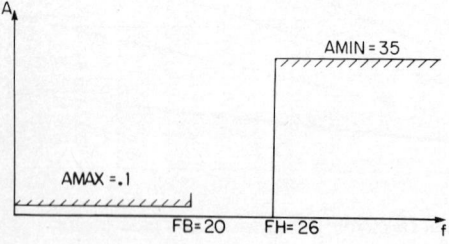

Fig. 14.6 Amplitude requirements for an elliptic lowpass filter.

These attenuation poles are such that the stopband attenuation is everywhere greater than 46.85 dB, which is larger than the requirement $A_{\min} = 35$ dB. It is possible to trade some of this "excess" stopband performance for better delay performance.

Before we can investigate the effects of modifying the approximation in (14.10), we must find the delay for this approximation, which requires knowing the numerator of $H(s)$. The program in Fig. 10.3 can be used to find the numerator of $H(s)$; the result is

$$e(s) = \prod_{i=1}^{3} \left[s^2 + \frac{2\pi ZF_i}{Q_i} s + (2\pi ZF_i)^2 \right]$$

where $ZQ_1 = 7.8805$ $ZF_1 = 20.827$
       $ZQ_2 = 1.7888$ $ZF_2 = 18.331$
       $ZQ_3 = 0.6250$ $ZF_3 = 12.975$

Figure 14.3 can now be applied to find the delay. The results are plotted in Fig. 14.7, where the curve is identified as approximation 1.

From Fig. 14.7 it can be seen that the delay for approximation 1 has a maximum at about $f = 20.5$. If the location of this maximum is increased, then the delay at the passband edge ($f = 20$) should decrease. This suggests one way to modify approximation 1: increase F_B so that the delay maximum is shifted to the right. Increasing F_B will decrease

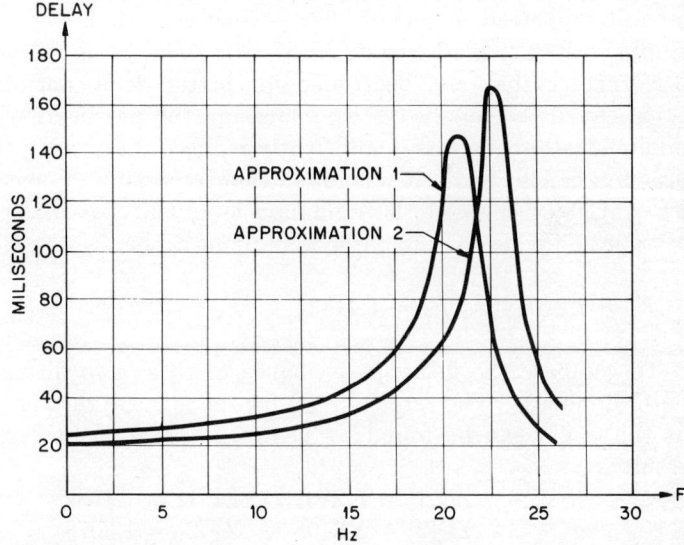

Fig. 14.7 Illustration of the fact that delay performance can be improved by increasing the passband edge.

the minimum stopband attenuation, but the minimum stopband attenuation of approximation 1 was greater than $A_{\min} = 35$ dB. Thus, consider approximation 2, which follows:

APPROXIMATION 2

$$A_{\max} = 0.1 \qquad FB = 22 \qquad FH = 26$$

The elliptic filter for this approximation has loss peaks at

$$F_1 = 26.472 \qquad F_2 = 32.191 \qquad F_3 = 76.886$$

The minimum stopband loss is 38.1 dB, which is greater than $A_{\min} = 35$ dB.

The delay for approximation 2 is shown in Fig. 14.7. Notice that the delay variation in the passband is much less than for approximation 1.

We have just seen that it is possible to trade excess stopband performance for better delay performance by increasing the bandwidth. We can apply this procedure to demonstrate that a lowpass that is maximally flat beyond the origin can have better delay performance than one that is maximally flat at the origin. To show this, we shall first reconsider the filter in Fig. 11.10 that is maximally flat at the origin. This has a passband edge of $FB = 1$, where the attenuation is $A_{\max} = 0.1$ dB. The attenuation poles were chosen so that the stopband is equiminimum and has minimum loss of 30.22 dB. We saw in Fig. 11.12 that if the lowpass is instead maximally flat beyond the origin, then the minimum stopband loss is increased to 53.87 dB; thus we have excess stopband performance that can be traded for better delay performance. This trade can be accomplished by increasing the passband width until the minimum stopband loss is sufficiently reduced. By using the pole-placer program of Fig. 11.11, it was found that increasing the passband edge to $FB = 1.185$ decreased the minimum stopband loss to 30.48 dB. The attenuation poles for this filter were found to be

$$F_1 = 1.31054 \qquad F_2 = 1.41434 \qquad F_3 = 1.80193 \qquad F_4 = 4.38846$$

To compare the delay performance of these two filters we need the natural modes. The zeros of $H(s)$ for the lowpass that is maximally flat at the origin can be found by using the program in Fig. 11.14. The result is

$$\begin{aligned}
ZQ[1] &= 5.2732 & ZF[1] &= 1.1584 \\
ZQ[2] &= 1.6408 & ZF[2] &= 1.3073 \\
ZQ[3] &= 0.8584 & ZF[3] &= 1.6697 \\
ZQ[4] &= 0.5414 & ZF[4] &= 2.2442
\end{aligned}$$

Similarly, the natural modes for the lowpass that is maximally flat beyond the origin can be found by using the program in Fig. 11.15. The result is

$$ZQ[1] = 14.7245 \quad ZF[1] = 1.2403$$
$$ZQ[2] = 3.9267 \quad ZF[2] = 1.2710$$
$$ZQ[3] = 1.5145 \quad ZF[3] = 1.3159$$
$$ZQ[4] = 0.6203 \quad ZF[4] = 1.2285$$

The delay performance can now be found by using the program in Fig. 14.3. The results shown in Fig. 14.8 demonstrate that, within the original passband of FB = 1, the filter that is maximally flat beyond the origin has better delay performance.

It has just been demonstrated that one way to improve delay performance is to increase the passband width. Another way to improve delay performance is to decrease the maximum passband loss A_{max}. This again reduces the stopband loss, which makes the transition between passband and stopband more gradual. A gradual transition region implies good delay performance.

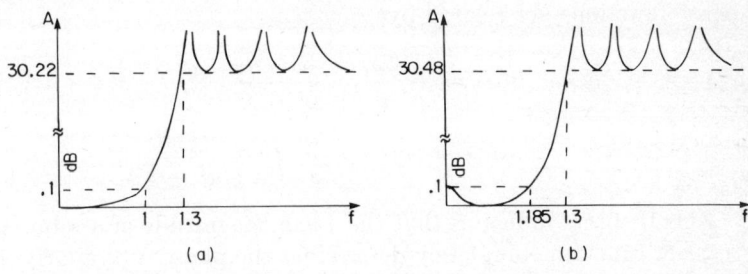

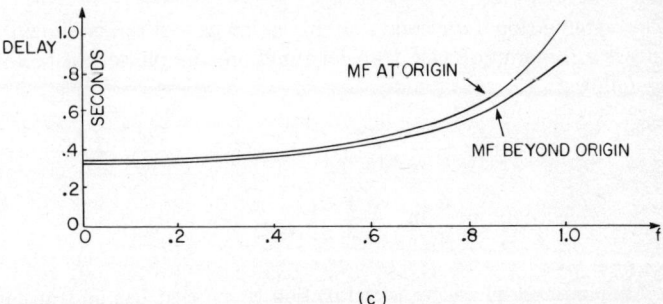

Fig. 14.8 A lowpass filter maximally flat beyond the origin can have better delay performance than one which is maximally flat at the origin.

14.6 THE HILBERT TRANSFORMATIONS[1]

Previous sections have shown that the magnitude and delay responses are related: if we change the magnitude response, we also alter the delay. In fact, for a minimum-phase transfer function* the magnitude response uniquely determines the phase response, and vice versa! This section presents some equations that relate the phase and magnitude. These equations, known as *Hilbert transformation pairs*, will then be applied to specific examples.

The equations are stated in terms of functions α and β, defined by

$$H(j\omega) = e^{\alpha + j\beta} \qquad (14.11)$$

From this it follows that

$$\alpha = \ln |H(j\omega)| \qquad \beta = \arg H(j\omega)$$

α is called the *attenuation function*, and β the *phase function*. The following theorem indicates the relation between the attenuation and phase functions.

Theorem 14.4†

If $H(s)$ is a minimum-phase transfer function, then its attenuation and phase functions are related by

$$\beta(\omega) = -\frac{2\omega}{\pi} \int_0^\infty \frac{\alpha(\xi)\, d\xi}{\omega^2 - \xi^2} \qquad (14.12)$$

$$\alpha(\omega) = \alpha(0) + \frac{2\omega^2}{\pi} \int_0^\infty \frac{\beta(\xi)\, d\xi}{\xi(\omega^2 - \xi^2)} \qquad (14.13)$$

This theorem indicates that the j-axis magnitude of a minimum-phase transfer function completely determines the phase. Similarly, the phase determines the amplitude to within a constant multiplier.

Example 14.1

The attenuation function $\alpha(\omega)$ for a lowpass filter is shown in Fig. 14.9. If this is a minimum-phase transfer function, the phase can be found from (14.12) as follows:

$$\beta(\omega) = -\frac{2\omega}{\pi} \int_0^\infty \frac{\alpha(\xi)}{\omega^2 - \xi^2}\, d\xi = -\frac{2\omega A}{\pi} \int_{\omega_1}^\infty \frac{1}{\omega^2 - \xi^2}\, d\xi$$

$$= \frac{A}{\pi} \ln \left| \frac{\omega + \omega_1}{\omega - \omega_1} \right|$$

---

* A minimum-phase transfer function is one that has no transmission zeros in the right half-plane.

† This theorem comes from Eqs. 82 (page 554) and 89 (page 558) of Guillemin.[1]

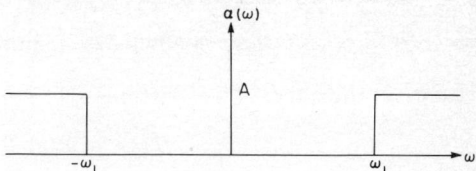

Fig. 14.9 Example of the attenuation function $\alpha(\omega)$ for a lowpass filter.

The delay for Example 14.1 can be found by differentiating the phase. The result is

$$|D(\omega)| = \frac{2A\omega_1}{\pi|\omega + \omega_1||\omega - \omega_1|} \quad (14.14)$$

This equation demonstrates that an ideal lowpass filter (i.e., one in which the constant A is infinite) would have infinite delay, which implies that an ideal lowpass filter is physically unrealizable.

Equation (14.14) demonstrates that at $\omega = \omega_1$, where the amplitude changes rapidly, the delay becomes infinite. This implies that if one wants a fairly constant delay in the passband, the attenuation must have a smooth transition region. The next few sections present approximations that have gradual transition regions and thus good delay properties. These sections describe various polynomials that can be used for lowpass filter approximations. We would use such polynomials if we were interested in a lowpass filter that has good delay performance. By definition, a lowpass filter attenuates high frequencies. If we want to pass all frequencies, but still wish to provide a certain amount of delay, we can instead use allpass networks. Section 14.15 demonstrates that allpass networks can be derived from lowpass networks in such a way that their delay performance is very similar.

14.7 THE BESSEL APPROXIMATION

Section 14.2 demonstrated that if an output pulse is a delayed replica of the input pulse, then

$$H(s) = e^{sT} \quad (14.15)$$

where T indicates the amount by which the pulse is delayed. If the delay is normalized to unity, then (14.15) can be written as

$$H(s) = e^s = \cosh s + \sinh s \quad (14.16)$$

We would like to approximate this with an nth-order polynomial

$H(s) = \sum_{i=0}^{n} h_i s^i$. The coefficients h_i must be chosen such that $H(s)$ is Hurwitz; thus, if $H(s)$ is written as the sum of an even and odd polynomial

$$H(s) = m(s) + n(s) \qquad (14.17)$$

then $m(s)/n(s)$ must be a reactance function.

We can find $m(s)$ and $n(s)$ such that $m(s)/n(s)$ is a reactance function by considering the following series expansions:

$$\cosh s = 1 + \frac{s^2}{2!} + \frac{s^4}{4!} + \cdots$$
$$\sinh s = s + \frac{s^3}{3!} + \frac{s^5}{5!} + \cdots \qquad (14.18)$$

These expansions demonstrate that $\cosh s$ is an even polynomial and $\sinh s$ is an odd polynomial. It can be shown[2] that a continued-fraction expansion of $(\cosh s)/\sinh s$ yields

$$\frac{\cosh s}{\sinh s} = \frac{1}{s} + \cfrac{1}{\frac{3}{s} + \cfrac{1}{\frac{5}{s} + \cfrac{1}{\frac{7}{s} + \cdots}}}$$

If this function is truncated at the Nth term, we obtain

$$\frac{\cosh s}{\sinh s} = \frac{1}{s} + \cfrac{1}{\frac{3}{s} + \cfrac{1}{\frac{5}{s} + \cdots + \cfrac{1}{\frac{2N-1}{s}}}} = \frac{m(s)}{n(s)} \qquad (14.19)$$

Since all the terms in the continued-fraction expansion are positive, $m(s)/n(s)$ as given in (14.19) must be a reactance function. Thus, if we approximate e^s by $H(s) = m(s) + n(s)$, then $H(s)$ will be Hurwitz, as desired.

Example 14.2

Approximate e^s by $H(s) = \sum_{i=0}^{4} h_i s^i$.

Solution
From (14.19),

$$\frac{m(s)}{n(s)} = \frac{1}{s} + \cfrac{1}{\cfrac{3}{s} + \cfrac{1}{\cfrac{5}{s} + \cfrac{1}{\cfrac{7}{s}}}} = \frac{105 + 45s^2 + s^4}{105s + 10s^3}$$

Thus
$$m(s) = C(105 + 45s^2 + s^4)$$
$$n(s) = C(105s + 10s^3)$$

Substituting into (14.17) yields

$$H(s) = C(105 + 105s + 45s^2 + 10s^3 + s^4) \approx e^s$$

The constant C should be chosen such that $H(0)$ is unity, which implies that $C = 1/105$.

Example 14.2 found a fourth-degree $H(s)$ which approximates e^s; the same approach could be used for any degree approximation. Storch[2] has shown that the general result can be written in terms of Bessel polynomials, which are defined next.

Definition 14.2
The nth-order Bessel polynomial $B_n(s)$ is given by the following recursion formula:

$$B_n = (2n - 1)B_{n-1} + s^2 B_{n-2} \qquad (14.20)$$

where the initial values are $B_0 = 1$, $B_1 = s + 1$.

The following theorem summarizes Storch's results.

Theorem 14.5
An nth-order Hurwitz approximation for e^s is given by

$$H(s) = \frac{B_n(s)}{b_0}$$

where $B_n(s)$ is the nth-order Bessel polynomial, and $b_0 = B_n(0)$.

Storch showed that, instead of using the recursive relation in Definition 14.2, one can find Bessel polynomials by using

$$B_n(s) = \sum_{i=0}^{n} b_i s^i$$

where
$$b_i = \frac{(2n - i)!}{2^{n-i} i! (n - i)!} \qquad (14.21)$$

```
1; DETERMINATIØN ØF DELAY FØR BESSEL FILTERS
1.1 DEMAND N,DO,W
1.2 TYPE #
1.3 DØ PART 2
1.4 K1=N/2,K2=N/2-1 IF N=2*IP(N/2)
1.41 K2=(N-1)/2,K2=K1 IF N>2*IP(N/2)
1.5 DØ PART 6

2; DETERMINATIØN ØF H(S) FRØM THEØREM 14.5
2.1 B[I]=1 FØR I=0:1:N
2.2 B[0]=B[0]*K FØR K=3:2:2*N-1
2.3 I=1,II=1
2.5 B[I]=B[I]*K FØR K=N-I+1:1:2*N-I
2.6 II=I*II,B[I]=B[I]/2^(N-I)/II
2.7 I=I+1
2.8 TØ PART 2.5 IF I<N
2.9 H[I]=B[I]*DO^I/B[0] FØR I=0:1:N

6; DETERMINATIØN ØF D(W) BY THEØREM 14.3
6.1 X1=0,X2=0,X3=0,X4=0
6.2 Y1=2*K,X1=X1+(-1)^K*H[Y1]*W^Y1 FØR K=0:1:K1
6.3 Y1=2*K,Y2=Y1+1,X2=X2+(-1)^K*Y2*H[Y2]*W^Y1 FØR K=0:1:K2
6.4 Y1=2*K,Y2=Y1-1,X3=X3+(-1)^K*Y1*H[Y1]*W^Y2 FØR K=1:1:K1
6.5 Y1=2*K+1,X4=X4+(-1)^K*H[Y1]*W^Y1 FØR K=0:1:K2
6.6 D=(X1*X2-X3*X4)/(X1*X1+X4*X4)
6.7 A=10*LØG(X1*X1+X4*X4)
6.8 TYPE W,D,A IN FØRM 6

>
```

Fig. 14.10 Program for delay calculation of Bessel filters.

Filters designed according to Theorem 14.5 are known by many equivalent terms. They were first made popular by W. E. Thomson[3,4] and are thus sometimes called *Thomson filters*. The first n derivatives of the delay are zero at the origin,[2] so they are also called *maximally flat delay filters*. However, they are most frequently called *Bessel filters*, as the recursion relation in (14.20) is the same as for Bessel polynomials.

A program that can be used to determine the attenuation and delay of a Bessel filter is given in Fig. 14.10. To understand the program, recall from Theorem 14.5 that

$$e^s \approx \sum_{i=0}^{n} b_i s^i$$

where the b_i are given in (14.21). The b_i are calculated in Part 2 of the program. If the zero-frequency delay is not unity, then

$$e^{sD_0} \approx \sum_{i=0}^{n} b_i (sD_0)^i = \sum_{i=0}^{n} H_i s^i$$

Step 2.9 finds H_i. The delay and attenuation are then found in Part 6.

The curves in Fig. 14.11 were found by using the Bessel delay program. These curves demonstrate that for delay approximations an important parameter is the delay-bandwidth product. For example, assume we want to design a lowpass filter that has no greater than a 1 percent delay error for $\omega \leq 3$ rad/s. From Fig. 14.11, a fourth-degree Bessel filter has a delay error of 1 percent at $\omega D_0 = 1.9$. Thus the requirements could be met by a fourth-degree Bessel filter if $D_0 < 1.9/3$ s.

When one is designing a delay filter, the delay and bandwidth will often both be specified. If this is the case, one can use the curves in Fig. 14.11 to find the degree of the Bessel filter. For example, assume we want to design a filter that has a zero-frequency delay of $D_0 = 20$ ms and a delay error of less than 1 percent for $\omega \leq 120$ rad/s. Examination of Fig. 14.11 at

$$\omega D_0 = 120 \times 0.02 = 2.4$$

indicates that a fifth-degree Bessel filter will meet the requirements.

The program in Fig. 14.10 can also be used to find the attenuation of a Bessel filter as a function of the delay-bandwidth product. Some results are plotted in Fig. 14.12. The following example demonstrates one use of these curves.

Example 14.3
What degree Bessel filter is needed for the following requirements:
  a. Zero-frequency delay of 20 ms
  b. Less than 1 percent delay error for $\omega \leq 120$ rad/s
  c. Less than 2.0 dB variation in loss for $\omega \leq 120$ rad/s

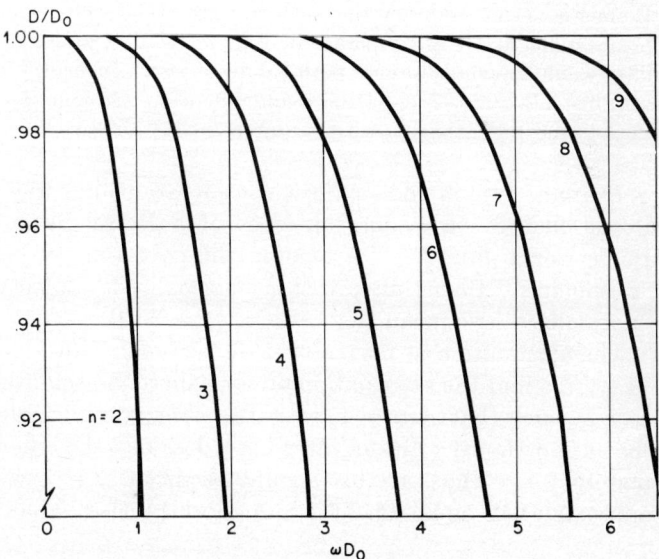

Fig. 14.11 Normalized delay of Bessel filters.

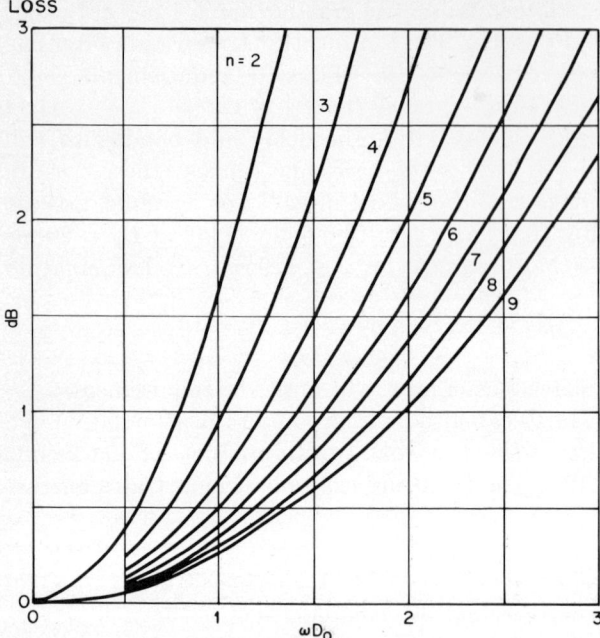

Fig. 14.12 Loss of Bessel filters.

Solution

From Fig. 14.11 we know that a fifth-degree Bessel filter would satisfy requirements a and b. However, Fig. 14.12 indicates that at $\omega D_0 = 2.4$ a fifth-degree filter would have much more than 2.0 dB of loss. Instead, Fig. 14.12 shows that we must have $n = 7$ to satisfy requirement c. Of course, this seventh-degree Bessel filter more than satisfies requirement b.

Example 14.3 introduced loss considerations into the requirements for Bessel filters. How does the loss of a Bessel filter compare with a Butterworth filter?* To answer this question, we shall compare a "normalized" Bessel filter with a "normalized" Butterworth filter. By "normalized" we mean that $A(\omega = 1) = 3$ dB.

The attenuation of normalized Butterworth filters was plotted in Fig. 14.4. To find the attenuation of normalized Bessel filters, one must first find D_0 such that $A(\omega = 1) = 3$ dB. For example, Fig. 14.12 indicates that a fourth-order Bessel filter has a loss of 3 dB when ωD_0 is approximately 2.1. Thus, if this should occur at $\omega = 1$ we must have D_0 approximately equal to 2.1 s. A more precise value can be obtained

---

* We shall compare Bessel and Butterworth filters, as they both have gradual transition regions.

TABLE 14.1 DC Delay for Normalized Bessel Lowpass Filters

n	2	3	4	5	6	7	8	9
D_0	1.359	1.753	2.111	2.424	2.699	2.947	3.174	3.386

by using the program in Fig. 14.10 iteratively. The results are shown in Table 14.1.

The values of n and D_0 in Table 14.1 can be used as inputs for the program in Fig. 14.10. This yields the attenuation performance shown in Fig. 14.13 and the delay performance in Fig. 14.14.

To compare the attenuation of a normalized Butterworth filter with that of a normalized Bessel filter, we can compare Figs. 14.4 and 14.13. These indicate that (for the same degree) the stopband attenuation of a Bessel filter is less than that of a Butterworth. But recall that when we studied lowpass filters we concluded that a Butterworth filter has poor* attenuation performance, so a Bessel filter must have horrible lowpass properties! This does not imply that Bessel filters are useless, but rather that they should not be used when one wants a steep transition region. Bessel filters have very gradual transition regions, but this gives good delay performance.

To be fair to the Bessel filter, we should compare its delay performance with the Butterworth delay performance. Comparing Figs. 14.14 and 14.5 indicates that a Bessel filter shows much flatter delay performance than does a Butterworth filter.

* A very high degree is needed to give a steep transition region.

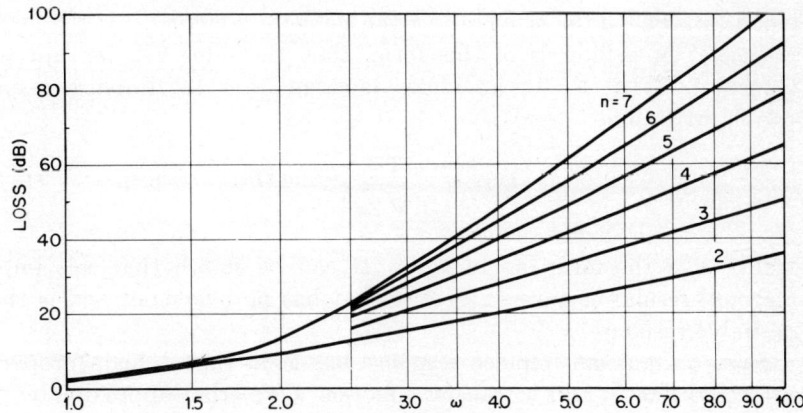

Fig. 14.13 Attenuation performance of normalized Bessel filters.

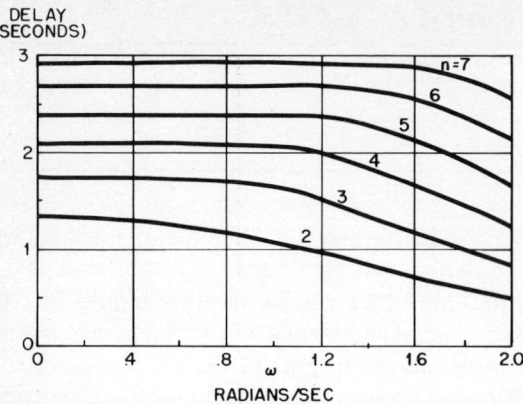

Fig. 14.14 Delay performance of normalized Bessel filters.

14.8 THE GAUSSIAN MAGNITUDE APPROXIMATION[5]

The Bessel approximation has very good delay performance because it has a gradual transition between the passband and stopband. Another approximation that has a gradual transition is the Gaussian approximation. This has good delay performance, although not as good as the Bessel approximation; however, the Gaussian approximation does have better step response than the Bessel. This section will investigate the response of Gaussian filters.

By definition, a Gaussian function has the form

$$g(x) = \frac{1}{\sigma \sqrt{2\pi}} e^{-\frac{1}{2}[(x-m)/\sigma]^2} \qquad (14.22a)$$

where m is called the *mean*, and σ the *standard deviation*. If the impulse response* of a filter is of this form, then the filter will be said to be Gaussian. That is, for an ideal Gaussian filter the impulse response can be written as

$$f_\delta(t) = \frac{1}{\sigma \sqrt{2\pi}} e^{-\frac{1}{2}[(t-T)/\sigma]^2} \qquad (14.22b)$$

which is of the form of (14.22a). It can be shown that this impulse response results in a step response that has no overshoot and is thus a desirable response.

Since a Gaussian impulse response has good time-domain properties, we would like to find a transfer function $T(j\omega)$ that approximates this.

---
* Impulse response is discussed in Chap. 15.

By definition, $T(j\omega)$ is the Fourier transformation of the impulse response; thus,

$$T(j\omega) = \int_{-\infty}^{\infty} f_\delta(t) e^{-j\omega t}\, dt$$

$$= \frac{1}{\sigma\sqrt{2\pi}} \int_{-\infty}^{\infty} e^{-\frac{1}{2}[(t-T)/\sigma]^2} e^{-j\omega t}\, dt$$

$$= e^{-\sigma^2\omega^2/2 - j\omega T}$$

The input/output transfer function $H(j\omega)$ is the reciprocal of $T(j\omega)$; thus, we can write

$$H(j\omega) = e^{(\omega/\omega_0)^2 + j\omega T} \qquad (14.23)$$

where $\omega_0^2 = 2/\sigma^2$. The frequency ω_0 is a normalizing frequency; it can be related to the 3-dB point by noting from (14.23) that the ideal Gaussian-magnitude shape can be written as

$$|H(j\omega)| = e^{(\omega/\omega_0)^2} \qquad (14.24)$$

Thus it follows that the 3-dB point $\omega_{3\text{dB}}$ is the solution to

$$|H(j\omega_{3\text{dB}})| = \sqrt{2} = e^{(\omega_{3\text{dB}}/\omega_0)^2}$$

so that

$$\omega_{3\text{dB}} = \left(\frac{\ln 2}{2}\right)^{1/2} \omega_0$$

Equation (14.23) indicates that $H(j\omega)$ has a linear phase; thus if the impulse response of a filter is Gaussian, then its delay will be constant. From another viewpoint we can instead conclude that if a transfer function $H(s)$ has approximately constant delay, then its impulse response will be approximately Gaussian.

The ideal Gaussian-magnitude shape given in (14.24) is unrealizable. This can be proved by applying the theorem of Paley and Wiener[6] that states that if $A(\omega)$ is realizable, then

$$\int_{-\infty}^{\infty} \frac{|\log A(\omega)|}{1 + \omega^2}\, d\omega \qquad (14.25)$$

must be finite. This integral is not finite for the Gaussian-magnitude shape.

Realizable approximations to the ideal Gaussian-magnitude shape can be obtained by using the following series expansion:

$$|H(j\omega)|^2 = e^{2(\omega/\omega_0)^2} = \sum_{i=0}^{\infty} \frac{2^i}{i!} \left(\frac{\omega}{\omega_0}\right)^{2i} \qquad (14.26)$$

An nth-order approximation to the ideal Gaussian-magnitude shape uses the first $2n$ powers in the series. For example, the third-order approximation is

$$|H(j\omega)|^2 = 1 + 2\left(\frac{\omega}{\omega_0}\right)^2 + \frac{2^2}{2!}\left(\frac{\omega}{\omega_0}\right)^4 + \frac{2^3}{3!}\left(\frac{\omega}{\omega_0}\right)^6 \quad (14.27)$$

The parameter ω_0 can be chosen so as to give a required passband width. For example, by definition a normalized Gaussian filter should have $|H(j)|^2 = 2$. Thus, for a third-order normalized Gaussian filter, (14.27) implies

$$2 = 1 + 2\left(\frac{1}{\omega_0}\right)^2 + \frac{2^2}{2!}\left(\frac{1}{\omega_0}\right)^4 + \frac{2^3}{3!}\left(\frac{1}{\omega_0}\right)^6$$

The solution to this is $\omega_0 = 1.692$. Similarly, one can show that $\omega_0 = 1.698$ for $n = 4$. These two examples indicate that as n increases, ω_0 is approaching an asymptotic value. This value can be determined by realizing that as n increases we are approaching the ideal Gaussian-magnitude shape described by (14.22). Thus the asymptotic value of ω_0 for the normalized Gaussian filter is the solution to

$$2 = e^{2(1/\omega_0)^2}$$

This solution is $\omega_0 = \sqrt{2/\ln 2}$.

The stopband losses of some normalized Gaussian filters are plotted in Fig. 14.15. These curves were found by using the first $2n$ powers in the series (14.26). The values for ω_0 were chosen such that $|H(j)|^2 = 2$. Comparing Fig. 14.15 with Fig. 14.13 indicates that the stopband loss of an nth-order Gaussian filter is less than the loss of an nth-order Bessel filter.

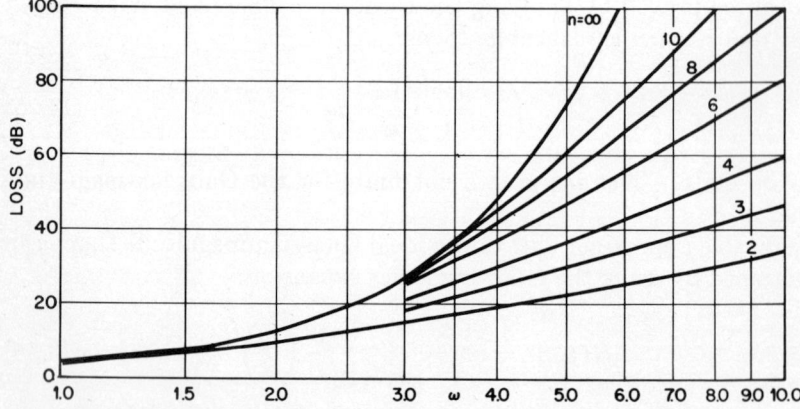

Fig. 14.15 Attenuation performance of normalized Gaussian filters.

In order to find the delay of an nth-order Gaussian filter, we need an equation for $H(s)$. This can be found from

$$H(s)H(-s) = \sum_{i=0}^{n} \frac{(-2)^i}{i!} \left(\frac{s}{\omega_0}\right)^{2i} \qquad (14.28)$$

$H(s)$ is determined by assigning to it the left half-plane roots of (14.28). M. Dishal[5] has done this in his article, and his Table I lists the roots for the first nine normalized Gaussian filters. Knowing the roots of $H(s)$, we can then apply Theorem 14.2 to find the delay. Zverev[7] plots delay versus frequency for Gaussian filters. As is to be expected, the delay performance is good because of the smooth Gaussian-magnitude shape; however, the delay performance is not as good as the Bessel delay plotted in Fig. 14.14.

Since the Gaussian-magnitude approximation has poorer amplitude and delay response than Bessel filters, why even mention Gaussian filters? The reason they are considered is that they have a very good step response, as will be noted in the next chapter.

14.9 TRANSITIONAL BUTTERWORTH–THOMSON FILTERS[8]

We have examined two different types of maximally flat filters: the Butterworth filter had a maximally flat amplitude response, whereas the Bessel (also called the Thomson) filter had a maximally flat delay response. In the next chapter we shall see that a high-degree Butterworth filter has excessive transient overshoot. On the other hand, the Thomson filter has small overshoot but relatively poor rise time.

A class of filters having characteristics lying between the Butterworth and Thomson approximations can be obtained by choosing the zeros of $H(s)$ correctly. One way of doing this was introduced by Y. Peless,[8] who called the resulting filters *transitional Butterworth-Thomson* filters (TBT). To understand how he chose the zeros of $H(s)$, consider Fig. 14.16. Z_B indicates a typical Butterworth root. It is located on the unit circle and at an angle ϕ_B. Z_T indicates a typical Thomson root which is located a distance r_T from the origin at an angle ϕ_T. Z indicates the TBT zero which is defined by

$$Z = re^{j(\pi - \phi)} \qquad (14.29)$$

where $\qquad r = r_T{}^m \qquad \phi = \phi_B - m(\phi_B - \phi_T) \qquad (14.30)$

The parameter m allows the designer to place the TBT zero between the Butterworth zero and the Thomson zero. In fact, if $m = 0$ the TBT zero coincides with the Butterworth zero; if $m = 1$, it coincides with the

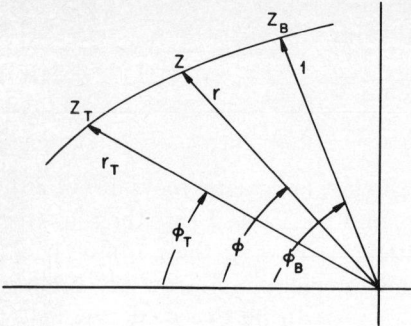

Fig. 14.16 Determination of natural modes for transitional Butterworth-Thomson filters.

Thomson zero. The article by Y. Peless and T. Murakami[8] contains many detailed graphs that can be used to help choose the proper TBT filter for a specific design.

It should be obvious that other "transitional" filters can be derived—for example, a transitional Butterworth-Tschebycheff filter.[35]

14.10 TSCHEBYCHEFF APPROXIMATION OF CONSTANT DELAY

The Bessel polynomials are used to provide a maximally flat approximation of constant delay. They approximate a constant delay very well at the origin, but at the expense of the rest of the frequency band. If one wants to approximate a constant delay over a wider frequency range, one can instead use a Tschebycheff type of approximation. That is, the zeros of $H(s)$ can be chosen in such a way that a constant delay is approximated in an equiripple sense.

A closed-form expression for a Tschebycheff approximation of constant delay does not exist; however, a computer can be used iteratively to find the roots of $H(s)$ so that an equiripple approximation results.[9-13] For that matter, the roots of $H(s)$ can be chosen so as to yield a Tschebycheff approximation of linear phase instead of constant delay. D. S. Humpherys[32] described a program that can be used to yield a Tschebycheff approximation of linear phase or of constant delay; but he noted that for small amounts of ripple, Tschebycheff delay implies Tschebycheff phase, so that the two approaches are equivalent for small ripples.

Some curves for equiripple approximations to linear phase are given in Zverev.[7] It is noted there that the attenuation characteristics become rather "lumpy" as the ripple factor ϵ is increased. Also, by comparing different delay curves, one can observe that the equiripple delay filters approximate a constant delay for a broader frequency band than do the flat delay approximations such as Bessel or Gaussian.

14.11 DELAY CONSIDERATIONS FOR BANDPASS FILTERS

The approximations in the previous sections are for lowpass filters; however, in designing bandpass filters one often wants the delay to be "as flat as possible" in the passband. With this in mind, we shall now investigate the effect on delay of a lowpass-to-bandpass transformation.

As in Chap. 6, lowercase letters will denote bandpass variables, and uppercase letters will denote normalized lowpass variables. For example, $d(\omega)$ will be bandpass delay, while $D(\Omega)$ will be lowpass delay. From Theorem 6.1, a normalized lowpass-to-bandpass transformation is

$$\Omega = \frac{\omega}{\omega_B - \omega_A} - \frac{\omega_A \omega_B}{\omega_B - \omega_A} \frac{1}{\omega} \tag{14.31}$$

This transforms the lowpass frequencies $\Omega = -1, 0, +1$ to the bandpass frequencies $\omega = \omega_A, \sqrt{\omega_A \omega_B}, \omega_B$. Thus it is geometrically symmetrical.

We are assuming that we have a lowpass delay function $D(\Omega)$ which is constant in the lowpass passband, and we want to know whether or not the transformed bandpass delay function $d(\omega)$ is constant in its passband. To answer this question we have to examine the input/output transfer functions

$$h(j\omega) = H(j\Omega)\Big|_{\Omega = \Omega(\omega)} \tag{14.32}$$

where $\Omega(\omega)$ is given by (14.31). Thus the phase can be written as

$$\arg h(j\omega) = \arg H[j\Omega(\omega)]$$

Since the delay is the derivative of the phase, it follows that

$$\begin{aligned} d(\omega) &= \frac{d}{d\omega} \arg h(j\omega) = \frac{d}{d\omega} \arg H[j\Omega(\omega)] \\ &= \frac{d}{d\Omega} \arg H(j\Omega) \frac{d\Omega}{d\omega} \\ &= D(\Omega) \frac{d\Omega}{d\omega} \end{aligned}$$

Thus, from (14.31), the bandpass delay $d(\omega)$ is related to the normalized lowpass delay by

$$d(\omega) = \left[\frac{1}{\omega_B - \omega_A} \left(1 + \frac{\omega_A \omega_B}{\omega^2} \right) \right] D(\Omega) \tag{14.33}$$

The term in brackets represents distortion; that is, even if $D(\Omega)$ is flat in its passband, $d(\omega)$ will not be flat in the passband because of the dis-

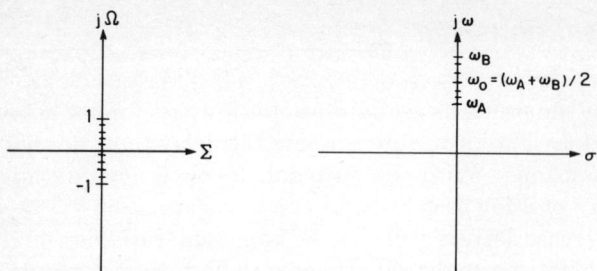

Fig. 14.17 Ideal lowpass–to–arithmetically symmetrical bandpass transformation.

tortion term. However, for a narrow passband, ω will not vary much ($\omega^2 \approx \omega_A \omega_B$); then the distortion will be quite small. That is, if a constant-delay lowpass filter is transformed to a narrow-bandpass filter, then the bandpass filter has approximately constant delay.

The lowpass-to-bandpass transformation introduces delay distortion because it distorts the frequency scale: it is a geometrically symmetrical transformation, not an arithmetically symmetrical one. Szentirmai[27] discusses a way to approximate arithmetically symmetrical bandpass filters which leads to bandpass filters that have approximately constant delay. The method consists of a periodic frequency transformation which is then truncated. Another approach, described by Geffe,[33] will be given next.

We would like to obtain an arithmetically symmetrical bandpass filter from a normalized lowpass as illustrated in Fig. 14.17. A transformation that changes the frequencies $\Omega = -1, 0, 1$ to $\omega = \omega_A, (\omega_A + \omega_B)/2, \omega_B$ is

$$\Omega = \frac{2}{\omega_B - \omega_A} (\omega - \omega_0)$$

where $\omega_0 = (\omega_A + \omega_B)/2$. This transformation implies that

$$S = \frac{2}{\omega_B - \omega_A} (s - j\omega_0) \qquad (14.34)$$

There is a problem with this transformation: it only yields the passband region $\omega_A \leq \omega \leq \omega_B$; it does not yield the passband region $-\omega_B \leq \omega \leq -\omega_A$. To transform the frequencies $\Omega = -1, 0, 1$ to $\omega = -\omega_B, -(\omega_A + \omega_B)/2, -\omega_A$ we instead need the transformation

$$S = \frac{2}{\omega_B - \omega_A} (s + j\omega_0) \qquad (14.35)$$

Geffe[33] combined the transformations in (14.34) and (14.35) to yield a lowpass-to-bandpass transformation that is approximately arithmetically symmetrical. His work is the basis of the following theorem.

Theorem 14.6

If $H(s)$ is the transfer function of a normalized lowpass filter, then a bandpass transfer function $h(s)$ that is approximately arithmetically symmetrical can be obtained by

$$h(s) = H(S_+)H(S_-) \tag{14.36}$$

where
$$S_+ = \frac{2}{\omega_B - \omega_A}(s - j\omega_0) \tag{14.37}$$

$$S_- = \frac{2}{\omega_B - \omega_A}(s + j\omega_0) \tag{14.38}$$

Figure 14.18 illustrates that the transformation in (14.37) transforms the lowpass roots to the upper s plane; the transformation in (14.38) transforms the lowpass roots to the lower s plane. Recall that we wanted to use just the transformation in (14.37), as this would yield an arithmetically symmetrical bandpass filter. However, we cannot use just that transformation, as it would yield an $h(s)$ that has imaginary coefficients. If $h(s)$ is formed as in Theorem 14.6, then all its coefficients are real, but the transformation results in a bandpass that is only approximately arithmetically symmetrical.

Blinchikoff[34] demonstrated that the approximation is very nearly arithmetically symmetrical. He did this by noting that the delay $d(\omega)$ of the bandpass function $h(s)$ can be written as

$$d(\omega) = \frac{2}{\omega_B - \omega_A}\left\{D\left[\frac{2(\omega - \omega_0)}{\omega_B - \omega_A}\right] + D\left[\frac{2(\omega + \omega_0)}{\omega_B - \omega_A}\right]\right\} \tag{14.39}$$

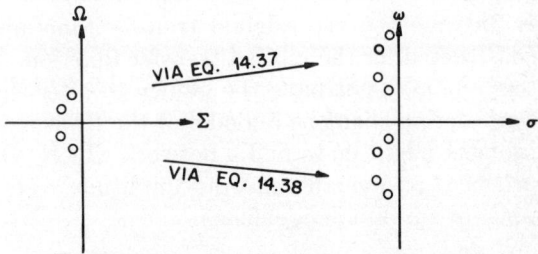

Fig. 14.18 Illustration of Geffe's transformation for arithmetically symmetrical bandpass filters.

$D[\Omega]$ is the lowpass delay; thus, the first term,

$$\frac{2}{\omega_B - \omega_A} D\left[\frac{2(\omega - \omega_0)}{\omega_B - \omega_A}\right]$$

represents the transformed lowpass delay. The second term is the error due to the approximation. It can usually be neglected in the passband so that (in the passband) the delay $d(\omega)$ is just a scaled version of the lowpass delay.

14.12 ADDITION OF ATTENUATION POLES

We have examined various approximations to a constant delay: Bessel, Gaussian, transitional Butterworth-Thomson, Tschebycheff. All these approximations have a common feature—the input/output transfer function $H(s)$ is a polynomial. Because there are no finite attenuation poles, the transition region is very gradual. With this as motivation, consider the following transfer function:

$$H(s) = \frac{N(s)}{\prod_{i=1}^{n} (s^2 + \omega_i^2)} \qquad (14.40)$$

The numerator polynomial $N(s)$ will be assumed to be specified as a Bessel polynomial or any other polynomial that gives satisfactory delay response. The attenuation poles ω_i do not affect the phase.*

Since the attenuation poles do not change the delay performance, their locations can be chosen solely by considering the amplitude response. For instance, the attenuation poles can be chosen such that the stopband has equal minimums of attenuation. Or, in general, the attenuation poles can be chosen so that the loss approximates an arbitrary shape.[14,15]

The addition of attenuation poles to a prescribed delay polynomial—such as a Bessel function—is one method of amplitude equalization.[20,21] In this method the original transfer function is altered, and a filter is constructed for the altered transfer function. In a communications system one may not have the ability to alter the "original" transfer function, as this might be a model of the transmission line. In this case the problem might be to find a network which, when cascaded with the line, produces an overall constant-amplitude response. This introduces the topic of amplitude equalizers.

* Except at the pole, where the phase changes by 180°; however, the pole will be in the stopband so that passband delay is unaltered.

14.13 AMPLITUDE EQUALIZERS

In the present chapter we have been concerned with delay, as this quantity gives an indication of the distortion a pulse will encounter. If a filter has excess delay distortion it will require delay equalization, which is the subject of the next section. However, we shall first consider amplitude equalizers, as they affect the delay response; in practice, one first equalizes the amplitude and then the delay.

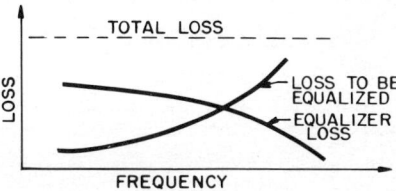

Fig. 14.19 Typical loss shape that requires equalization.

It is frequently desirable to preserve the wave shape of a signal in a transmission system; thus amplitude equalizers are often needed. These equalizers may provide equalization for filter characteristics, cable loss, amplifier performance, etc. Assume we have a loss that we want to equalize, as illustrated in Fig. 14.19. One way to design an equalizer is to choose the coefficients of the loss function

$$|H(j\omega)|^2 = \frac{a_0 + a_1\omega^2 + a_2\omega^4 + \cdots}{b_0 + b_1\omega^2 + b_2\omega^4 + \cdots} \qquad (14.41)$$

such that $|H(j\omega)|^2$ matches the loss characteristic at n specified frequencies. This yields a set of n equations for the n unknowns a_i, b_i. If certain conditions are fulfilled, a network can then be synthesized. This is essentially the method described by Zobel;[19] it designs the amplitude equalizer as one network. The resulting network can be quite involved and difficult to adjust; in practice, it is usually easier to design an amplitude equalizer as a cascade of simple networks.

An amplitude equalizer is frequently realized by cascading networks of the type shown in Fig. 14.20. If the impedances Z_1 and Z_2 are chosen such that

$$Z_1 = \frac{R^2}{Z_2} \qquad (14.42)$$

then the network in Fig. 14.20 is a constant-resistance bridged T. That is, if the output is terminated in a resistor R, then the input impedance is also R; this implies that the bridged T's can be cascaded without inter-

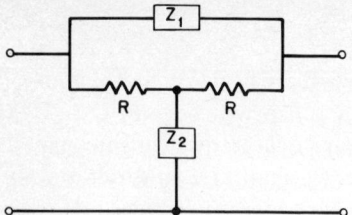

Fig. 14.20 Bridged-T network.

action. The total loss is thus simply the sum of the losses of the individual sections.

Assuming that $Z_1 = R^2/Z_2$, the transfer function of a terminated bridged T is*

$$\frac{V_{\text{in}}}{V_{\text{out}}} = H(s) = 1 + \frac{Z_1}{R} \qquad (14.43)$$

This formula can be used to find the loss of a bridged-T section. To simplify the design procedure R is usually normalized to unity, and Z_1 is assumed to consist of a conductance G in parallel with a susceptance B. Thus (14.43) becomes

$$H(j\omega) = 1 + Z_1 = 1 + \frac{1}{G + jB} \qquad (14.44)$$

so that
$$A(\omega) = 10 \log \frac{(1+G)^2 + B^2}{G^2 + B^2} \qquad (14.45)$$

Some of the more common forms for Z_1 and the resulting losses are shown in Fig. 14.21.

In practice, loss equalizers are designed by breaking the desired loss characteristic into parts and assigning one of the simple sections of Fig. 14.21 to each part. As an example, consider the equalization of a lowpass filter.† Figure 14.22 shows the ideal characteristics (curve 1) and the dissipative computed loss (curve 2). The needed equalizer is clearly the type shown in the third row of Fig. 14.21 that can, for instance, have the loss shape indicated by curve 3 in Fig. 14.22. Curve 4 shows the total loss, which is the sum of curves 2 and 3.

The example in Fig. 14.22 is fairly simple; practical problems will usually require a cascade of equalizer sections. Iterative computer programs[17,26] can help design these equalizers, but the engineer must choose the initial circuit configurations; he must also choose initial parameter

---

* See Problem 14.14.
† This example and Figs. 14.21 and 14.22 were supplied by G. Szentirmai.

values. Since the computer cannot change the circuit configuration (only the element values), the engineer has a great influence on the outcome of the optimization. Of course, the initial guess as to the actual element values will also influence how rapidly the iterations converge. There exist many charts that can help one choose the initial design.

Unlike previous topics in this book, the material on amplitude equalizers was discussed with an actual circuit realization assumed; that is,

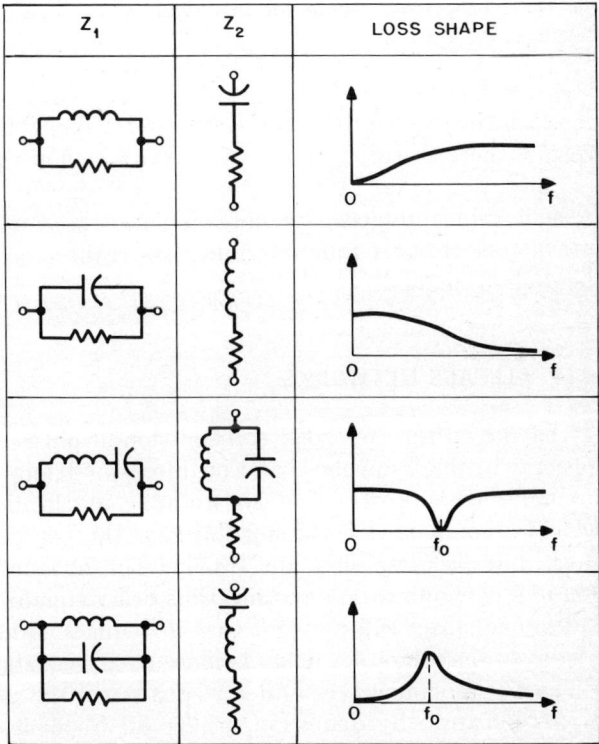

Fig. 14.21 Common networks used in constant-resistance bridged-T networks.

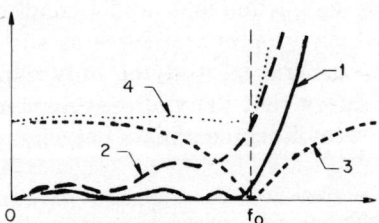

Fig. 14.22 Example of loss equalization.

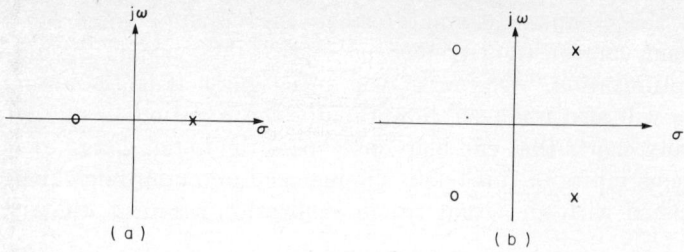

Fig. 14.23 Pole-zero locations for first- and second-degree allpass functions.

we assumed a constant-resistance bridged-T realization.* This was done because there already exist many curves which are very useful in designing such networks. If one wanted to design an amplitude equalizer without using bridged T's, one could use the same approach to find a satisfactory transfer function, and then realize the transfer function by the desired method.

14.14 ALLPASS NETWORKS

We have seen that for good transmission of pulses, the delay should be constant in the frequency band of interest. If one designs a filter from an amplitude approximation (for example, an elliptic filter), there will be a large amount of delay distortion near the passband edge. Not only filters, but also amplifiers and transmission lines introduce delay distortion in a communication system; thus delay equalizers are often needed. A delay equalizer is a network that introduces additional delay into the system so that the total delay is more nearly constant.

The most common type of delay equalizer is the allpass network. The allpass network, by definition, passes all frequencies without affecting their amplitudes—it just affects the delay. If the allpass network is to be able to produce an arbitrary delay without changing amplitude response, it must be a nonminimum phase network.† With this as motivation, we shall define first- and second-order allpass networks to be networks that have poles and zeros as shown in Fig. 14.23. We shall be completely general in studying only allpass sections of this type, as it has been shown that "any allpass network is equivalent to a number of first- and second-degree allpass networks in tandem."[16]

---

* Of course there are other equivalent realizations, e.g., a lattice network.

† If it were minimum phase, the amplitude and phase response would be interrelated by the Hilbert transformation.

The transfer function for a first-degree allpass function can be written as

$$H_1(s) = \frac{\omega_0 + s}{\omega_0 - s} \qquad (14.46)$$

Its phase is given by

$$\arg H_1(j\omega) = 2 \tan^{-1} \frac{\omega}{\omega_0} \qquad (14.47)$$

from which it follows that the delay is given by

$$D_1(\omega) = \frac{2/\omega_0}{1 + (\omega/\omega_0)^2} \qquad (14.48)$$

The phase and delay for a first-degree allpass function are sketched in Fig. 14.24.

The transfer function for a second-order allpass function is usually written as

$$H_2(s) = \frac{s^2 + (2\omega_0/b)s + \omega_0^2}{s^2 - (2\omega_0/b)s + \omega_0^2} \qquad (14.49)$$

Its phase is given by

$$\arg H_2(j\omega) = 2 \operatorname{ctn}^{-1}\left[\frac{b}{2}\left(\frac{\omega_0}{\omega} - \frac{\omega}{\omega_0}\right)\right] \qquad (14.50)$$

from which it follows that the delay is given by

$$D_2(\omega) = \frac{b}{\omega_0} \frac{1 + (\omega_0/\omega)^2}{1 + (b^2/4)(\omega/\omega_0 - \omega_0/\omega)^2} \qquad (14.51)$$

The above formulas indicate that there are two parameters that characterize a second-order allpass function: ω_0 and b. The parameter ω_0 is called the *critical frequency* and is the frequency at which there is 180°

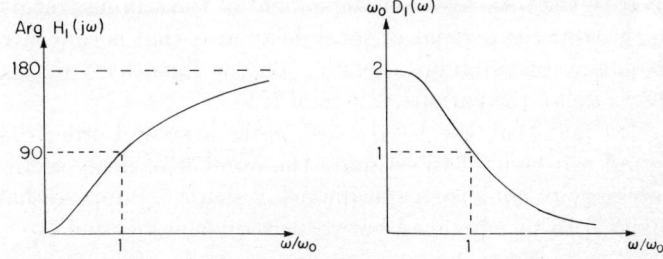

Fig. 14.24 Phase and delay of a first-order allpass function.

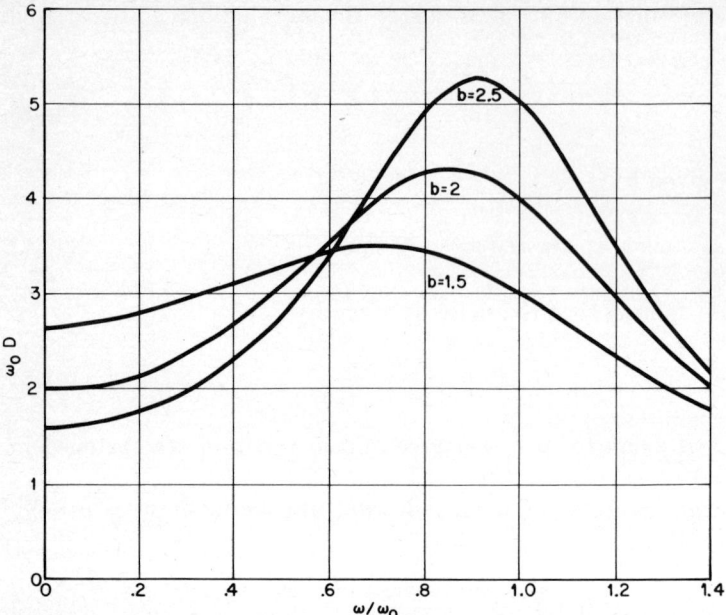

Fig. 14.25 Delay of various second-order allpass functions.

of phase shift. ω_0 is very close to the frequency of maximum delay ω_m. The parameter b is called the *stiffness ratio:* it determines the amount of concentration of delay around ω_m.

It follows from (14.50) that the phase of a second-order allpass function is zero at $\omega = 0$ and increases monotonically as frequency increases. As ω approaches infinity, the phase approaches 2π rad. Since delay is the derivative of phase, this implies that the total area under the delay curve is*

$$\int_0^\infty D(\omega)\, d\omega = 2\pi$$

That is, the total area is independent of the stiffness ratio b: the higher b, the greater the percent of total delay area that is concentrated about the frequency of maximum delay. This is illustrated in Fig. 14.25, which shows delay for various values of b.

The fact that the total area under a second-order delay function is 2π rad can be used to estimate the number of delay sections that will be necessary to equalize a specific delay shape. Suppose that a given delay shape is to be equalized between frequencies ω_1 and ω_2. If the amount

---

* Similarly, the total area under a first-order allpass function is π rad.

of "equalization area" that is required is Ar, then the equalizer will need at least $Ar/2\pi$ sections. This minimum number of sections is possible only if the sections produce no delay outside the range ω_1 and ω_2. In actual practice more sections will be needed, as the networks do produce delay outside the passband. Also, a greater number of sections provides flexibility in matching the required delay shape.

To obtain an intuitive appreciation of how one can produce a desired delay response by cascading allpass sections, consider Fig. 14.26. This figure assumes that we want to produce a negative delay slope for a certain frequency region. As illustrated, this can be done by properly choosing the ω_0 and b of each of the delay sections.

In practice, if one wants to delay equalize a given shape, one usually just makes an initial guess as to the ω_0 and b of each allpass section. An optimization program* can then be used to move the parameters to better locations.[17,18]

In concluding this introduction to allpass functions, it might be of interest to present typical passive and active realizations of a second-order allpass function. The passive network[25] in Fig. 14.27 has the property (for proper element values) that the input and output impedances are not a function of frequency; thus such sections can be cascaded

---

* If one does not have access to a computer, then the book by Skwirzynski[36] should be consulted for its discussion of delay equalization.

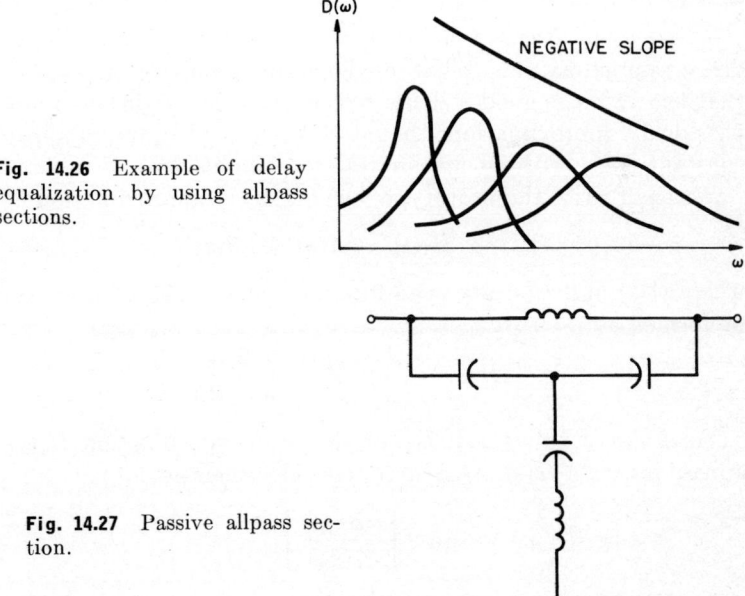

Fig. 14.26 Example of delay equalization by using allpass sections.

Fig. 14.27 Passive allpass section.

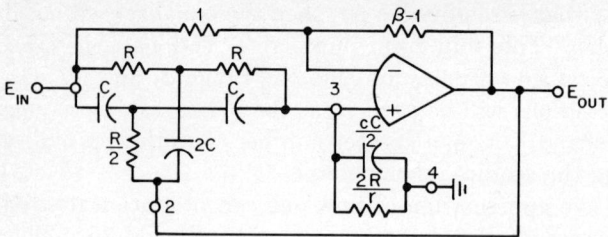

Fig. 14.28 Active allpass section.

without interaction. The active allpass in Fig. 14.28 can be used for second-order functions that have complex roots.[23] These sections can be cascaded without interactions because the output is a voltage source. Similar circuits exist for first-order allpass functions and second-order allpass functions which have real roots.[24]

14.15 ALLPASS NETWORKS DERIVED FROM LOWPASS DELAY APPROXIMATIONS

In previous sections we encountered many lowpass delay approximations—for example, the Bessel approximation. These polynomial lowpass approximations give good delay performance but poor attenuation performance, as they have very gradual transition regions. If we want a sharper transition region, attenuation poles can be added as described in Sec. 14.12.

This section discusses another way to modify the polynomial lowpass delay approximations. The modification produces an allpass network that has twice as good a delay response as the original polynomial lowpass delay approximation; thus it is useful if we want to approximate a constant delay without attenuating the amplitude.

We shall write the prototype lowpass delay approximation as

$$H_1(s) = A(s) + sB(s)$$

where $A(s)$ and $B(s)$ are even functions of s. The allpass function will be formed as

$$H(s) = \frac{A(s/2) + (s/2)B(s/2)}{A(s/2) - (s/2)B(s/2)}$$

To discover how the delay of the prototype function $H_1(s)$ and the allpass function $H(s)$ are related, consider their angles

$$\arg H_1(j\omega) = \tan^{-1}\left[\frac{\omega B(j\omega)}{A(j\omega)}\right]$$

$$\arg H(j\omega) = 2\tan^{-1}\left[\frac{\omega}{2}\frac{B(j\omega/2)}{A(j\omega/2)}\right] = 2\arg H_1\left(\frac{j\omega}{2}\right)$$

Differentiating with respect to ω yields the delay:

$$D(\omega) = \frac{d}{d\omega} \arg H(j\omega) = 2\frac{d}{d\omega} \arg H_1\left(\frac{j\omega}{2}\right) = D_1\left(\frac{\omega}{2}\right)$$

This result is summarized in the next theorem.

Theorem 14.7

If a prototype polynomial transfer function $H_1(s)$ is written in terms of even polynomials $A(s)$ and $B(s)$ as

$$H_1(s) = A(s) + sB(s)$$

and a related allpass function is given by

$$H(s) = \frac{A(s/2) + (s/2)B(s/2)}{A(s/2) - (s/2)B(s/2)}$$

then the delays are related by

$$D(\omega) = D_1\left(\frac{\omega}{2}\right)$$

This theorem indicates that the delay bandwidth of the allpass function is twice that of the prototype function. This is demonstrated in the following example.

Example 14.4

From Theorem 14.5 a third-order polynomial approximation of e^s can be written by using third-order Bessel polynomials. This yields

$$H_1(s) = \frac{s^3 + 6s^2 + 15s + 15}{15}$$

The related allpass is

$$H(s) = \frac{(s/2)^3 + 6(s/2)^2 + 15(s/2) + 15}{-(s/2)^3 + 6(s/2)^2 - 15(s/2) + 15}$$

$$= \frac{s^3 + 12s^2 + 60s + 120}{-s^3 + 12s^2 - 60s + 120}$$

Both these functions have unity delay at zero frequency. At $\omega = 1.2$ the delay of the polynomial approximation has fallen by 1 percent; but not until $\omega = 2.4$ has the allpass delay fallen by 1 percent. That is,

$$D(2.4) = 0.99 = D_1(1.2)$$

In the above example we approximated e^s by

$$e^s = \frac{s^3 + 12s^2 + 60s + 120}{-s^3 + 12s^2 - 60s + 120}$$

This is a special case of a Padé[29] approximation. A Padé approximation of a function $f(s)$ is obtained by expanding $f(s)$ in a power series:

$$f(s) = \sum_{i=0}^{\infty} a_i s^i$$

Definition 14.3

$F_{m,n} = P_m(s)/Q_n(s)$ is a Padé approximation of $f(s) = \sum_{i=0}^{\infty} a_i s^i$ if

1. $P_m(s)$ is a polynomial of degree m.
2. $Q_n(s)$ is a polynomial of degree n.
3. $\dfrac{P_m(s)}{Q_n(s)} = \sum_{i=0}^{m+n} a_i s^i + \sum_{i=m+n+1}^{\infty} b_i s^i$

It can be shown that the Padé approximation $F_{m,n}$ is unique. For $f(s) = e^s$, G. Nielsen[30] gives the unique solution as

$$P_m(s) = 1 + \frac{n}{n+m}\frac{s}{1!} + \frac{n(n-1)}{(n+m)(n+m-1)}\frac{s^2}{2!} + \cdots$$

$$+ \frac{n(n-1)\cdots 2\cdot 1}{(n+m)\cdots(m+1)}\frac{s^n}{n!}$$

$$Q_n(s) = 1 - \frac{m}{n+m}\frac{s}{1!} + \frac{m(m-1)}{(n+m)(n+m-1)}\frac{s^2}{2!} + \cdots$$

$$+ (-1)^m \frac{m(m-1)\cdots 2\cdot 1}{(n+m)\cdots(n+1)}\frac{s^m}{m!}$$

This set of equations can be used to obtain a Padé table for the function e^s (see Table 14.2).

TABLE 14.2 A Pade Table for Approximations to e^s

m \ n	1	2	3
1	$\dfrac{s+2}{-s+2}$	$\dfrac{2s+6}{s^2-4s+6}$	$\dfrac{6s+24}{-s^3+6s^2-18s+24}$
2	$\dfrac{s^2+4s+6}{-2s+2}$	$\dfrac{s^2+6s+12}{s^2-6s+12}$	$\dfrac{3s^2+24s+60}{-s^3+9s^2-36s+60}$
3	$\dfrac{s^3+6s^2+18s+24}{-6s+24}$	$\dfrac{s^3+9s^2+36s+60}{3s^2-24s+60}$	$\dfrac{s^3+12s^2+60s+120}{-s^3+12s^2-60s+120}$

The Padé approximation to e^s results in delays that decrease monotonically with frequency. Hepner[31] has described rational approximations that produce a ripple in the delay, which results in a larger bandwidth.

14.16 CONCLUSIONS

Delay is a frequency-domain quantity that can be used to obtain insight into a system's transient response—the flatter the delay, the better the step response. Often one wants to design a lowpass filter that has a constant delay. This can be done by using polynomial delay approximations, such as the Bessel, which have very gradual transition regions. If one wants a more abrupt transition region, attenuation poles can be added. Another way to obtain a filter with good delay response is to design the filter according to amplitude approximations (e.g., an elliptic filter), and then to delay equalize it.

Instead of designing a lowpass filter that has constant delay, one often wants to design a delay line. Unlike a lowpass filter, a delay line should not attenuate high frequencies—it should present negligible amplitude distortion at all frequencies of interest. Thus, delay lines do not usually use polynomial approximations; instead, they use allpass approximations. Section 14.15 demonstrates that allpass approximations can be derived from polynomial delay approximations.

Bandpass filters with good delay performance can be derived from lowpass delay networks by using proper frequency transformations. Another method of obtaining the desired amplitude and delay response is to select a general transfer function such as

$$H(s) = \frac{a_6 s^6 + a_5 s^5 + \cdots + a_0}{b_6 s^6 + b_5 s^5 + \cdots + b_0}$$

and then let a computer choose the coefficients such that the amplitude and delay performance are approximated in an optimal manner.[28,29] That is, the optimization techniques described in Chap. 13 can be used to optimize amplitude and delay performance simultaneously.

REFERENCES

1. E. A. Guillemin, *Theory of Linear Physical Systems*, John Wiley & Sons, Inc., New York, 1963, chap. 18.
2. L. Storch, "Synthesis of Constant-Time-Delay Ladder Networks Using Bessel Polynomials," *Proc. IRE*, November 1954, pp. 1666–1675.
3. W. E. Thomson, "Networks with Maximally Flat Delay," *Wireless Eng.*, October 1952, pp. 255–263.
4. W. E. Thomson, "Delay Networks Having Maximally Flat Frequency Characteristics," *Proc. IEE*, November 1949, pp. 487–490.

5. M. Dishal, "Gaussian-Response Filter Design," *Electr. Commun.*, vol. 36, no. 1, pp. 3–29.
6. G. E. Valley and H. Wallman, *Vacuum Tube Amplifiers*, M.I.T. Radiation Laboratory Series, vol. 18, McGraw-Hill Book Company, New York, 1948, app. A.
7. A. I. Zverev, *Handbook of Filter Synthesis*, John Wiley & Sons, Inc., New York, 1967.
8. Y. Peless and T. Murakami, "Analysis and Synthesis of Transitional Butterworth-Thomson Filters and Bandpass Amplifiers," *RCA Rev.*, March 1957, pp. 60–94.
9. A. B. MacNee, "Chebychev Approximation of a Constant Group Delay," *IEEE Trans. on Circuit Theory*, June 1963, pp. 284–285.
10. E. Ulbrich and H. Piloty, "Uber don Entwirf von Allpässen, Tiefpässen und Bandpässen mit einer im Tschebyscheffschen Sinne approximierten konstanten Gruppenlaufzeit," *Arch. Elektr. Ubertragung*, October 1960, pp. 451–467.
11. J. J. Neirnck, "Transient Behavior of Systems with Equal-ripple Delay," *IEEE Trans. on Circuit Theory*, June 1964, pp. 302, 303.
12. W. M. Bucker, "Symmetrical Equal-ripple Delay and Symmetrical Equal-ripple Phase Filters," *IEEE Trans. on Circuit Theory*, August 1970, pp. 455–458.
13. T. A. Abele, "Ubertragungsfaktoren mit Tschebyscheffscher Approximierten Konstanter Gruppenlaufzeit," *Arch. Elektr. Ubertragung*, January 1962, pp. 9–17.
14. F. F. Kuo and M. Karnaugh, "Approximation of Linear Phase Filters with Gaussian Damped Impulse Response," *IEEE Trans. on Circuit Theory*, June 1964, pp. 255–260.
15. A. F. Beletskiy, "Synthesis of Filters with Linear Phase Characteristics," *Telecommunication*, April 1961, pp. 39–48.
16. H. W. Bode, *Network Analysis and Feedback Amplifier Design*, D. Van Nostrand Company, Inc., New York, 1945, pp. 239–242.
17. F. F. Kuo and J. F. Kaiser, *System Analysis by Digital Computer*, John Wiley & Sons, Inc., New York, 1966, chap. 6.
18. S. Hellerstein, "Synthesis of All-pass Delay Equalizers," *IRE Trans. on Circuit Theory*, September 1961, pp. 215–222.
19. O. J. Zobel, "Distortion Correction in Electrical Circuits with Constant-Resistance Networks," *Bell Syst. Tech. J.*, July 1928, pp. 438–534.
20. A. Budak, "A Maximally Flat Phase and Controllable Magnitude Approximation," *IEEE Trans. on Circuit Theory*, June 1965, p. 279.
21. T. Deliyannis, "Six New Delay Functions and their Realization Using Active RC Networks," *Radio & Electron. Eng.*, March 1970, pp. 139–144.
22. E. A. Guillemin, *Theory of Linear Physical Systems*, John Wiley & Sons, Inc., New York, 1963, pp. 486–490.
23. G. S. Moschytz, "A General All-pass Network Based on the Sallen-Key Circuit Topology," private correspondence.
24. S. C. Dutta Roy, "RC Active All-Pass Networks Using a Differential-Input Operational Amplifier," *Proc. IEEE*, November 1969, pp. 2055–2056.
25. V. R. Cunningham, "Pick a Delay Equalizer," *Electron. Des.*, May 1966, pp. 62–66.
26. D. A. Calahan, *Modern Network Synthesis*, Hayden Book Company, Inc., New York, 1964, chap. 5.
27. G. Szentirmai, "The Design of Arithmetically Symmetrical Band-Pass Filters," *IEEE Trans. on Circuit Theory*, September 1963, pp. 367–375.
28. J. G. Linvill, "The Approximation with Rational Functions of Prescribed Magnitude and Phase Characteristics," *Proc. IRE*, June 1952, pp. 711–721.
29. J. E. Storer, *Passive Network Synthesis*, McGraw-Hill Book Company, New York, 1957.

30. G. Nielsen, "Time Delay Approximations," *Servolaboratoriet*, Denmarks tekniske Højskole, Lyngby, 1968.
31. C. F. Hepner, "Improved Methods of Simulating Time Delays," *IEEE Trans. on Electron. Comput.*, April 1965, pp. 239–243.
32. D. S. Humpherys, "Equiripple Network Approximations Using Iteration Techniques," Natl. Electron. Conf., 1964, pp. 753–758.
33. P. R. Geffe, "On the Approximation Problem for Band-Pass Delay Lines," *Proc. IRE*, September 1962, pp. 1986–1987.
34. H. Blinchikoff, "A Note on Wide-Band Group Delay," *IEEE Trans. on Circuit Theory*, September 1971, pp. 577–578.
35. A. Budak and P. Aronhime, "Transitional Butterworth-Chebyshev Filters," *IEEE Trans. on Circuit Theory*, May 1971, pp. 413–415.
36. J. K. Skwirzynski, *Design Theory and Data for Electrical Filters*, D. Van Nostrand Company, Inc., Princeton, N.J., 1965.

PROBLEMS

14.1 This problem establishes Theorem 14.3. We assume that the denominator of $H(s)$ does not influence the delay; thus we are only interested in the numerator $e(s) = \sum_{i=0}^{n} h_i s^i$.

a. Show that $e(j\omega)$ can be written as

$$e(j\omega) = \sum_{k=0}^{k_1} (-1)^k h_{2k} \omega^{2k} + j \sum_{k=0}^{k_2} (-1)^k h_{2k+1} \omega^{2k+1}$$

where k_1 = integer part of $n/2$
k_2 = integer part of $(n-1)/2$

b. If the parameter u is defined as

$$u = \frac{\sum_{k=0}^{k_2} (-1)^k h_{2k+1} \omega^{2k+1}}{\sum_{k=0}^{k_1} (-1)^k h_{2k} \omega^{2k}}$$

then it follows that $\arg e(j\omega) = \tan^{-1} u$. Find $du/d\omega$.

c. Prove Theorem 14.3 by using

$$D(\omega) = \frac{d}{d\omega} \arg e(j\omega) = \frac{1}{1+u^2} \frac{du}{d\omega}$$

14.2 Many of the delay approximations in this chapter were normalized to unity delay; that is, we usually approximated e^s instead of e^{sT}. For example, the third-order Bessel approximation of e^s is $H(s) = (15 + 15s + 6s^2 + s^3)/15$.

a. What is the third-order Bessel approximation of e^{sT}?
b. For the approximation in a, what is $D(0)$?

14.3* This problem demonstrates that if a signal is composed of two frequencies (ω_1 and ω_2) and the delay is constant between these frequencies, then there still might be delay distortion.

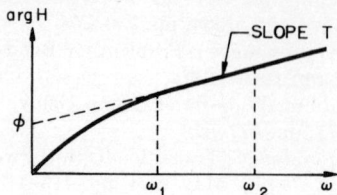

a. Assume $H(j\omega)$ has a linear phase between ω_1 and ω_2 as shown above: That is,
$$\arg H = \phi + \omega T \quad \text{for} \quad \omega_1 \leq \omega \leq \omega_2$$
If the input is
$$V_{in} = \cos \omega_1 t + \cos \omega_2 t$$
what is the output (assuming no attenuation of the amplitude)?

b. Show that the answer to a can be written as
$$V_{out} = \cos \phi \, [\cos \omega_1(t - T) + \cos \omega_2(t - T)]$$
$$+ \sin \phi \left[\cos \omega_1 \left(t - T - \frac{1}{4f_1} \right) + \cos \omega_2 \left(t - T - \frac{1}{4f_2} \right) \right]$$

Note that in the first expression both waveforms are delayed by the same amount T. However, in the second expression one is delayed by $T + \frac{1}{4}f_1$, and the other is delayed by $T + \frac{1}{4}f_2$. Since there are different amounts of delay, the output is not a replica of the input—there is delay distortion unless $\sin \phi = 0$.

14.4 a. If $H(s) = \sum_{i=0}^{n} h_i s^i / q(s)$, where the zeros of $q(s)$ have quadrantal symmetry, what is the delay at $\omega = 0$?

b. The nth-order Bessel lowpass filter is given by $H(s) = B_n(s)/b_0$, where $B_n(s)$ is the nth-order Bessel polynomial, and $b_0 = B_n(0)$. Show that this nth-order filter has unity delay at $\omega = 0$.

14.5 a. Find the fifth-order Bessel polynomial by using the recursion formula in (14.20).

b. Check the result in part a by using the relation in (14.21).

c. Approximate e^s by using the fifth-order Bessel polynomial.

d. What is the delay at $\omega = 3$?

14.6 The Bessel lowpass filter was obtained as follows:

a. e^s was written as $e^s = \cosh s + \sinh s$.

b. $\cosh s$ and $\sinh s$ were expanded in power series.

c. A continued-fraction expansion was formed and then truncated.

* A. E. Devletoglou, private correspondence.

A simpler way to approximate e^s would be to expand e^s in a power series and truncate the power series. Show that a fifth-order approximation that is done in this manner is not Hurwitz.

14.7 Assume that we want to design a lowpass filter that has a 3-dB point at $\omega = 1$. Also, the filter should have greater than 42 dB of suppression at $\omega = 4$.

  a. What is the minimum-degree Bessel filter that can be used?

  b. At approximately what frequency has the dc delay changed by 5 percent?

  c. Repeat parts a and b for a Butterworth filter.

14.8 Find the transfer function $H(s)$ for the lowest-degree Bessel filter that has a zero-frequency delay of 10 ms and a delay error of less than 2 percent for $f \leq 100/\pi$ Hz.

14.9 What is $H(s)$ for the normalized second-order Gaussian filter?

14.10 If the zeros of $q(s)$ have quadrantal symmetry, then the program in Fig. 14.3 finds the delay of the transfer function

$$H(s) = C_H \frac{\sum_{i=1}^{K}(s + \sigma_i) \prod_{i=1}^{L_1}[s^2 + (2\pi ZF_i/ZQ_i)s + (2\pi ZF_i)^2]}{q(s)}$$

A more general denominator polynomial $q(s)$ can be written as

$$q(s) = \prod_{i=1}^{N}[s^2 - 2\pi R_i s + (2\pi)^2(R_i^2 + F_i^2)]$$

The program in Fig. 14.3 can also be used for this more general transfer function. To do this we can define the parameter L of Step 1.1 to be $L_1 + N$, where L_1 and N are as identified above. Then the first L_1 values of ZF[I] and ZQ[I] describe the numerator of $H(s)$. Show that to describe the denominator, the next N values of ZF[I] and ZQ[I] should be

$$ZF[I] = SQRT(R_i^2 + F_i^2)$$
$$ZQ[I] = ZF[I]/R_i$$

14.11 From Example 14.2, a fourth-degree Bessel lowpass approximation of unity delay is

$$H(S) = \frac{105 + 105S + 45S^2 + 10S^3 + S^4}{105}$$

  a. What is the loss at $\Omega = 1$?

  b. Define the passband edge to be at $\Omega = 1$. Use the transformation in (14.31) to find a bandpass function $h(s)$ that has transformed passband edges $\omega_A = 3$, $\omega_B = 4$.

  c. What is the delay of $h(s)$ at ω_A, $\sqrt{\omega_A \omega_B}$, and ω_B?

14.12 As in Problem 14.11, let the prototype filter be defined as

$$H(S) = \frac{105 + 105S + 45S^2 + 105S^3 + S^4}{105}$$

a. Define the passband edge to be $\Omega = 1$. Use the transformation in (14.36) to find a bandpass function $h(s)$ that has transformed passband edges $\omega_A = 3$, $\omega_B = 4$.

b. What is the delay of $h(s)$ at ω_A, $(\omega_A + \omega_B)/2$, and ω_B?
Hint: $D(13) = 0.060$, $D(14) = 0.052$, and $D(15) = 0.045$.

14.13 Theorem 14.6 describes a lowpass-to-bandpass transformation that results in negligible delay distortion. This problem illustrates how the transformation can be modified so that there is also negligible amplitude distortion

a. Show that when $S_+ = j$, $s = j\omega_B$; and when $S_+ = -j$, $s = j\omega_A$.

b. Show that

$$h(j\omega_A) = H(S_+ = -j) H\left(S_- = j\frac{3\omega_A + \omega_B}{\omega_B - \omega_A}\right)$$

$$h(j\omega_B) = H(S_+ = j) H\left(S_- = j\frac{3\omega_B + \omega_A}{\omega_B - \omega_A}\right)$$

c. Let $H(S)$ represent an nth-degree polynomial filter that has passband edges at $S = \pm j$. Assume that the roots of $H(S)$ have magnitudes less than unity. Use the result in part *b* to show that if $h(s)$ has a relatively narrow passband [that is, $(\omega_A + \omega_B)/(\omega_B - \omega_A) \gg 1$], then

$$\frac{|h(j\omega_A)|}{|h(j\omega_B)|} \approx \left(\frac{3\omega_A + \omega_B}{3\omega_B + \omega_A}\right)^n$$

d. Part *c* implies that $|h(j\omega_A)|$ is not equal to $|h(j\omega_B)|$; thus the transformation has resulted in amplitude distortion. Show that if $\omega_B = \omega_A(1 + \epsilon)$, where $\epsilon \ll 1$, then

$$\left|\frac{h(j\omega_A)}{h(j\omega_B)}\right| \approx \left(1 - \frac{\epsilon}{2}\right)^n$$

e. If $h(s)$ is formed as in Theorem 14.6, the delay performance is good but there is some amplitude distortion as indicated in part *d*. A. Anuff* has suggested adding $n/2$ poles at the origin to $h(s)$, as this will not affect the delay and will improve the amplitude performance. What is $|h(j\omega_A)|/|h(j\omega_B)|$ for this case?

14.14 If the bridged T in Fig. 14.20 has $Z_1 = R^2/Z_2$ and is terminated in a resistor of value R,

a. Show that the input impedance is equal to R.

b. Show that

$$\frac{V_{\text{in}}}{V_{\text{out}}} = 1 + \frac{Z_1}{R}$$

14.15 Figure 14.20 shows a bridged-T network. For this to be a constant-resistance network, we must have $Z_1 = R^2/Z_2$. It was mentioned that Z_1 is usually constructed as a parallel combination of a resistor and a reactance net-

---
* Private correspondence.

work; that is, $1/Z_1 = 1/R_1 + 1/(jx_1)$. If Z_1 is of this form, what is the form of Z_2?

14.16 Table 14.2 shows a Padé table for e^s. One of the entries is

$$F_{m,n} = \frac{2s + 6}{s^2 - 4s + 6}$$

Show that the first four terms in the Taylor-series expansion of $F_{m,n}$ are identical with the first four terms in the power-series expansion of e^s.

CHAPTER FIFTEEN

Time-Domain Response

15.1 INTRODUCTION

Most of the chapters in this book investigate the amplitude responses of various filter approximations. The amplitude of a transfer function indicates the steady-state response to sinusoids of various frequencies; however, there are many applications for which one is interested in the transient response of a filter. For example, in a digital communication system one would want to know the response to a pulse. In the previous chapter we discussed the delays of different filter approximations, since the delay gives a good indication of the transient response of a filter. This is because a pulse can be considered to be comprised of many different frequencies—if the filter delays each by the same amount of time, then the response will be a perfect replica of the input; if not, there will be distortion.

If one is interested in the transient response of a specific filter, why consider the intermediate quantity delay; why not attempt to solve the approximation problem in the time domain? Some do prefer to work in the time domain; for example, Guillemin[1] devotes a chapter to time-domain approximation. However, the majority of filter designers prefer to investigate the approximation problem in the frequency domain.

Since most filter data are given in the frequency domain, this chapter is limited to a discussion of the transient responses of filters that are specified in the frequency domain.

15.2 DEFINITION OF TERMS

There are two idealized input signals that are usually used in a mathematical investigation of transient response: the unit impulse function and the unit step function. The *unit impulse function* $\delta(t)$ is zero for all time except $t = 0$, and then it is infinite. Furthermore, the area under the impulse function is unity; i.e.,

$$\int_{-\infty}^{\infty} \delta(t)\, dt = 1 \tag{15.1}$$

The *unit step function* $u(t)$ is defined such that

$$u(t) = \begin{cases} 0 & t < 0 \\ 1 & t \geq 0 \end{cases} \tag{15.2}$$

Quite logically, the responses to these inputs are termed the *impulse response* and *step response*. If one knows the impulse response $f_\delta(t)$, one can determine the step response by integration:*

$$f_s(t) = \int_{-\infty}^{t} f_\delta(\tau)\, d\tau \tag{15.3}$$

In fact, the superposition integral can be used to find the response $y(t)$ to any input $x(t)$. Expressed in terms of the impulse response, it is

$$y(t) = \int_{-\infty}^{t} x(\tau) f_\delta(t - \tau)\, d\tau \tag{15.4}$$

In terms of the step response we instead have

$$y(t) = \int_{-\infty}^{t} x'(\tau) f_s(t - \tau)\, d\tau \tag{15.5}$$

where the prime denotes differentiation.

Since the step response (or impulse response) can be used to determine the response to any input, we shall be content in this chapter to compare only the step responses of various filters. The step response of a typical filter is shown in Fig. 15.1.† As indicated in this figure, the *delay time* t_d is defined to be the time necessary to reach 50 percent of the final value. The *rise time* t_r is the time required to go from 10 percent to 90 percent of

---

* This formula assumes a linear system.

† The step response in Fig. 15.1 has been normalized to a steady-state value of unity.

the final value. Also of interest is the *percent overshoot*, which is the difference between the peak value and final value (expressed as a percent).

15.3 TRANSIENT RESPONSE AND THE LAPLACE TRANSFORMATION

The filter transfer functions we found in previous chapters were expressed in terms of the Laplace transformation. For example, for a second-order Butterworth filter we had

$$\frac{V_{\text{out}}(s)}{V_{\text{in}}(s)} = T(s) = \frac{1}{s^2 + \sqrt{2}\,s + 1} \qquad (15.6)$$

$V_{\text{out}}(s)$ is the Laplace transformation of the output waveform, and $V_{\text{in}}(s)$ is the Laplace transformation of the input waveform. In general, the Laplace transformation is defined as

$$\mathcal{L}[v(t)] = V(s) = \int_0^\infty v(t) e^{-st}\, dt \qquad (15.7)$$

The inverse transformation can be found by using

$$\mathcal{L}^{-1}[V(s)] = v(t) = \frac{1}{2\pi j} \int_{\sigma_0 - j\infty}^{\sigma_0 + j\infty} V(s) e^{st}\, ds \qquad (15.8)$$

To find the transient response of a filter, one must solve

$$v_{\text{out}}(t) = \mathcal{L}^{-1}[V_{\text{in}}(s) T(s)] \qquad (15.9)$$

In particular, to find the step response, we can use the fact that

$$\mathcal{L}[u(t)] = \frac{1}{s}$$

Thus
$$f_s(t) = \mathcal{L}^{-1}\left[\frac{T(s)}{s}\right] \qquad (15.10)$$

The next several sections discuss various ways of determining the inverse Laplace transformation.

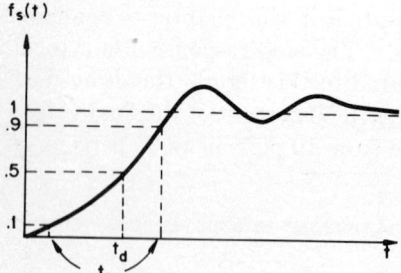

Fig. 15.1 Step response that illustrates overshoot.

15.4 PARTIAL FRACTIONS AND THE INVERSE LAPLACE TRANSFORMATION[2]

If one wants an explicit form for the inverse Laplace transformation, one usually does not use the integral relation in (15.8). Instead, one usually expands the function in terms of partial fractions and then writes down (by inspection) the inverse transformation of each term. The inverse transformations that are usually needed are

$$\mathcal{L}^{-1}\left[\frac{1}{s+a}\right] = e^{-\sigma t} \qquad t \geq 0 \qquad (15.11)$$

and

$$\mathcal{L}^{-1}\left[\frac{C}{s+\sigma_0-j\omega_0} + \frac{C^*}{s+\sigma_0+j\omega_0}\right] = 2|C|e^{-\sigma_0 t}\cos(\omega_0 t + \sphericalangle C) \quad (15.12)$$

where $\sphericalangle C$ is the angle of C.

Example 15.1
Find the step response of the second-order Butterworth filter.

Solution

$$V_{\text{out}}(s) = V_{\text{in}}(s)T(s) = \frac{1}{s}\frac{1}{s^2 + \sqrt{2}s + 1}$$

The partial-fraction expansion of this is

$$V_{\text{out}}(s) = \frac{1}{s} - \frac{1}{2}\frac{1+j}{s+1/\sqrt{2}+j/\sqrt{2}} - \frac{1}{2}\frac{1-j}{s+1/\sqrt{2}-j/\sqrt{2}}$$

Applying the inverse transformations of (15.11) and (15.12) yields

$$v_{\text{out}}(t) = 1 + \sqrt{2}\, e^{-t/\sqrt{2}}\cos\left(\frac{t}{\sqrt{2}} + \frac{3\pi}{4}\right) \qquad \text{for} \qquad t \geq 0$$

The partial-fraction method has been used by many people to investigate transient response; a particularly detailed investigation was done by Henderson and Kautz.[3] However, there is a major disadvantage to the partial-fraction method: One must determine the roots of the denominator before making a partial-fraction expansion. In the previous example the roots were easy to find, as the denominator was essentially a quadratic polynomial. In a more complicated filter one would need to use a root-finding technique before doing a partial-fraction expansion. The next section describes some inverse Laplace transformation methods that do not require root finding.

15.5 TAYLOR SERIES AND THE INVERSE LAPLACE TRANSFORMATION

Assume that we want to find the inverse Laplace transformation of

$$X(s) = \frac{a_{m-1}s^{m-1} + a_{m-2}s^{m-2} + \cdots + a_1 s + a_0}{s^m + b_{m-1}s^{m-1} + \cdots + b_1 s + b_0} \quad (15.13)$$

Corrington[4] introduced a method that avoided the necessity of finding the denominator's roots. He expanded the ratio of polynomials in a Taylor series and used this to obtain a linear difference equation. The linear difference equation was then solved recursively.

M. L. Liou[5] described a simpler method of finding the inverse Laplace transformation, based on the state-space approach: a recursive formula was derived from the exact solution of the state-space equation. This led to a numerical solution that could be obtained to any desired accuracy. One of the calculations that consumed a substantial amount of computer time was the calculation of the transition matrix. W. E. Thomson[6] published some relations that made it necessary to evaluate only some of the terms of the transition matrix. Liou[7] then further simplified the calculations used in evaluating the transition matrix.

An alternative method for the solution of the inverse Laplace transformation was presented by R. I. Ross.[8] A computer program based on his method is described in this section. His method was chosen because it is very simple to program.

Equation (15.13) is equivalent to a differential equation with a certain set of initial conditions. The differential equation is

$$\frac{d^m x}{dt^m} = -b_0 x - b_1 \frac{dx}{dt} - \cdots - b_{m-1} \frac{d^{m-1} x}{dt^{m-1}} \quad (15.14)$$

which can be rewritten using simpler notation as

$$x^{(m)} = -\sum_{j=0}^{m-1} b_j x^{(j)}$$

where the superscript denotes the order of the differentiation. The initial conditions are given by

$$x(0+) = a_{m-1}$$
$$x^{(i)}(0+) = a_{m-1-i} - \sum_{j=0}^{i-1} b_{m-i-j} x^{(j)}(0+) \quad i = 1, \ldots, m-1 \quad (15.15)$$

If (15.14) is further differentiated, we can write

$$x^{(i)}(t) = -\sum_{j=i-m}^{i-1} b_{m-i+j} x^{(j)}(t) \quad i = m, \ldots, v \quad (15.16)$$

where $v - m$ is the number of additional differentiations. This equation is valid for $t \geq 0$. In particular, it can be evaluated at $t = 0+$ to give the initial conditions for these higher-order derivatives.

We shall now show that the derivatives in (15.16) can be used in a Taylor series to find an approximate value for the transient response. The accuracy of the approximation is improved by increasing the parameter v. A Taylor-series approximation of $x(t + T)$ expanded at time t is

$$x(t + T) \approx \sum_{j=0}^{v} x^{(j)}(t) \frac{T_j}{j!}$$

Differentiating with respect to time (and keeping terms only to the vth derivative) yields

$$x^{(i)}(t + T) \approx \sum_{j=0}^{v-i} x^{(i+j)}(t) \frac{T_j}{j!} \qquad i = 0, \ldots, m - 1 \qquad (15.17)$$

Equations (15.16) and (15.17) can be used recursively to solve for the transient response. To be specific, (15.16) can be used to find $x^{(i)}(0+)$ for $i = m, \ldots, v$. Then (15.17) can be used to solve for $x^{(i)}(T)$ for $i = 0, \ldots, m - 1$. Next (15.16) can be used to find $x^{(i)}(T)$ for $i = m$, $\ldots, v$. Then (15.17) can be used to find $x^{(i)}(2T)$ for $i = 0, \ldots$, $m - 1$; etc.

A program based on this recursive procedure is given in Fig. 15.2; a sample use of the program is shown in Fig. 15.3. The only input parameter not identified in the previous equations is NI, which is the number of increments for which the transient response should be calculated.

The filter analyzed in Fig. 15.3 is the same one that was encountered in Example 15.1. As a check on the accuracy of the program, we can

```
1; TRANSIENT RESPØNSE EVALUATIØN
1.1 DEMAND M,V,T,NI
1.11 DEMAND A[I],B[I] FØR I=0:1:M-1
1.12 TYPE #,#,#
1.14 T[0]=1,X=0,N=0
1.2 T[J]=T[J-1]*T/J FØR J=1:1:V
1.3; CALCULATIØN ØF EQ. 15.15
1.31 X[I]=A[M-1-I] FØR I=0:1:M-1
1.32 X[I]=X[I]-B[M-I+J]*X[J] FØR J=0:1:I-1 FØR I=1:1:M-1
1.4; CALCULATIØN ØF EQ. 15.16
1.41 X[I]=0 FØR I=M:1:V
1.42 X[I]=X[I]-B[M-I+J]*X[J] FØR J=I-M:1:I-1 FØR I=M:1:V
1.5; CALCULATIØN ØF EQ. 15.17
1.51 X[I]=X[I]+X[I+J]*T[J] FØR J=1:1:V-I FØR I=0:1:M-1
1.6 N=N+1
1.7 TYPE N*T,X[0] IN FØRM 1
1.8 TØ PART 1.4 IF N<NI
```

Fig. 15.2 Program for calculating transient response.

288   Approximation Methods for Electronic Filter Design

```
DO PART 1
        M=3
        V=6
        T=.1
        NI=80
     A[0]=1
     B[0]=0
     A[1]=0
     B[1]=1
     A[2]=0
     B[2]=SQRT(2)
```

.100	.0048	4.000	1.0380
.200	.0182	4.100	1.0402
.300	.0390	4.200	1.0418
.400	.0660	4.300	1.0427
.500	.0981	4.400	1.0432
.600	.1344	4.500	1.0431
.700	.1740	4.600	1.0427
.800	.2161	4.700	1.0419
.900	.2599	4.800	1.0409
1.000	.3048	4.900	1.0396
1.100	.3503	5.000	1.0381
1.200	.3959	5.100	1.0364
1.300	.4410	5.200	1.0347
1.400	.4854	5.300	1.0328
1.500	.5288	5.400	1.0309
1.600	.5708	5.500	1.0289
1.700	.6113	5.600	1.0269
1.800	.6501	5.700	1.0250
1.900	.6870	5.800	1.0231
2.000	.7219	5.900	1.0212
2.100	.7549	6.000	1.0193
2.200	.7858	6.100	1.0175
2.300	.8146	6.200	1.0158
2.400	.8413	6.300	1.0142
2.500	.8660	6.400	1.0127
2.600	.8887	6.500	1.0112
2.700	.9094	6.600	1.0098
2.800	.9282	6.700	1.0085
2.900	.9453	6.800	1.0073
3.000	.9605	6.900	1.0062
3.100	.9742	7.000	1.0052
3.200	.9863	7.100	1.0043
3.300	.9969	7.200	1.0034
3.400	1.0061	7.300	1.0027
3.500	1.0141	7.400	1.0020
3.600	1.0209	7.500	1.0014
3.700	1.0266	7.600	1.0008
3.800	1.0313	7.700	1.0003
3.900	1.0351	7.800	.9999
		7.900	.9995
		8.000	.9992

Fig. 15.3 Sample use of the transient-response program.

compare the results at $t = 1$. For $t = 1$, the formula in Example 15.1 yields $v_{out}(1) = 0.3048316$. This compares favorably with the result in Fig. 15.3.

Usually one will not have an exact formula for the transient response (if one did, a recursive relationship would not be needed). In this case the accuracy of the program can be checked by increasing the parameter

v and determining how much the results change. If the results vary significantly, v should be increased further.

Another way to check the accuracy of the program is to examine the asymptotic value approached by the transient response. This can be compared with that predicted by the *final-value theorem*

$$v(t \to \infty) = \lim_{s \to 0} [sV(s)] \tag{15.18}$$

15.6 COMPARISON OF TRANSIENT RESPONSES

The program in the previous section can be used to calculate the impulse or step response of any filter. However, we shall not use it to generate data for a catalog of filters, as references[3,9] already exist that examine the transient responses of many conventional filters. This section merely summarizes some of the properties of these filters.

Many lowpass filters have a step response similar to that shown in Fig. 15.1. That is, the steady-state value is approached in an oscillatory manner. Butterworth, Tschebycheff, elliptic, and Legendre filters all have transient responses of this type. For these filters, as the degree is increased, the amount of ringing increases; that is, there is more overshoot, and the oscillations take longer to die out.

Gaussian and Bessel filters have a step response of the type shown in Fig. 15.4. The Gaussian filters have no overshoot, and the Bessel filters have negligible overshoot (less than 1 percent). If the filters are normalized so that the 3-dB points are located at $\omega = 1$, then Zverev's figures[9] indicate that the delay time t_d increases as the degree of the filter is increased; however, the rise time stays approximately constant. Thus, compared with the filters that have an oscillatory type of response, the Gaussian and Bessel filters have better time-domain response; however, their magnitude response is poorer.

The Butterworth-Thomson[10] filters offer a compromise between the responses shown in Figs. 15.1 and 15.4. The amount of overshoot can be controlled by properly choosing the poles.

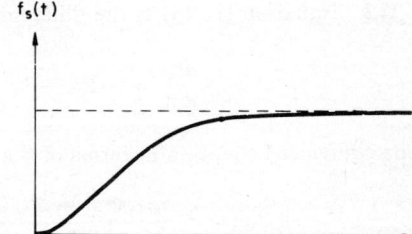

Fig. 15.4 Illustration of a non-oscillatory step response.

REFERENCES

1. E. A. Guillemin, *Synthesis of Passive Networks,* John Wiley & Sons, Inc., 1957, chap. 15.
2. M. F. Gardner and J. L. Barnes, *Transients in Linear Systems,* vol. 1, John Wiley & Sons, Inc., New York, 1942.
3. K. W. Henderson and W. H. Kautz, "Transient Responses of Conventional Filters," *IRE Trans. on Circuit Theory,* December 1958, pp. 333–347.
4. M. S. Corrington, "Simplified Calculation of Transient Response," *Proc. IEEE,* March 1965, pp. 287–292.
5. M. L. Liou, "A Novel Method of Evaluating Transient Response," *Proc. IEEE,* January 1966, pp. 20–23.
6. W. E. Thomson, "Evaluation of Transient Response," *Proc. IEEE,* November 1966, p. 1584.
7. M. L. Liou, "Evaluation of the Transition Matrix," *Proc. IEEE,* February 1967, pp. 228–229.
8. R. I. Ross, "Evaluating the Transient Response of a Network Function," *Proc. IEEE,* May 1967, pp. 693–694.
9. A. I. Zverev, *Handbook of Filter Synthesis,* John Wiley & Sons, Inc., New York, 1967.
10. Y. Peless and T. Murakami, "Analysis and Synthesis of Transitional Butterworth-Thomson Filters and Band-pass Amplifiers," *RCA Rev.,* March 1957, pp. 60–94.

PROBLEMS

15.1 *a.* The final-value theorem states that $v(t \to \infty) = \lim_{s \to 0} [sV(s)]$. It was stated in Sec. 15.5 that this could be used to check the accuracy of the transient-response program. Use the final-value theorem to find the steady-state value of the function described in Fig. 15.3.

b. What is the initial-value theorem? Use this to check the initial response in Example 15.1.

15.2 Equation (15.13) gives a general form for the step response in terms of the Laplace variable s as

$$X(s) = \frac{a_{m-1}s^{m-1} + a_{m-2}s^{m-2} + \cdots + a_1 s + a_0}{s^m + b_{m-1}s^{m-1} + \cdots + b_1 s + b_0}$$

Why was the numerator chosen to be of lower degree than the denominator?

15.3 Equation (15.14) is the differential equation

$$\frac{d^m x}{dt^m} = -b_0 x - b_1 \frac{dx}{dt} - b_2 \frac{dx}{dt^2} - \cdots$$

The equivalent equation in terms of the Laplace transformation is

$$X(s) = \frac{a_{m-1}s^{m-1} + a_{m-2}s^{m-2} + \cdots + a_1 s + a_0}{s^m + b_{m-1}s^{m-1} + \cdots + b_1 s + b_0}$$

Show that the initial conditions are related to the parameters a_i and b_i as in (15.15).

15.4 Use the method due to R. I. Ross to find the inverse Laplace transformation of

$$X(s) = \frac{1}{s(s^2 + \sqrt{2}s + 1)}$$

at time $t = 0.1$. Check with the result in Fig. 15.3.

Hint: Because time $t = 0.1$ is very near the initial time ($t = 0$), the parameter v may be quite small; in fact, $v = 4$ will yield accurate results. Also, the increment step may be chosen as $T = 0.1$.

15.5 A very simple circuit is shown below:

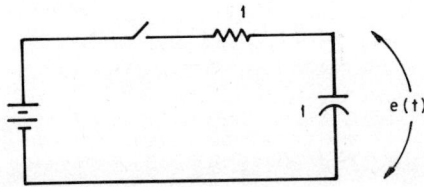

a. Use the partial-fraction method to find $e(t)$.

b. Use the method described in Sec. 15.5 to find $e(t)$ for $t = 0.1, 0.2$, and 0.3. In the solution choose $T = 0.1$ and $v = 3$. Check the results with the answer to part a.

15.6 The program in Fig. 15.2 can be used to find the transient response of a filter. Describe the input for the program if one wants to find the step response of a third-order Bessel filter. Assume that the Bessel filter is normalized so that its 3-dB point is at $\omega = 1$. For the program assume that we want the response for the first 2 s in 0.05-s increments.

CHAPTER SIXTEEN

Approximation Methods and Passive Network Synthesis

16.1 INTRODUCTION

The first part of this book concentrated on the approximation problem. An understanding of that material will allow a filter designer to obtain a transfer function that is suitable for his particular requirements. There are many different ways in which the transfer function can be realized. This chapter discusses one of the "classical" ways: insertion loss synthesis. The next chapter will discuss some active network realizations, and the last chapter will deal with digital filters. The treatment in each of these three chapters will not be comprehensive; there are many books devoted to each of these synthesis techniques. The purpose of discussing the various synthesis techniques is to help the reader become acquainted with different methods of realizing the transfer functions we have derived. For those who have no synthesis background, this should help show that the transfer functions are not just mathematical abstractions; networks can be constructed to perform these filtering operations. For those who are familiar with one of the synthesis techniques, it can help put that knowledge in broader perspective. Also, an introduction to these filtering techniques will enable us to show how our notation is related to the jargon of those who synthesize filters. This should greatly facilitate the reading of synthesis texts.[1-4]

This chapter discusses the realization of transfer functions via insertion loss theory. We shall first define various terms and discuss some properties that these functions have. We shall then demonstrate how the theory can be used to synthesize passive networks. The synthesis will be performed in terms of the transformed variable Z because this will help reduce inaccuracies due to numerical calculations.

16.2 INSERTION LOSS

In this chapter we assume that we have a voltage source that has a source resistance R_1. We want the source to be coupled to a load resistance R_2 so as to generate a certain transfer function $H(s)$. This situation is illustrated in Fig. 16.1.

When we introduced the input/output transfer function $H(s)$ in Chap. 1, $H(s)$ was a ratio of voltages. In this chapter we introduce a constant that modifies the voltage ratio. The effect of a constant multiplier is simply to add a constant loss to $A(\omega)$. That is, the constant does not change the shape of the loss curve; it merely changes the overall level. To see why passive-filter designers modify a voltage ratio with a multiplicative constant, we must first define insertion loss.

Definition 16.1

The insertion loss IL of the coupling network in Fig. 16.1 is defined as

$$\text{IL} = 20 \log \left(\frac{R_2}{R_1 + R_2} \left| \frac{V_0}{V_2} \right| \right) \qquad (16.1)$$

In order to obtain a physical interpretation of insertion loss, first let the coupling network in Fig. 16.1 be replaced by a short circuit. Then the power dissipated in the load is

$$P_0 = \frac{|V_0|^2 R_2}{(R_1 + R_2)^2} \qquad (16.2)$$

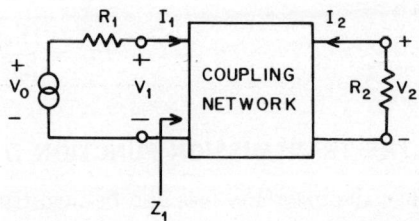

Fig. 16.1 Standard configuration used in insertion loss theory to synthesize a transfer function.

294   Approximation Methods for Electronic Filter Design

With the coupling network in place, the power dissipated in the load is $P_2 = |V_2|^2/R_2$. Equation (16.1) indicates that the insertion loss is given by

$$\text{IL} = 10 \log \frac{P_0}{P_2} \qquad (16.3)$$

That is, the insertion loss is the loss due to the insertion of the coupling network.

Example 16.1

Find the insertion loss due to the coupling networks in the following figures. Assume that the source and load resistors are $R_1 = 300$, $R_2 = 1{,}200$.

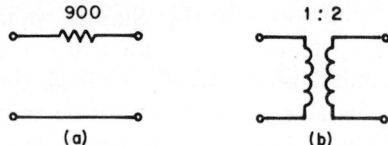

(a)   (b)

Solution

For circuit (a), $V_0/V_2 = 2$. Applying (16.1) thus yields

$$\text{IL}_a = 20 \log \left(\frac{1{,}200}{1{,}500} \times 2 \right) = 4.08 \text{ dB}$$

For circuit (b), $V_0 = V_2$; thus

$$\text{IL}_b = 20 \log \frac{1{,}200}{1{,}500} = -1.94 \text{ dB}$$

The above example demonstrates that a passive network can have a negative insertion loss, that is, an insertion gain! The reason for this is that P_0 as given by (16.2) is not the maximum power that can be delivered by the source. The source delivers the maximum power P_m when it is matched to the load. This maximum power is

$$P_m = \frac{|V_0|^2}{4R_1} \qquad (16.4)$$

16.3 THE TRANSMISSION FUNCTION $H(s)$

Because the insertion loss can be negative, it is more convenient to work with a quantity termed the *transmission function*, which will be denoted as $H(s)$. We shall define $H(s)$ in terms of P_m (the maximum power that can be delivered by the source) and P_2 (the power dissipated in the load).

Approximation Methods and Passive Network Synthesis

In order that H be a function of the complex frequency s, P_m and P_2 must also be functions of s. This is possible if we extend our previous definitions as below:

$$P_m(s) = \frac{V_0(s)V_0(-s)}{4R_1} \qquad P_2(s) = \frac{V_2(s)V_2(-s)}{R_2} \qquad (16.5)$$

The transmission function $H(s)$ is defined as*

$$H(s) = \left(\frac{R_2}{4R_1}\right)^{1/2} \frac{V_0(s)}{V_2(s)} = \frac{e(s)}{q(s)} \qquad (16.6)$$

It thus follows that

$$H(s)H(-s) = \frac{P_m(s)}{P_2(s)} \qquad (16.7)$$

Example 16.2

In this example we find the transmission function $H(s)$ for a third-degree elliptic lowpass filter that has a passband ripple of 0.25 dB, a passband edge of 10 Hz, and a stopband edge of 20 Hz.

A third-degree filter will have an attenuation pole located at F_1 and an attenuation pole located at infinity. The frequency F_1 can be determined by using elliptic functions as described in Chap. 5. However, it is easier to use the pole-placer program in Fig. 8.15. With the input data $S = 1$, FS[1] = 20, A[1] = 10, $A_{\max} = 0.25$, FB = 10, NIN = 1, and $N = 1$, the program was used iteratively to find $F_1 = 22.7007$. This yields a minimum stopband loss of 28.05 dB.

The zeros of $H(s)$ can be found by using the zero-finder program of Fig. 10.3. The results of this program and the pole placer let us write $H(s)$ as

$$H(s) = \frac{e(s)}{q(s)}$$
$$q(s) = s^2 + (2\pi \times 22.7007)^2$$
$$e(s) = 0.0719(s + \sigma)\left(s^2 + \frac{\omega Z_1}{ZQ_1}s + \omega Z_1^2\right)$$

where $\sigma = 54.0546$
$\omega Z_1 = 2\pi \times 11.515$
$ZQ_1 = 1.8021$

We define the attenuation $A(\omega)$ exactly as we did earlier:

$$A(\omega) = 20 \log |H(j\omega)| \qquad (16.8)$$

---

* Because $H(s)$ is an input/output transfer function, it might be more appropriate to call this a loss function; it is, however, conventional to refer to it as the transmission function.

It thus follows that

$$A(\omega) = 20 \log\left[\left(\frac{R_2}{4R_1}\right)^{1/2} \left|\frac{V_0}{V_2}\right|\right] \qquad (16.9)$$

Comparing (16.9) with (16.1), we see that the attenuation $A(\omega)$ is related to the insertion loss IL via

$$A(\omega) = \text{IL} + 10 \log \frac{(R_1 + R_2)^2}{4R_1 R_2} \qquad (16.10)$$

Recall that the insertion loss could be a negative number. However, the attenuation $A(\omega)$ cannot be negative, because it can be written in terms of the maximum available power P_m as

$$A(\omega) = 10 \log \frac{P_m}{P_2} \qquad (16.11)$$

Since the coupling network is assumed to be passive, the power P_2 that is delivered to the load must be less than P_m, which implies that $A(\omega)$ must be nonnegative.

We can interpret (16.10) as implying that adding the proper constant to the insertion loss IL guarantees that the attenuation $A(\omega)$ is always positive. Also, if the source resistance is equal to the load resistance, then $A(\omega)$ is equal to the insertion loss.

Further examination of (16.10) yields another very important practical constraint. For a lossless lowpass coupling network, the low-frequency insertion loss approaches 0 dB because inductors behave as short circuits, and capacitors as open circuits. Thus, for a lowpass filter,

$$A(0) = 10 \log \frac{(R_1 + R_2)^2}{4R_1 R_2}$$

which implies that only if the source resistance is equal to the load impedance can a lowpass filter have zero loss at dc. Therefore an odd-order Tschebycheff or elliptic filter must have the source resistance equal to the load resistance. On the other hand, an even-order Tschebycheff or elliptic filter cannot have the source resistance equal to the load. If either of these restrictions is objectionable, the frequency transformation of Sec. 6.8 can be used to alter the dc loss.

In conclusion, in this section we have defined the transmission function $H(s)$ as an input/output transfer function. It has the constant multiplier $(R_2/4R_1)^{1/2}$ so that the attenuation function $A(\omega)$ is nonnegative. This implies that $|H(j\omega)|$ has a minimum of unity; thus, it will again be useful to introduce the characteristic function $K(s)$:

$$H(s)H(-s) = 1 + K(s)K(-s) \qquad (16.12)$$

Approximation Methods and Passive Network Synthesis

If we want to find a function $H(s)$ that will yield a certain loss shape $A(\omega)$, we can use the various approximation methods described in the first part of this book. We can then find a passive network that realizes $H(s)$ by using the insertion loss method described in the rest of this chapter.

Example 16.3

Example 16.2 found the transmission function $H(s)$ for a specific third-order elliptic lowpass filter. This example continues that work and finds the characteristic function $K(s)$. Actually, the denominator of $K(s)$ is the same as the denominator of $H(s)$; thus we need only determine the numerator $f(s)$.

Part 50 of the zero-finder program in Fig. 10.3 finds $F(Z)$, which is the transformed version of $f(s)$. A step can be added to the program so that the coefficients A_i are printed. The result is

$$F(Z) = A_0 + A_2 Z^2$$
$$= C(0.19618 + 0.68048 Z^2)$$

The constant C must be included because the program only determines $F(Z)$ to within a constant multiplier. The function $f(s)$ can be found by using the fact that $F(Z) = f(s)/(s^2 + \omega_A^2)^{m/2}$. Since this is a lowpass filter, $\omega_A = 0$, and we thus have

$$f(s) = s^m F(Z)$$
$$= s^3 C(0.19618 + 0.68048 Z^2)$$

where $Z^2 = \dfrac{s^2 + \omega_B^2}{s^2} = \dfrac{s^2 + (20\pi)^2}{s^2}$

This yields

$$f(s) = C(0.87666s)(s^2 + 3064.4)$$

The constant C can be evaluated by using the fact that since there is a pole at infinity, the constant multiplier of $H(s)$ and $K(s)$ must be the same. From Example 16.2 the constant multiplier of $H(s)$ is 0.0719; thus

$$f(s) = 0.0719s(s^2 + 3064.4)$$

Before proceeding with the insertion loss method we shall give an expression for $H(s)$ which will be useful in the sections that follow. This expression will be given in terms of the A, B, C, D parameters for the network in Fig. 16.1. By definition, these parameters satisfy the relations

$$V_1 = AV_2 - BI_2$$
$$I_1 = CV_2 - DI_2 \quad (16.13)$$

Since

$$H(s) = \left(\frac{R_2}{4R_1}\right)^{1/2} \frac{V_0}{V_2}$$

it follows that the transmission function can be expressed as

$$H(s) = \frac{1}{2}\left[\left(\frac{R_2}{R_1}\right)^{1/2} A + \left(\frac{1}{R_1 R_2}\right)^{1/2} B + (R_1 R_2)^{1/2} C + \left(\frac{R_1}{R_2}\right)^{1/2} D\right] \quad (16.14)$$

16.4 THE REFLECTION FUNCTION $T_1(s)$

The source can deliver a maximum power $P_m = |V_0|^2/4R_1$. However, in general, not all of this power is delivered because the source resistance R_1 is not matched to the input impedance Z_1 (Z_1 is identified in Fig. 16.1). The power that is delivered to the coupling network will be denoted as*

$$P_1(s) \triangleq \frac{V_1(s)I_1(-s) + V_1(-s)I_1(s)}{2}$$

$$= \frac{Z_1(s) + Z_1(-s)}{2Z_1(s)Z_1(-s)} V_1(s)V_1(-s) \quad (16.15)$$

Because the power delivered to the coupling network is in general less than the maximum available power P_m, it is a common practice to define the *reflected power* P_r as

$$P_r = P_m - P_1 \quad (16.16)$$

The following theorem, which relates the reflected power to the maximum available power, is established in Problem 16.1.

Theorem 16.1

$$\frac{P_m(s)}{P_r(s)} = \frac{Z_1(s) + R_1}{Z_1(s) - R_1} \frac{Z_1(-s) + R_1}{Z_1(-s) - R_1} \quad (16.17)$$

This theorem motivates us to define a *reflection function* $T_1(s)$ as follows:

$$T_1(s) \triangleq \frac{Z_1(s) + R_1}{Z_1(s) - R_1} \quad (16.18)$$

Some people prefer instead to deal with the reciprocal of this and term the function the *reflection coefficient*.

In the previous section we gave an expression for the transmission function $H(s)$ in terms of the A, B, C, D parameters. A similiar expression can be given for the reflection function $T_1(s)$:

$$T_1(s) = \frac{(R_2/R_1)^{1/2} A + (1/R_1 R_2)^{1/2} B + (R_1 R_2)^{1/2} C + (R_1/R_2)^{1/2} D}{(R_2/R_1)^{1/2} A + (1/R_1 R_2)^{1/2} B - (R_1 R_2)^{1/2} C - (R_1/R_2)^{1/2} D} \quad (16.19)$$

* This is a meaningful definition because it implies that $P_1(j\omega)$ is the real part of $V_1(j\omega)I_1(-j\omega)$.

In the material that follows we shall assume that the coupling network is lossless; that is, the power delivered to the coupling network is all dissipated in the load, so that $P_1 = P_2$. This is a reasonable assumption because insertion loss theory is based on using capacitors, inductors, and transformers to form the coupling network. Since these "ideal" elements dissipate no real power, they form a lossless coupling network.

Theorem 16.2

For a lossless coupling network,

$$K(s)K(-s) = \frac{P_r}{P_2} \tag{16.20}$$

Proof

$$H(s)H(-s) = \frac{P_m}{P_2} = \frac{P_1 + P_r}{P_2} = 1 + K(s)K(-s)$$

Using the fact that $P_1 = P_2$ for a lossless coupling network yields the theorem.

This theorem has an interesting implication for the reflection function $T_1(s)$. First note from (16.17) and (16.18) that

$$T_1(s)T_1(-s) = \frac{P_m(s)}{P_r(s)}$$

so that
$$T_1(s)T_1(-s) = \frac{P_m}{P_2} \frac{P_2}{P_r} = \frac{H(s)H(-s)}{K(s)K(-s)}$$

Since $H(s) = e(s)/q(s)$ and $K(s) = f(s)/q(s)$, we can combine these results to yield

Theorem 16.3

$$T_1(s)T_1(-s) = \frac{P_m}{P_r} = \frac{H(s)H(-s)}{K(s)K(-s)} = \frac{e(s)e(-s)}{f(s)f(-s)} \tag{16.21}$$

This theorem gives an expression for $T_1(s)T_1(-s)$ but does not uniquely determine $T_1(s)$. In most of the work that follows we shall assume that

$$T_1(s) = \frac{e(s)}{f(s)} \tag{16.22}$$

However, an equally valid relation is

$$T_1(s) = \frac{e(s)}{-f(-s)} \tag{16.23}$$

Using a different expression for $T_1(s)$ would result in a different coupling network.

300 Approximation Methods for Electronic Filter Design

If we use Eq. (16.22) as the defining relation for $T_1(s)$, it follows that $T_1(s) = H(s)/K(s)$. We can then express the characteristic function $K(s)$ in terms of the A, B, C, D parameters by using (16.14) and (16.19):

$$K(s) = \frac{1}{2}\left[\left(\frac{R_2}{R_1}\right)^{1/2} A + \left(\frac{1}{R_1 R_2}\right)^{1/2} B - (R_1 R_2)^{1/2} C - \left(\frac{R_1}{R_2}\right)^{1/2} D\right] \quad (16.24)$$

16.5 THE LOSSLESS COUPLING NETWORK

We are assuming that we have solved the approximation problem and thus have a transmission function $H(s)$ that we want to synthesize. The most common passive realizations use a lossless coupling network between the source and load. This section develops equations that describe this lossless coupling network. These equations will be used in later sections to synthesize the network.

The transmission function and characteristic function can be written as

$$H(s) = \frac{e(s)}{q(s)} \qquad K(s) = \frac{f(s)}{q(s)}$$

Using (16.14) and (16.24) to express these functions in terms of the A, B, C, D parameters leads to

$$e(s) = \frac{q(s)}{2}\left[\left(\frac{R_2}{R_1}\right)^{1/2} A + \left(\frac{1}{R_1 R_2}\right)^{1/2} B + (R_1 R_2)^{1/2} C + \left(\frac{R_1}{R_2}\right)^{1/2} D\right] \quad (16.25)$$

$$f(s) = \frac{q(s)}{2}\left[\left(\frac{R_2}{R_1}\right)^{1/2} A + \left(\frac{1}{R_1 R_2}\right)^{1/2} B - (R_1 R_2)^{1/2} C - \left(\frac{R_1}{R_2}\right)^{1/2} D\right] \quad (16.26)$$

We are assuming that the coupling network will be realized with inductors, capacitors, and ideal transformers; it will thus be lossless. For this case, the parameters A and D are even whereas B and C are odd. Also, the polynomial $q(s)$ must be even or odd. These facts allow us to write the A, B, C, D parameters in terms of the even and odd parts of $e(s)$ and $f(s)$. Using the notation

$$f_e = \frac{f(s) + f(-s)}{2} \qquad f_o = \frac{f(s) - f(-s)}{2}$$

and similarly for e_e and e_o leads to

CASE A $q(s)$ an even polynomial:

$$A = \left(\frac{R_1}{R_2}\right)^{1/2} \frac{e_e + f_e}{q(s)} \qquad B = (R_1 R_2)^{1/2} \frac{e_o + f_o}{q(s)}$$

$$C = \frac{1}{(R_1 R_2)^{1/2}} \frac{e_o - f_o}{q(s)} \qquad D = \left(\frac{R_2}{R_1}\right)^{1/2} \frac{e_e - f_e}{q(s)} \quad (16.27a)$$

CASE B $q(s)$ an odd polynomial:

$$A = \left(\frac{R_1}{R_2}\right)^{1/2} \frac{e_o + f_o}{q(s)} \qquad B = (R_1 R_2)^{1/2} \frac{e_e + f_e}{q(s)}$$
$$C = \frac{1}{(R_1 R_2)^{1/2}} \frac{e_e - f_e}{q(s)} \qquad D = \left(\frac{R_2}{R_1}\right)^{1/2} \frac{e_o - f_o}{q(s)} \tag{16.27b}$$

Equations (16.27) completely describe the lossless coupling network in terms of the A, B, C, D parameters. However, we shall find it more convenient to use the open-circuit–impedance description which is defined by*

$$V_1 = z_{11} I_1 + z_{12} I_2 \qquad V_2 = z_{12} I_1 + z_{22} I_2 \tag{16.28}$$

The impedance parameters can be written in terms of the A, B, C, D parameters as

$$z_{11} = \frac{A}{C} \qquad z_{12} = \frac{1}{C} \qquad z_{22} = \frac{D}{C} \tag{16.29}$$

These relations allow us to describe the lossless coupling network in terms of the impedance parameters. Equations (16.27) and (16.29) yield

CASE A $q(s)$ an even polynomial:

$$z_{11} = R_1 \frac{e_e + f_e}{e_o - f_o} \qquad z_{22} = R_2 \frac{e_e - f_e}{e_o - f_o}$$
$$z_{12} = (R_1 R_2)^{1/2} \frac{q(s)}{e_o - f_o} \tag{16.30a}$$

CASE B $q(s)$ an odd polynomial:

$$z_{11} = R_1 \frac{e_o + f_o}{e_e - f_e} \qquad z_{22} = R_2 \frac{e_o - f_o}{e_e - f_e}$$
$$z_{12} = (R_1 R_2)^{1/2} \frac{q(s)}{e_e - f_e} \tag{16.30b}$$

Similar equations can be derived for admittance parameters instead of impedance parameters (see Problem 16.4). The admittance relations are

CASE A $q(s)$ an even polynomial:

$$y_{11} = \frac{1}{R_1} \frac{e_e - f_e}{e_o + f_o} \qquad y_{22} = \frac{1}{R_2} \frac{e_e + f_e}{e_o + f_o}$$
$$y_{12} = \frac{1}{(R_1 R_2)^{1/2}} \frac{q(s)}{e_o + f_o} \tag{16.31a}$$

* This assumes a reciprocal network so that $z_{12} = z_{21}$.

CASE B $q(s)$ an odd polynomial:

$$y_{11} = \frac{1}{R_1}\frac{e_o - f_o}{e_e + f_e} \quad y_{22} = \frac{1}{R_2}\frac{e_o + f_o}{e_e + f_e} \quad y_{12} = \frac{1}{(R_1R_2)^{1/2}}\frac{q(s)}{e_e + f_e} \quad (16.31b)$$

Example 16.4

Examples 16.2 and 16.3 found the transmission function and characteristic function for a specific third-order elliptic lowpass filter. From these examples we have

$$e(s) = 0.0719(s^3 + 94.203s^2 + 7{,}404.8s + 282{,}956)$$
$$f(s) = 0.0719s(s^2 + 3{,}064.4)$$
$$q(s) = s^2 + 20{,}344$$

For simplicity we assume that the source and load resistances are both equal to R. Then (16.30a) yields

$$z_{11} = z_{22} = R\frac{94.203s^2 + 282{,}956}{4{,}340.4s}$$

$$z_{12} = K\frac{s^2 + 20{,}344}{s}$$

where the constant K is unimportant, as will be demonstrated in later examples.

16.6 SYNTHESIS OF THE LOSSLESS COUPLING NETWORK

We want to synthesize the transfer function $H(s)$ by using a lossless coupling network between the source and load. The previous section shows how the impedance or admittance parameters can be obtained for the coupling network. This section demonstrates how the coupling network can be synthesized.

We shall restrict the coupling network to be a lossless ladder. Furthermore, except for a transformer at the output, the ladder will contain only inductors and capacitors. Since we are not considering the most general type of coupling network (see, for example, Cauer's realization as discussed on page 220 of Ref. 2), we shall not be able to realize all types of transfer functions $H(s)$; however, most practical functions can be realized by this technique.

The lossless ladder will be realized by the *zero-shifting* technique, which can be performed on an impedance or admittance basis. We shall concentrate on the impedance formulation because the admittance formulas follow by analogy. The zero-shifting technique can be used to synthesize the impedance parameters z_{11} and z_{12} simultaneously. We now show that if z_{11} and z_{12} are properly synthesized, then the resulting ladder network automatically realizes z_{22}. That is, we can concentrate on synthesizing z_{11} and z_{12}, and z_{22} will take care of itself! Alternatively we could, of course, realize z_{22} and z_{12} and neglect z_{11}.

For the sake of discussion, assume that $q(s)$ is an even polynomial. Then Eq. (16.30) yields

$$z_{11} = R_1 \frac{e_e + f_e}{e_o - f_o} \qquad z_{12} = (R_1 R_2)^{1/2} \frac{q(s)}{e_o - f_o}$$

Problem 16.5 demonstrates that this implies

$$\frac{z_{12}^2}{z_{11}} = R_2 \left(\frac{e_e - f_e}{e_o - f_o} - \frac{e_o + f_o}{e_e + f_e} \right) \tag{16.32}$$

We are assuming that the coupling network is realized as a ladder. The poles of a ladder network are compact; that is, at any pole,

$$(\text{Residue } z_{11})(\text{residue } z_{22}) = (\text{residue } z_{12})^2 \tag{16.33}$$

Because the poles of the ladder network are the solutions to

$$e_o(s) - f_o(s) = 0$$

it follows that (16.32) and (16.33) imply

$$z_{22} = R_2 \frac{e_e - f_e}{e_o - f_o}$$

But this is the same as the expression for z_{22} in (16.30a).

We have just demonstrated that if z_{11} and z_{12} are realized as a ladder network, then the proper z_{22} is automatically realized. We shall now discuss the zero-shifting technique that can be used to realize z_{11} and z_{12} simultaneously.

16.7 THE ZERO-SHIFTING TECHNIQUE

Figure 16.2 shows a general ladder network. We want to use a lossless ladder to realize z_{11} and z_{12} simultaneously. Alternatively, we could realize y_{11} and y_{12}, but we shall only discuss the impedance case. Since only lossless elements will be used, z_{11} must be a reactance function; thus its poles and zeros must be on the $j\omega$ axis, where they alternate. But

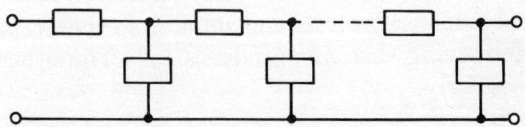

Fig. 16.2 General ladder network.

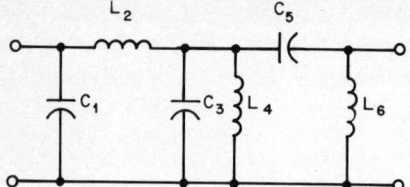

Fig. 16.3 Ladder network that has three attenuation poles at the origin and three at infinity.

what are the restrictions on z_{12}? Of course, its poles must be the same as the poles of z_{11}, but what about its zeros?

For a ladder network, the zeros of z_{12} are where the series branches are open circuits or where the shunt branches are shorts. Because the series and shunt branches are assumed to contain only inductors and capacitors, this implies that the zeros of z_{12} must be on the $j\omega$ axis. With this in mind, let us look at the following z_{11} and z_{12} to determine how they can be realized:*

$$z_{11} = \frac{s(s^2+2)(s^2+4)}{(s^2+1)(s^2+3)(s^2+5)} \qquad z_{12} = \frac{Ks^3}{(s^2+1)(s^2+3)(s^2+5)} \qquad (16.34)$$

First note that the poles and zeros of z_{11} are on the $j\omega$ axis, where they alternate. The zeros of z_{12} are also on the $j\omega$ axis: three are at the origin, and three are at infinity. We want to realize z_{11} in such a way that we simultaneously realize these zeros of z_{12}. One possible realization is shown in Fig. 16.3. The first three elements (C_1, L_2, C_3) produce transmission zeros (attenuation poles) at infinity, while the last three elements produce transmission zeros at the origin. The element values can be determined by properly expanding z_{11}. For example, C_1 can be found by writing $1/z_{11}$ as

$$\frac{1}{z_{11}} = \frac{(s^2+1)(s^2+3)(s^2+5)}{s(s^2+2)(s^2+4)}$$

$$= s + \frac{3s^4 + 15s^2 + 15}{s(s^2+2)(s^2+4)}$$

$$= C_1 s + y_r$$

This implies that the shunt capacitor has the value $C_1 = 1$. The remainder y_r has a zero at infinity so that $1/y_r = z_r$ has a pole at infinity. Removing this pole produces L_2. The other elements may be obtained similarly.

* This, and several other examples, come from a course taught at M.I.T. by Harry B. Lee.

The development just described realized z_{11} in such a way that the zeros of z_{12} were simultaneously synthesized. However, we had no control over the constant multiplier of z_{12}. That is, the synthesized network produces a specific constant multiplier. If it is not the proper value, any desired value can be obtained by placing a transformer at the output. This does not affect z_{11}, but it does multiply z_{12} by the reciprocal of the turns ratio.

In the example just discussed, we developed z_{11} in such a way that zeros of transmission were produced at the origin or at infinity. In general, we shall also have to produce transmission zeros on the $j\omega$ axis. For example, we may want to synthesize

$$z_{11} = \frac{s^2 + 1}{s} \qquad z_{12} = K \frac{s^2 + 2}{s} \tag{16.35}$$

The impedances in (16.35) can be realized by using the zero-shifting technique. To see this, we must first introduce the concept of a partial pole removal. In (16.35), z_{11} has a pole at infinity. This can be totally removed by extracting a series inductor of unity value. If an inductor of value less than unity is used, we say that the pole has been *partially removed*, because the remainder impedance still has a pole at infinity.

A general partial pole removal is shown in Fig. 16.4. Assume that the network Z_p produces a partial pole removal at $s = j\omega_p$. Thus, for s near this frequency we can write

$$Z_p \approx \frac{K_p}{s - j\omega_p}$$

where K_p is the residue of the partial pole. Because it was only a partial pole removal, for s near ω_p we can also write

$$Z_R \approx \frac{K_R}{s - j\omega_p}$$

This implies that $E_2 \approx K_R E_1/(K_p + K_R) \neq 0$. That is, if we perform a partial pole removal in producing z_{11}, it does not produce a transmission zero. Thus, if the transmission zeros are not included in the zeros of z_{11},

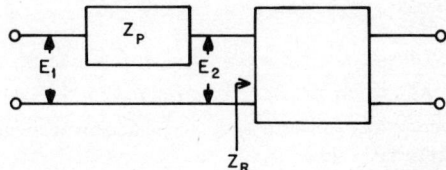

Fig. 16.4 Partial removal of a pole by the series element Z_p.

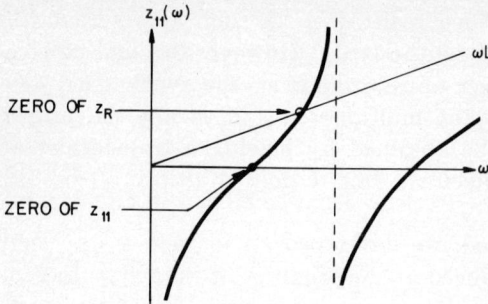

Fig. 16.5 Illustration of zero shifting by partial removal of a pole at infinity.

we may shift the zeros by making partial pole removals. The following theorem describes how a partial pole removal shifts the zeros of z_{11}.

Theorem 16.4

Let $z_{11} = z_p + z_R$, where z_p is used to produce a partial pole removal at ω_p, and z_R is the remainder function. If z_{11} has a zero at ω_0, then the zero of z_R corresponding to ω_0 is between ω_0 and ω_p.

Proof

The theorem will be proved for $z_p = sL$, but it can be similarly established for other types of partial pole removals. Figure 16.5 shows a typical form of $z_{11}(j\omega)$. The intersection of $z_{11}(j\omega)$ with the line ωL determines the zeros of z_R. From the figure it is obvious that the partial removal of a pole at infinity has shifted the zero to the right, which is in accord with the theorem.

Example 16.5

Examples 16.2 to 16.4 have been tracing through the realization of a third-order elliptic lowpass filter. We are finally at the stage where we can obtain element values. We want to synthesize*

$$z_{11} = z_{22} = \frac{94.203s^2 + 282{,}956}{4{,}340.4s}$$

$$= 0.0217 \frac{s^2 + 3{,}003.7}{s}$$

$$z_{12} = K \frac{s^2 + 20{,}344}{s}$$

We can shift the zero of z_{11} to be at the zero of z_{12} by doing a partial pole removal at infinity. That is,

$$z_{11} = sL + z_R$$

* The source and load resistors have been normalized to $R = 1$.

The value of L should be chosen such that z_R is zero at $s^2 = -20{,}344$. This implies $L = 0.0185$, so that we can solve for z_R and obtain

$$z_R = 0.0032 \frac{s^2 + 20{,}344}{s}$$

This remainder function has a zero at $s^2 = -20{,}344$, as desired. The realization of $z_{11} = 0.0217(s^2 + 3{,}003.7)/s$ is shown in Fig. 16.6a. Because of the zero shifting, we simultaneously realized $z_{12} = K(s^2 + 20{,}344)/s$. Now consider the network in Fig. 16.6b. This has the same open-circuit input impedance (z_{11}) as the circuit in Fig. 16.6a. Similarly, z_{12} is the same (except for a different constant multiplier). However, the open-circuit output impedances are different. In fact, only the circuit in Fig. 16.6b has $z_{11} = z_{22}$ as is required. In general, after realizing z_{11} and z_{12}, we must check to see whether or not we have also synthesized the proper z_{22}. If not, an extra element will have to be added. The complete realization of the transmission function $H(s)$ is shown in Fig. 16.6c.

16.8 TYPICAL NETWORK CONFIGURATIONS

The lossless coupling networks encountered in practice can be considered to be comprised of various building blocks. For example, the network in Fig. 16.6a can be considered to be one of these building blocks. It was obtained by using the series inductor to shift a transmission zero toward infinity; the capacitor-inductor combination then realized the transmission zero.

Other commonly encountered filter sections are shown in Fig. 16.7; these sections can be used to produce attenuation poles (transmission zeros). For example, circuit 7 can be used to produce a transmission

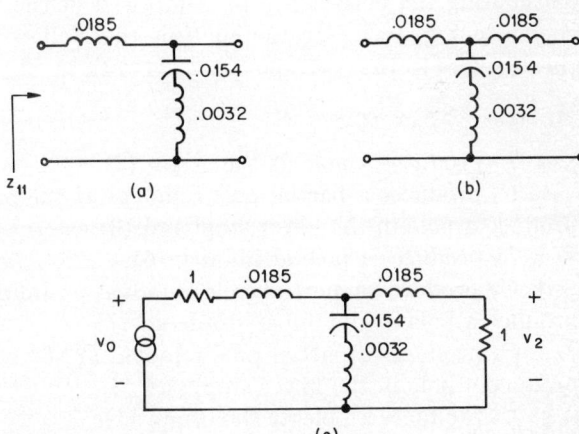

Fig. 16.6 Development of the lossless coupling network in Example 16.5.

308   Approximation Methods for Electronic Filter Design

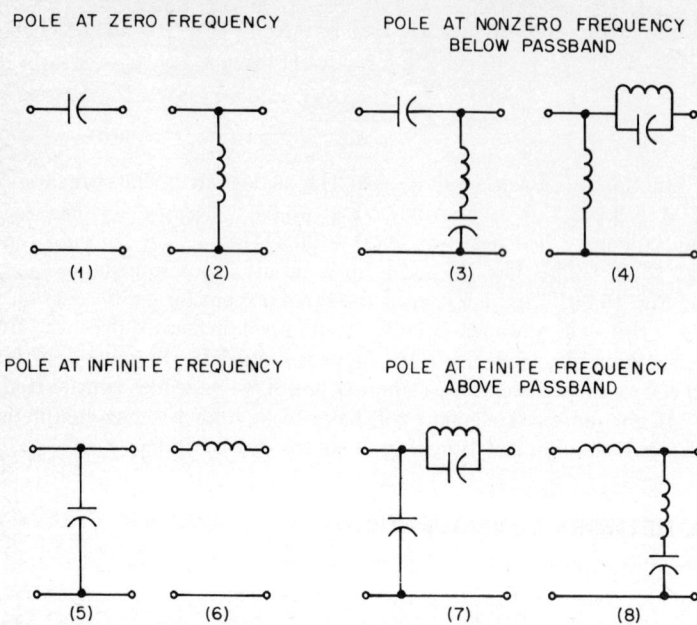

Fig. 16.7 Common building blocks used in insertion loss synthesis.

zero above the passband. The shunt capacitor shifts the zero toward infinity. If, instead, one wants to shift a zero toward the origin, then circuit 3 or 4 can be used.

A bandpass filter realized from the filter sections is shown in Fig. 16.8. A good insight into the performance of this filter can be obtained by considering the network to be comprised of the various building blocks shown in Fig. 16.7. In the analysis that follows, the number in parenthesis refers to the corresponding network in Fig. 16.7.

Analysis of Fig. 16.8

  a. L_1 produces a pole at the origin (2).
  b. C_1 produces a partial pole removal at the origin so that C_2, L_2 can produce a pole in the lower stopband (3).
  c. L_3 produces a pole at infinity (6).
  d. C_3 produces a partial pole removal at infinity so that C_4, L_4 can produce a pole in the upper stopband (7).
  e. C_5 produces a partial pole removal at the origin so that C_6, L_5 can produce a pole in the lower stopband (3).
  f. C_7 produces a pole at the origin (1).

From the above discussion it is obvious that there are two poles at finite frequencies below the passband, and one pole at a finite frequency above the passband; however, it is less obvious how many poles there are at the origin and at infinity. The low-frequency behavior can be analyzed with the help of Fig. 16.8b, which models the behavior of the circuit as $s \to 0$. The capacitor C corresponds to the series combination C_1, C_5, C_7. It follows from Fig. 16.8b that there are two poles at the origin; that is, $NZ = 2$. Similarly, Fig. 16.8c implies that $NIN = 2$.

Another bandpass filter realized from the filter sections is shown in Fig. 16.9. The realization of this filter is discussed in great detail in Saal and Ulbrich's classical paper,[5] and the reader is enthusiastically referred there if he wants to follow through the zero-shifting technique with a numerical example. Instead of paraphrasing Saal and Ulbrich's work, we shall limit ourselves to a cursory discussion of the network.

The numbers in parentheses in Fig. 16.9 refer to the filter sections identi-

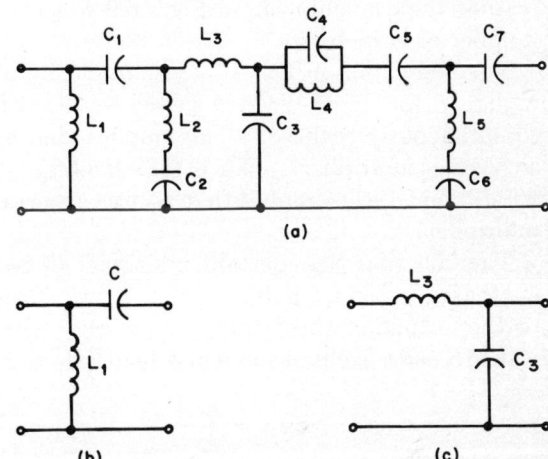

Fig. 16.8 Typical bandpass filter. (b) and (c) are useful for analyzing low- and high-frequency behavior.

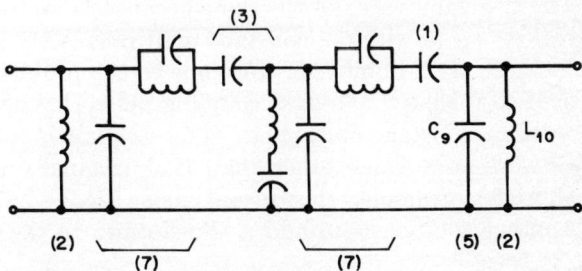

Fig. 16.9 Typical bandpass filter.

fied in Fig. 16.7. Thus, proceeding from left to right, we see that the filter realizes

   a. An attenuation pole at the origin (2)
   b. An attenuation pole in the upper stopband (7)
   c. An attenuation pole in the lower stopband (3)
   d. An attenuation pole in the upper stopband (7)
   e. An attenuation pole at the origin (1)
   f. An attenuation pole at infinity (5)
   g. An attenuation pole at the origin (2)

The filter just described had three attenuation poles at the origin, one in the lower stopband, two in the upper stopband, and one at infinity. Of course the realization was not unique; for example, the attenuation poles could have been removed in another order.* One of the major reasons the configuration in Fig. 16.9 was used is that it has a minimum number of inductors.

The realization in Fig. 16.9 was found by using the zero-shifting technique: z_{11} was synthesized in such a way that the transmission zeros were simultaneously realized. This implies that z_{12} was realized (except for a constant multiplier). Since, for a ladder realization z_{11} and z_{12} uniquely determine z_{22}, this implies that z_{22} was also realized (except for a constant multiplier).

The fact that the constant multiplier of z_{22} was not controlled by the synthesis procedure indicates that a transformer may have to be added to the output of the coupling network. For example, assume that we want to use a 1-ohm source and load ($R_1 = 1 = R_2$). From (16.30),

$$z_{11} = \frac{e_o + f_o}{e_e - f_e} \qquad z_{22} = \frac{e_o - f_o}{e_e - f_e}$$

Proceeding from left to right (i.e., synthesizing z_{11}) yielded that the last inductor was $L_{10} = 0.362307674$. However, proceeding from right to left (i.e., synthesizing z_{22}) yielded that this inductor was $L'_{10} = 0.342004574$. The two values are not the same because the zero-shifting technique only realizes z_{12} to within a constant multiplier.

There are a number of solutions to the problem that if z_{11} and z_{12} are used to realize the lossless coupling network, then z_{22} is only realized to within a constant multiplier. The simplest is to terminate the coupling network in a value other than that originally specified. In Saal and Ulbrich's example, this meant using $R_2 = 1.059$ instead of $R_2 = 1$. Another solution is to add a transformer to the output of the coupling

---

* It often happens that removing the attenuation poles in one order will produce negative elements; then, another order should be attempted.

network. For this example, the turns ratio is

$$t^2 = \frac{L_{10}}{L'_{10}} = 1.059365$$

Another possible solution is to perform some transformations on the network.[6] The Norton transformation is frequently used to shift impedance levels in networks. This can be used not only to yield a proper terminating impedance, but also to produce reasonable element values.

The zero-shifting procedure can produce errors due to numerical inaccuracies. As each successive element is computed, roundoff errors make the remaining ones more inaccurate. One can determine how accurate the results are by synthesizing the lossless coupling network from both directions. For example, in Fig. 16.9 synthesizing z_{11} yielded

$$C_9 = 2.530775564 \qquad L_{10} = 0.362307674$$

while synthesizing z_{22} yielded

$$C'_9 = 2.68098815 \qquad L'_{10} = 0.342004574$$

If there were no inaccuracies we should have

$$\frac{L_{10}}{L'_{10}} = \frac{C'_9}{C_9}$$

However, $\quad \dfrac{L_{10}}{L'_{10}} = 1.059365 \quad$ and $\quad \dfrac{C'_9}{C_9} = 1.059354$

16.9 SYNTHESIS IN TERMS OF THE TRANSFORMED VARIABLE

The previous section demonstrated that the zero-shifting technique can introduce numerical inaccuracies. In a high-degree filter this can result in erroneous element values. The improper values will manifest themselves most by the way they affect passband performance; for example, the passband may not be equiripple.

The transformed variable Z was introduced in Chap. 7 because it expanded the passband frequencies to the entire imaginary Z axis. This improved the computational accuracy when we found the natural modes of $H(s)$. The same transformation can be used to improve the accuracy of the zero-shifting technique. This section describes how it is possible to do the entire zero-shifting procedure in terms of the transformed variable. This is the method used by most, modern passive-synthesis computer programs.[7,8]

The zero-shifting technique uses as a starting point the fact that the lossless coupling network can be expressed in terms of $e(s)$ and $f(s)$. The equations we shall use in this section are given below.

CASE A $q(s)$ an even polynomial:

$$\frac{z_{11}}{R_1} = \frac{e_e + f_e}{e_o - f_o} \quad \frac{z_{22}}{R_2} = \frac{e_e - f_e}{e_o - f_o} \quad \frac{y_{11}}{G_1} = \frac{e_e - f_e}{e_o + f_o} \quad \frac{y_{22}}{G_2} = \frac{e_e + f_e}{e_o + f_o} \quad (16.36)$$

CASE B $q(s)$ an odd polynomial:

$$\frac{z_{11}}{R_1} = \frac{e_o + f_o}{e_e - f_e} \quad \frac{z_{22}}{R_2} = \frac{e_o - f_o}{e_e - f_e} \quad \frac{y_{11}}{G_1} = \frac{e_o - f_o}{e_e + f_e} \quad \frac{y_{22}}{G_2} = \frac{e_o + f_o}{e_e + f_e} \quad (16.37)$$

We want to express these equations in terms of the transformed variable, which requires finding $E(Z)$ and $F(Z)$, the transformed versions of $e(s)$ and $f(s)$. In our solution of the approximation problem we found expressions for $F(Z)$. For example, for an equiripple bandpass filter $F(Z)$ is given by

$$F(Z) = \epsilon \mathrm{Ev} \left[\left(Z + \frac{\omega_B}{\omega_A} \right)^{NZ} (Z+1)^{NIN} \prod_{i=1}^{N} (Z+Z_i)^2 \right] \quad (9.24)$$

We shall now discuss how the function $E(Z)$ can be determined. To simplify the discussion we assume that we are just considering equiripple or maximally flat bandpass filters.* This implies that we can write

$$E(Z)E^*(Z) = \prod_{i=1}^{m/2} (Z^4 + P_i Z^2 + Q_i) \quad (7.24)$$

where $E(Z)$ is the transformed version of $e(s)$, and $E^*(Z)$ is the transformed version of $e(-s)$.

In the solution of the approximation problem, the function $E(Z)E^*(Z)$ was found. In fact, the zero-finder program in Fig. 10.1 determines the factors of $Z^4 + P_i Z^2 + Q_i$. We would thus like to be able to find $E(Z)$ if the product $E(Z)E^*(Z)$ is as given by Eq. (7.24). The function $E(Z)$ can be found by determining $e(s)e(-s)$, allocating the left half-plane roots to $e(s)$, and then forming $E(Z)$. Problem 16.7 demonstrates that if $E(Z)E^*(Z)$ is as given above, then we can write

$$E(Z) = D \prod_{i=1}^{m/2} [U_i + V_i Z^2 + y_i \sqrt{(Z^2-1)(\omega_B{}^2 - Z^2 \omega_A{}^2)}] \quad (16.38)$$

where†

$$U_i = \omega_B{}^2 - a_i \quad V_i = a_i - \omega_A{}^2 \quad y_i = 2\left(\frac{a_i - b_i}{2}\right)^{1/2} \quad (16.39)$$

---

* Even-order lowpass filters can be treated by simply letting ω_A equal zero. Odd-degree lowpass filters are considered in Problems 16.12 to 16.16.

† The constants a_i and b_i are identified in Eq. (7.26).

The constant D can be evaluated by using the following result from Problem 16.8:

$$|E(0)| = \left(1 + \frac{1}{\epsilon^2}\right)^{1/2} |F(0)| \qquad (16.40)$$

The expression for $E(Z)$ in (16.38) can be expanded and written in simpler notation as

$$E(Z) = E_e(Z) + \sqrt{(Z^2 - 1)(\omega_B{}^2 - Z^2\omega_A{}^2)}\, E_o(Z) \qquad (16.41)$$

The functions $E_e(Z)$ and $E_o(Z)$ are both polynomials in Z^2. It is shown in Problem 16.9 that these polynomials are related to the even and odd parts of $e(s)$ as follows:

$$E_e(Z) = \frac{e_e(s)}{(s^2 + \omega_A{}^2)^{m/2}} \qquad (16.42)$$

$$\sqrt{(Z^2 - 1)(\omega_B{}^2 - Z^2\omega_A{}^2)}\, E_o(Z) = \frac{e_o(s)}{(s^2 + \omega_A{}^2)^{m/2}} \qquad (16.43)$$

We are now ready to write the impedance (or admittance) parameters of the lossless coupling network in terms of the transformed variable. Recall that we wanted to use the transformed variable because it will make the zero-shifting technique more accurate. By using the relations in (16.42) and (16.43), Problem 16.10 demonstrates that the impedance and admittance formulas in (16.36) and (16.37) can be rewritten as below, where $E_e(Z) = E_e$, etc.

CASE A $q(s)$ an even polynomial:

$$\begin{aligned}\frac{z_{11}(s)}{sR_1} &= \frac{1}{\omega_B{}^2 - Z^2\omega_A{}^2}\frac{E_e + F_e}{E_o - F_o} & \frac{z_{22}(s)}{sR_2} &= \frac{1}{\omega_B{}^2 - Z^2\omega_A{}^2}\frac{E_e - F_e}{E_o - F_o} \\ \frac{y_{11}(s)}{sG_1} &= \frac{1}{\omega_B{}^2 - Z^2\omega_A{}^2}\frac{E_e - F_e}{E_o + F_o} & \frac{y_{22}(s)}{sG_2} &= \frac{1}{\omega_B{}^2 - Z^2\omega_A{}^2}\frac{E_e + F_e}{E_o + F_o}\end{aligned} \qquad (16.44)$$

CASE B $q(s)$ an odd polynomial:

$$\begin{aligned}\frac{z_{11}(s)}{sR_1} &= (Z^2 - 1)\frac{E_o + F_o}{E_e - F_e} & \frac{z_{22}(s)}{sR_2} &= (Z^2 - 1)\frac{E_o - F_o}{E_e - F_e} \\ \frac{y_{11}(s)}{sG_1} &= (Z^2 - 1)\frac{E_o - F_o}{E_e + F_e} & \frac{y_{22}(s)}{sG_2} &= (Z^2 - 1)\frac{E_o + F_o}{E_e + F_e}\end{aligned} \qquad (16.45)$$

These relations allow us to find the impedance or admittance parameters of the lossless coupling network. The expressions are in terms of the transformed variable and thus help to reduce numerical inaccuracies which occur in the zero-shifting technique. To understand how the zero shifting can be performed in terms of the transformed variable, we shall reconsider a previous example.

In Example 16.5 we simultaneously realized z_{11} and z_{12}, which were given by

$$z_{11}(s) = 0.0217 \frac{s^2 + 3{,}003.7}{s} \qquad (16.46)$$

$$z_{12}(s) = K \frac{s^2 + 20{,}344}{s} \qquad (16.47)$$

This was for a third-degree lowpass filter that had a passband edge $F_B = 10$. Thus it follows that

$$Z^2 = \frac{s^2 + \omega_B^2}{s^2 + \omega_A^2}$$

where $\omega_B = 20\pi$
$\omega_A = 0$

We can now rewrite $z_{11}(s)$ in terms of the transformed variable by using the fact that $s^2 = (20\pi)^2/(Z^2 - 1)$:

$$\frac{z_{11}(s)}{s} = 0.0217 \frac{s^2 + 3{,}003.7}{s^2}$$

$$= 0.00519 + 0.01651 Z^2$$

$$= L + (0.00519 - L) + 0.01651 Z^2$$

The value of the inductor L should be chosen so that the remainder function

$$R = (0.00519 - L) + 0.01651 Z^2$$

is zero when $s^2 = -20{,}344$. This transmission zero corresponds to $Z^2 = 0.8059$, which implies that $L = 0.0185$. Thus, we can write

$$\frac{z_{11}(s)}{s} = 0.0185 + 0.01651(Z^2 - 0.8059)$$

The value of the inductor ($L = 0.0185$) is the same as the value that was found in Example 16.5. Similarly, if one transforms the function

$$0.01651 s (Z^2 - 0.8059)$$

back to the s domain, one finds the same RC series circuit as in Example 16.5.

16.10 CONCLUSIONS

The solution to the approximation problem can be stated in terms of the functions $e(s)$, $f(s)$, and $q(s)$. In this chapter we discussed the synthesis of the transmission function $H(s) = e(s)/q(s)$. This transmission function is related to an input/output voltage ratio via

$$H(s) = \left(\frac{R_2}{4R_1}\right)^{1/2} \frac{V_0(s)}{V_2(s)}$$

where R_1 and R_2 are the source and load resistors.*

It was shown that $H(s)$ could be realized by synthesizing a lossless coupling network that simultaneously realizes two impedance or admittance parameters, for example,

$$z_{11} = \frac{e_e + f_e}{e_o - f_o} \qquad z_{12} = \frac{q(s)}{e_o - f_o}$$

Thus the coupling network can be synthesized by using the zero-shifting technique. That is, partial poles are removed in such a manner that the zeros of transmission are shifted to the proper frequencies. Because the zero-shifting technique introduces numerical inaccuracies, practical computer programs perform the entire synthesis procedure in terms of the transformed variable.

REFERENCES

1. D. S. Humpherys, *The Analysis, Design, and Synthesis of Electrical Filters*, Prentice-Hall, Inc., Englewood Cliffs, N.J., 1970.
2. N. Balabanian, *Network Synthesis*, Prentice-Hall, Inc., Englewood Cliffs, N.J., 1958.
3. L. Weinberg, *Network Analysis and Synthesis*, McGraw-Hill Book Company, New York, 1962.
4. A. I. Zverev, *Handbook of Filter Synthesis*, John Wiley & Sons, Inc., New York, 1967.
5. R. Saal and E. Ulbrich, "On the Design of Filters by Synthesis," *IRE Trans. on Circuit Theory*, December 1958, pp. 284–327.
6. S. W. Conning, "A Survey of Network Equivalences," *Proc. IREE Aust.*, June 1969, pp. 166–184.
7. H. J. Orchard and G. C. Temes, "Filter Design Using Transformed Variables," *IEEE Trans. on Circuit Theory*, December 1968, pp. 385–408.
8. F. F. Kuo and W. G. Magnuson, eds., *Computer Oriented Circuit Design*, Prentice-Hall, Inc., Englewood Cliffs, N.J., 1969, chap. 6.

PROBLEMS

16.1 *a.* Using the expression for $P_1(s)$ in (16.15), show that

$$\frac{P_m(s)}{P_1(s)} = \frac{Z_1(s)Z_1(-s)}{2R_1[Z_1(s) + Z_1(-s)]} \frac{V_0(s)V_0(-s)}{V_1(s)V_1(-s)}$$

b. Establish the relation in (16.17).

* Problem 16.11 demonstrates how limiting cases such as $R_1 = 0$ can be treated.

16.2 Equation (16.14) expresses the transmission function $H(s)$ in terms of the A, B, C, D parameters. Establish this relation.

16.3 Equation (16.19) expresses the reflection function $T_1(s)$ in terms of the A, B, C, D parameters. Establish this relation.

16.4 *a.* For a reciprocal network, the short-circuit admittance parameters are defined by the relations

$$I_1 = y_{11}V_1 + y_{12}V_2 \qquad I_2 = y_{12}V_1 + y_{22}V_2$$

Show that they can be written in terms of the A, B, C, D parameters as

$$y_{11} = \frac{D}{B} \qquad y_{12} = -\frac{1}{B} \qquad y_{22} = \frac{A}{B}$$

b. Establish Eq. (16.31) by using the result of part *a*.

16.5 For $q(s)$ an even polynomial, show that if

$$z_{11} = R_1 \frac{e_e - f_e}{e_o - f_o} \qquad z_{12} = (R_1 R_2)^{\frac{1}{2}} \frac{q(s)}{e_o - f_o}$$

then

$$\frac{z^2{}_{12}}{z_{11}} = R_2 \left(\frac{e_e - f_o}{e_o - f_o} + \frac{e_o + f_o}{e_e + f_e} \right)$$

Hint: Use $q^2(s) = e(s)e(-s) - f(s)f(-s)$.

16.6 In most of this chapter we write the reflection function as

$$T_1(s) = \frac{e(s)}{f(s)} \tag{16.22}$$

However, we could instead write it as

$$T_1(s) = \frac{e(s)}{-f(-s)} \tag{16.23}$$

By using the expressions in (16.30) with $R_1 = R_2$ and (16.31), show that the two different expressions lead to two identical coupling networks—except that the input and output ports are interchanged.

16.7 *a.* If $e(s) = s^2 + As + B$, show that

$$E(Z) = L[(\omega_B{}^2 - B) + Z^2(B - \omega_A{}^2) + A\sqrt{(Z^2 - 1)(\omega_B{}^2 - Z^2\omega_A{}^2)}]$$

where L is an unimportant constant.

b. If $E(Z)E^*(Z) = C \prod_{i=1}^{m/2} (Z^4 + P_i Z^2 + Q_i)$, then

$$e(s) = K \prod_{i=1}^{m/2} \left[s^2 + 2\left(\frac{a_i - b_i}{2}\right)s + a_i \right] \tag{7.28}$$

Use this fact and the result of part *a* to establish Eq. (16.38).

16.8 The passband edge $s = j\omega_B$ corresponds to $Z = 0$. This implies that

$$\frac{F^2(0)}{Q^2(0)} = |K(j\omega_B)|^2 = \epsilon^2$$

$$\frac{E(0)E^*(0)}{Q^2(0)} = |H(j\omega_B)|^2 = 1 + \epsilon^2$$

Use these facts to establish (16.40).

16.9 For an even-degree filter we can write the transformed version of $e(s)$ as

$$E(Z) = E_e(Z) + \sqrt{(Z^2 - 1)(\omega_B^2 - Z^2\omega_A^2)}\, E_o(Z)$$

where $E_e(Z)$ and $E_o(Z)$ are even polynomials in the parameter Z^2. Using the fact that

$$E(Z) = \frac{e_e(s) + e_o(s)}{(s^2 + \omega_A^2)^{m/2}}$$

show that

$$E_e(Z) = \frac{e_e(s)}{(s^2 + \omega_A^2)^{m/2}}$$

Note that this result implies (16.43).

16.10 For $q(s)$ an even polynomial, $z_{11}/R_1 = (e_e + f_e)/(e_o - f_o)$. Divide both numerator and denominator by $(s^2 + \omega_A^2)^{m/2}$ and identify terms to:

a. Show that

$$\frac{z_{11}(s)}{R_1} = \frac{E_e(Z) + F_e(Z)}{E_o(Z) - F_o(Z)} \frac{1}{\sqrt{(Z^2-1)(\omega_B^2 - Z^2\omega_A^2)}}$$

b. Show that

$$\frac{z_{11}(s)}{sR_1} = \frac{E_e(Z) + F_e(Z)}{E_o(Z) - F_o(Z)} \frac{1}{(\omega_B^2 - Z^2\omega_A^2)}$$

16.11 The insertion loss synthesis method discussed in this chapter assumes that there are a source resistance R_1 and a load resistance R_2. This example demonstrates that we can also realize transfer functions for limiting cases of source and load. We shall consider the case of a source of $0\ \Omega$; for convenience, the load will be normalized to unity.

a. Show that

$$\frac{V_2}{I_1} = \frac{z_{12}}{z_{22} + 1}$$

where V_2 and I_1 are as identified in Fig. 16.1.

b. Show that

$$\frac{V_2}{V_1} = \frac{-y_{12}}{y_{22} + 1}$$

c. Assume we want to realize

$$\frac{V_2}{I_1} = \frac{q(s)}{e(s)} = \frac{q}{e_e + e_o}$$

Show that this can be realized by simultaneously realizing the following two parameters:

CASE A: $q(s)$ an even polynomial:

$$z_{22} = \frac{e_e}{e_o} \qquad z_{12} = \frac{q}{e_o}$$

CASE B: $q(s)$ an odd polynomial:

$$z_{22} = \frac{e_o}{e_e} \qquad z_{12} = \frac{q}{e_e}$$

16.12 In Sec. 16.9 we considered even-degree filters, but one may want to design odd-degree equiripple lowpass filters. Show that for an odd-degree lowpass filter we can write the transformed version of $e(s)$ as

$$E(Z) = \omega_B \sqrt{Z^2 - 1} E_e(Z) + E_o(Z)$$

where $E_e(Z)$ and $E_o(Z)$ are polynomials in Z^2. Furthermore, show that they are related to the even and odd parts of $e(s)$ via

$$\frac{e_e(s)}{s^m} = \omega_B \sqrt{Z^2 - 1} E_e(Z) \qquad \frac{e_o(s)}{s^m} = E_o(Z)$$

16.13 Example 16.4 considers a third-degree elliptic lowpass filter with $e(s) = 0.0719(s^3 + 94.203s^2 + 7{,}404.8s + 282{,}956)$. Using the notation in Problem 16.12, show that for $\omega_B = 20\pi$,

$$E_e(Z) = 0.0001305Z^2 + 0.0004103 \qquad E_o(Z) = 0.13468Z^2 - 0.06296$$

16.14 Example 16.3 considers a third-degree elliptic lowpass filter with $F(Z) = C(0.19618 + 0.68048Z^2)$. This filter has a passband ripple of 0.25 dB. Use this fact, the results of Problem 16.13, and Eq. (16.40) to show that $C = 0.082$.

16.15 Show that for an odd-order lowpass filter

$$\frac{z_{11}(s)}{sR_1} = \frac{(Z^2 - 1)E_e(Z)}{E_o(Z) - F_o(Z)}$$

where $E_o(Z) = \dfrac{e_o(s)}{s^m} \qquad F_o(Z) = \dfrac{e_o(s)}{s^m} \omega_B \sqrt{Z^2 - 1} \qquad E_e(Z) = \dfrac{e_e(s)}{s^m}$

16.16 For the filter described in Problems 16.13 and 16.14, show that

$$\frac{z_{11}(s)}{sR_1} = 0.01651Z^2 + 0.00519$$

CHAPTER SEVENTEEN

Approximation Methods and Active Filter Synthesis

17.1 INTRODUCTION

The previous chapter discussed transfer-function realizations that use passive elements; to be more specific, it discussed how inductors and capacitors are used in insertion loss theory. This chapter discusses how the same transfer functions can be realized by active circuits. Again, the major purpose of this dissertation is to demonstrate practical methods that can be used to synthesize the transfer functions found in the first part of this book.

Before embarking on our discussion of active filters, we should emphasize that they will never totally replace passive filters. When something new is introduced, there is a tendency to assume that the old will disappear; in fact, it often becomes fashionable to use the newer item just because it is modern. However, in industry one eventually encounters reality in the form of economics. If two items perform the same task equally well, the less expensive one will be produced. With this in mind, let us examine the areas where passive and active filters compete.

Active filters have the best chance of replacing passive filters at low frequencies (below about 100 kHz). As the frequency range decreases, the size of inductors increases. Not only do they become bulky, they also are of poorer quality—they have more loss. Active filters offer the possibility of replacing the bulky, poor-quality inductor with a small integrable device.

Practical active filters are often realized in hybrid form; that is, they use thin-film circuitry to realize resistors and capacitors while they use integrated circuits to realize the active device. In modern thin-film technology it is possible to have resistors and capacitors that compensate each other's temperature coefficient. Also, because the resistors (and capacitors) can be on the same substrate, they tend to track. That is, if one resistor increases by 1 percent, then all resistors tend to increase by 1 percent because they have been made by the same manufacturing process. The tracking property of thin-film circuits has been utilized by Bell Telephone Laboratories in the design of practical active filters.[1]

As the frequency range increases, inductors become smaller, have higher quality, and are less expensive, so that the competition they offer active filters becomes greater. Also, most active filters do not perform well at high frequencies. Thus, for economical and technological reasons, in industry active filters will be limited to frequencies below 100 kHz for some time to come. This is not meant to indicate that no research is being performed at higher frequencies, but instead is meant as a realistic appraisal of the competition—the passive filters.

Because the emphasis of this chapter is on practical active filters, very little time will be spent on one of the original ways of realizing active filters: the negative-impedance converter. This device has attracted much attention in theoretical papers, but in practice it is a poor solution to filtering problems. The next section indicates why very few people actually use negative-impedance converters.

Gyrators are more practical than negative-impedance converters; in fact, they are superior in many aspects. However, in any decision as to the type of active device to use in an active filter, the gyrator has a tough foe: the operational amplifier. This chapter discusses the pros and cons of using gyrators instead of operational amplifiers but, because the operational amplifier has more than its share of favorable arguments, most of the chapter is devoted to active filters that use operational amplifiers.

17.2 NEGATIVE-IMPEDANCE-CONVERTER ACTIVE FILTERS

Negative-impedance converters (NICs) have been discussed for some time; in fact, the widely referenced article by J. G. Linvill[2] was written in 1953. However, NICs have not found widespread practical acceptance. This section gives a brief description of the properties and applications of NICs wherein we shall see why NICs are currently held in disfavor.

The negative-impedance converters that are commonly encountered can be divided into two classes: current NICs and voltage NICs. Both circuits are two-ports; the current NIC can be described by

$$V_1 = V_2 \qquad kI_1 = I_2 \tag{17.1}$$

whereas the voltage NIC can be described by

$$V_1 = -kV_2 \qquad I_1 = -I_2 \tag{17.2}$$

The most important property of these two-ports is that they can be used to produce negative impedances. As illustrated in Fig. 17.1, if the output is terminated in an impedance Z_2, then the input impedance is

$$Z_1 = -kZ_2 \tag{17.3}$$

The constant k is a positive real number which can be greater than or less than unity.

Negative-impedance converters can be used to make RC circuits (circuits that contain only resistors and capacitors) behave like RL circuits. One circuit[3] that does this is shown in Fig. 17.2. For a current NIC it can be shown that

$$\left.\frac{V_2}{V_1}\right|_{I_2=0} = \frac{Y_2 - kY_1}{Y_2 + Y_4 - k(Y_1 + Y_3)} \tag{17.4}$$

The admittances Y_1, Y_2, Y_3, Y_4 are assumed to be made up of resistors and capacitors; however, because of the minus sign produced by the NIC, the overall transfer function can model the behavior of any circuit that uses resistors, capacitors, and inductors.

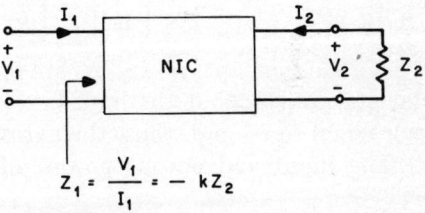

Fig. 17.1 Negative-impedance converter.

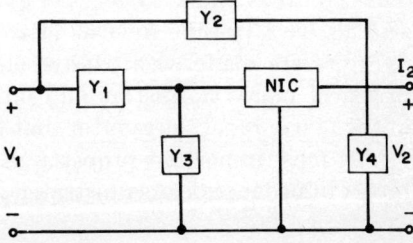

Fig. 17.2 Yanagisawa's circuit for synthesizing transfer functions with NICs.

Equation (17.4) embodies the two main reasons why negative-impedance-converter realizations are not very practical. The first is a consequence of the minus sign in the denominator. In a highly selective circuit, the denominator will have to be very small at the critical frequencies. If a small number is produced by the difference of two large numbers, the result is very sensitive to variations in the large numbers. This implies that NIC filters are very sensitive to element variations. A solution to this sensitivity problem is possible, but it demonstrates the other reason negative-impedance-converter realizations are not very practical.

One solution to the sensitivity problem is to realize a high-degree transfer function as a cascade of lower-order transfer functions. This can be symbolized as

$$T(s) = T_1(s)T_2(s)T_3(s) \cdots \quad (17.5)$$

This approach will be used in our discussions of operational-amplifier realizations; however, it is not a practical solution for NIC circuits because they cannot readily be cascaded. For example, the transfer function in (17.4) is only valid for $I_2 = 0$; that is, there can be no finite load impedance. Thus, to cascade such circuits, a buffer stage must be used. This buffering requirement is more than enough reason to make operational-amplifier active circuits more attractive than NICs.

17.3 GYRATOR ACTIVE FILTERS

We define a gyrator to be any two-port that satisfies the matrix equation

$$\begin{bmatrix} V_1 \\ V_2 \end{bmatrix} = \begin{bmatrix} 0 & -R_1 \\ R_2 & 0 \end{bmatrix} \begin{bmatrix} I_1 \\ I_2 \end{bmatrix} \quad (17.6)$$

where the constants R_1 and R_2 are positive real numbers. The important fact to notice about the impedance matrix for the gyrator is that z_{12} is not equal to z_{21} and, thus, the gyrator is a nonreciprocal two-port.

The input and output powers of the gyrator described by (17.6) are

$$\begin{aligned} P_1 &= V_1 I_1^* = -R_1 I_2 I_1^* \\ P_2 &= V_2 I_2^* = R_2 I_1 I_2^* \end{aligned} \quad (17.7)$$

Thus, if R_1 is equal to R_2, the gyrator dissipates no real power and is said to be a passive device. However, it should not be inferred that gyrators are made with passive elements. Practical gyrators are comprised of many transistors and do dissipate power internally; it is just at the two ports of the gyrator that the sum of the real power is zero.

The most important property of the gyrator is that it can be used to invert impedances. As illustrated in Fig. 17.3, if the output is terminated in an impedance Z_2, then the input impedance is

$$Z_1 = \frac{R_1 R_2}{Z_2} \tag{17.8}$$

The fact that a gyrator can be used as an impedance inverter implies that if the output is terminated in a capacitor, then the input impedance is that of an inductor. This leads to one simple way of designing active filters with gyrators: First design a prototype passive filter, and then replace each inductor with a gyrator terminated in a capacitor. D. F. Sheahan[4] has shown that insertion loss filters are very insensitive to element variation in their passband. If inductors are replaced with a gyrator-capacitor combination, the resulting active filter is insensitive, which is in sharp contrast to the highly sensitive NIC filters.

Active filters can also be constructed with gyrators by using the RC-RL partitioning method.[5,6] This is somewhat analogous to the NIC synthesis procedure discussed in the previous section. The NIC synthesis procedure was based on writing a transfer function in terms of the difference of two RC impedances. The NIC was then used to produce the negative of an RC impedance. In the RC-RL partitioning method the transfer function is written in terms of the sum of an RC and an RL impedance. The gyrator is used to produce the RL impedance from an RC impedance. Because the RC-RL partitioning method is based on producing the sum of two terms, it is less sensitive than the NIC procedure that produces the difference of two terms.

Even though active filters synthesized with gyrators are less sensitive than active filters synthesized with NICs, they still have some undesirable characteristics in comparison with operational-amplifier active filters. The main drawback is economical. Operational amplifiers are produced by the millions and are thus very inexpensive. Even before active filters became popular, operational amplifiers were produced in large quantities, and thus active filters have been able to exploit the fact that operational amplifiers are inexpensive. At the present time, gyrators have not yet been mass produced and are still expensive.

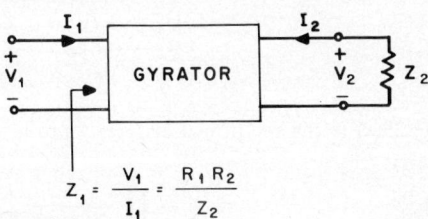

Fig. 17.3 A gyrator can be used as an impedance inverter.

324 Approximation Methods for Electronic Filter Design

Besides being less expensive, operational-amplifier active filters have other advantages: They can readily be cascaded without buffering, and they can easily provide gain at the same time they provide filtering. In special applications (such as when one may want a bilateral filter) the gyrator may prove superior; but for the foreseeable future the operational amplifier will see much more use in active filters. Thus the rest of this chapter concentrates on material that pertains to operational-amplifier active filters. For more information on gyrator active filters, one can consult any of the numerous available texts.[5–7] Also, J. A. Miller and R. W. Newcomb[8] have recently made available a very comprehensive bibliography.

17.4 SECOND-ORDER TRANSFER FUNCTIONS

Active-filter designers usually prefer to use the output/input transfer function $T(s)$ instead of the input/output transfer function $H(s)$ used by passive-filter designers. Since $T(s)$ is just the reciprocal of $H(s)$, this presents no difficulty. The following example demonstrates that we can easily express $T(s)$ in terms of our solution to the approximation problem.

Example 17.1
Find the output/input transfer function $T(s)$ for the lowpass elliptic filter described by $A_{max} = 0.1$ dB, $A_{min} = 40$ dB, FB = 20, FH = 26.

Solution
We could use the material in Chap. 5 to determine the attenuation poles, but it is just as easy to use the pole-placer program. As demonstrated in Fig. 8.16, choosing the attenuation poles as below yields an elliptic filter:

$$F_1 = 26.58 \quad F_2 = 33.28 \quad F_3 = 82.57$$

The natural modes [zeros of $H(s)$] can be found by using the zero-finder program as demonstrated in Fig. 10.5; this yields

$$ZQ_1 = 7.88 \quad ZF_1 = 20.83$$
$$ZQ_2 = 1.79 \quad ZF_2 = 18.33$$
$$ZQ_3 = 0.625 \quad ZF_3 = 12.98$$
$$C_H = 220.4$$

Since $T(s)$ is the reciprocal of $H(s)$, it follows that

$$T(s) = \frac{1}{220.4} \prod_{i=1}^{3} \frac{s^2 + (2\pi F_i)^2}{s^2 + (2\pi ZF_i/ZQ_i)s + (2\pi ZF_i)^2} \quad (17.9)$$

where F_i, ZF_i, and ZQ_i are as given above.

The form of $T(s)$ that was given in Example 17.1 is very convenient if we want to synthesize $T(s)$ as an active filter. Active filters are usually synthesized by cascading second-order transfer functions because this reduces the sensitivity problem. That is, if one realized the sixth-order filter of Example 17.1 with only one operational amplifier, the resulting network would be very sensitive to element variations; for example, if the characteristics of the operational amplifier changed only slightly, the filter characteristic could change drastically. To make the filter less sensitive, one would probably use three operational amplifiers for this elliptic filter: one for each second-order transfer function. In later sections we shall examine actual circuits that can be used for the second-order filter sections, and we shall see that the filter-section outputs are voltage sources. Since the output of a voltage source is not affected by its load, second-order sections can be cascaded without interaction problems.

Because active filters are usually comprised of various second-order sections, it will be helpful to study the different possibilities. The remaining part of this section examines the common second-order sections.

Lowpass

The lowpass second-order transfer function is defined to be of the form

$$T(s) = \frac{C}{s^2 + (\omega_p/Q)s + \omega_p^2} \tag{17.10}$$

The constant C determines the level (gain or loss) of the second-order section. As mentioned in Sec. 2.8, ω_p is called the *undamped natural frequency* of the pole, and Q is the *quality* of the pole. To see how these affect the transfer function $T(s)$, we can examine its magnitude

$$|T(j\omega)|^2 = \frac{C^2}{(\omega^2 - \omega_p^2)^2 + (\omega\omega_p/Q)^2} \tag{17.11}$$

The undamped natural frequency ω_p helps to determine where the maximum of $|T(j\omega)|^2$ occurs. The magnitude of this maximum is determined by the quality of the pole Q_p. This is best illustrated by considering the normalized function

$$\left|\frac{T(j\omega)}{T(0)}\right|^2 = \frac{\omega_p^4}{(\omega^2 - \omega_p^2)^2 + (\omega\omega_p/Q)^2} \tag{17.12}$$

which is plotted in Fig. 17.4.* From the figure it is obvious why we call it a lowpass function: it passes low frequencies (the normalized dc attenua-

---

* In Fig. 17.4 the frequency scale has been normalized so that $\omega_p = 1$.

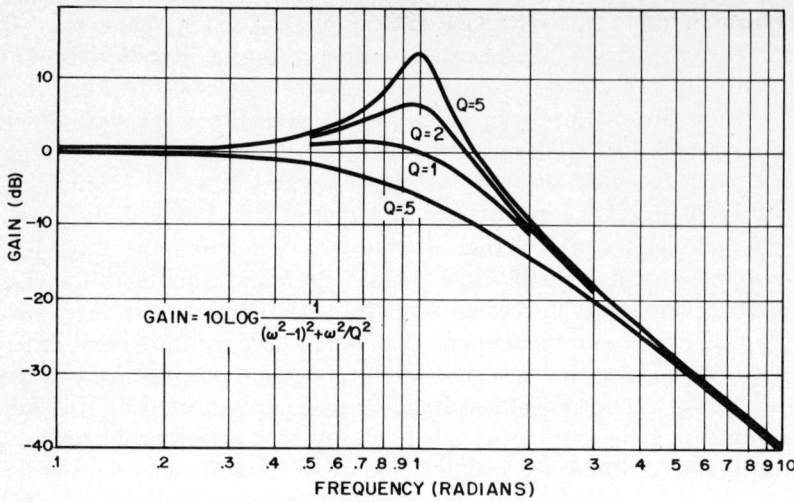

Fig. 17.4 Gain of normalized second-order lowpass functions.

tion is 0 dB) and attenuates high frequencies (at an asymptotic slope of 40 dB/decade).

Bandpass

The bandpass second-order transfer function is defined to be of the form

$$T(s) = \frac{Cs}{s^2 + (\omega_p/Q)s + \omega_p^2} \qquad (17.13)$$

This function was discussed in some detail in Sec. 2.8, where it was noted that the magnitude of $T(j\omega)$ has a maximum value at ω_p. If we normalize with respect to this maximum value, we can write

$$\left| \frac{T(j\omega)}{T(j\omega_p)} \right|^2 = \frac{(\omega\omega_p/Q)^2}{(\omega^2 - \omega_p^2)^2 + (\omega\omega_p/Q)^2} \qquad (17.14)$$

This normalized function is plotted (for $\omega_p = 1$) in Fig. 17.5. From the figure, it is obvious why it is called a bandpass function: it passes a band of frequencies centered about ω_p and attenuates the low and high frequencies (with an asymptotic slope of 20 dB/decade).

Highpass

The highpass second-order transfer function is defined to be

$$T(s) = \frac{Cs^2}{s^2 + (\omega_p/Q)s + \omega_p^2} \qquad (17.15)$$

This is very similar to the lowpass second-order transfer function: the low-frequency and high-frequency behaviors have been interchanged. Thus, for the highpass section, we normalize the magnitude response with respect to the infinite-frequency response:

$$\left|\frac{T(j\omega)}{T(\infty)}\right|^2 = \frac{\omega^4}{(\omega^2 - \omega_p^2)^2 + (\omega\omega_p/Q)^2} \qquad (17.16)$$

The behavior of this function (for $\omega_p = 1$) can be obtained from Fig. 17.4 by relabeling the abscissa to be $1/\omega$. This indicates that the highpass function does not attenuate high frequencies but does attenuate low frequencies with an asymptotic slope of 40 dB/decade.

Frequency Rejection

Another very important second-order transfer function is the frequency-rejection function

$$T(s) = C \frac{s^2 + \omega_0^2}{s^2 + (\omega_p/Q)s + \omega_p^2} \qquad (17.17)$$

The frequency ω_0 determines the location of the transmission zero; at ω_0 the frequency is completely "rejected." In terms of the notation used earlier in this book, ω_0 is an attenuation pole.

The magnitude of this transfer function is given by

$$|T(j\omega)|^2 = C^2 \frac{(\omega^2 - \omega_0^2)^2}{(\omega^2 - \omega_p^2)^2 + (\omega\omega_p/Q)^2} \qquad (17.18)$$

This function has a maximum at

$$\omega_m = \omega_p \left[\frac{1 - (\omega_0/\omega_p)^2(1 - 1/2Q^2)}{(1 - 1/2Q^2) - (\omega_0/\omega_p)^2}\right]^{1/2} \qquad (17.19)$$

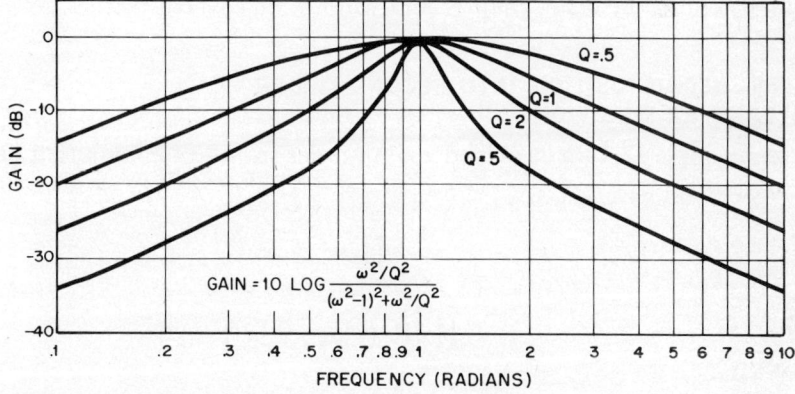

Fig. 17.5 Gain of normalized second-order bandpass functions.

328 Approximation Methods for Electronic Filter Design

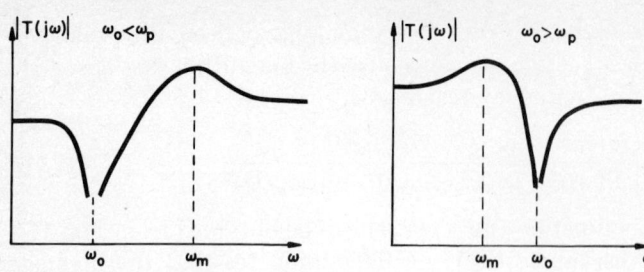

Fig. 17.6 Two different types of frequency-rejection functions.

There are two cases to be considered: $\omega_0 > \omega_p$ and $\omega_0 < \omega_p$. These cases are sketched in Fig. 17.6. The case $\omega_0 < \omega_p$ would be used if we wanted the transmission zero ω_0 to be in the lower stopband of a bandpass filter. The transmission zero produces attenuation in the passband; the transmission pole ω_p is in the passband and thus (in the passband) offsets the effect of the transmission zero.

The case $\omega_0 > \omega_p$ would be used if we wanted the attenuation pole to be in the upper stopband. Again this produces attenuation in the passband which is offset in the passband by the "gain bump." The magnitude of the gain bump is influenced by two items: (1) the quality Q of the transmission pole, and (2) the distance between ω_0 and ω_p. For a given distance between ω_0 and ω_p, the larger the quality Q the larger is the gain bump. Similarly, for a given quality Q, the greater the distance between ω_0 and ω_p the larger is the gain bump.

We have just examined four second-order transfer functions that are commonly encountered in the synthesis of active filters: the lowpass, bandpass, highpass, and frequency-rejection functions. In Chap. 14 we examined another function, the allpass function. That was adequately discussed in the delay chapter and will not be covered further here.

17.5 DECOMPOSITION INTO SECOND-ORDER TRANSFER FUNCTIONS[15,16]

In Example 17.1 we examined a sixth-order transfer function that can be written as

$$T(s) = K \frac{\prod_{i=1}^{3} (s^2 + \omega_{0i}^2)}{\prod_{j=1}^{3} [s^2 + (\omega_{pj}/Q_j)s + \omega_{pj}^2]} \qquad (17.20)$$

It was indicated that this will be realized as a cascade connection of second-order circuits. One of the second-order functions will be denoted as

$$T_j(s) = K_j \frac{s^2 + \omega_{0i}^2}{s^2 + (\omega_{pj}/Q_j)s + \omega_{pj}^2} \tag{17.21}$$

This innocuous-seeming equation has some profound implications.

First we should note that the pole ω_{pj} and zero ω_{0i} do not have the same subscript. This implies that we can group the poles and zeros in many different ways. For example, the poles ω_{p1}, ω_{p2}, ω_{p3} and the zeros ω_{01}, ω_{02}, ω_{03} could be grouped as

$$(\omega_{p1}, \omega_{01}) \quad (\omega_{p2}, \omega_{02}) \quad (\omega_{p3}, \omega_{03})$$
or as
$$(\omega_{p1}, \omega_{02}) \quad (\omega_{p2}, \omega_{01}) \quad (\omega_{p3}, \omega_{03})$$

just to demonstrate two different possibilities.

The constants K_j in (17.21) are subject to the constraint that

$$K_1 K_2 K_3 = K$$

but other than that the individual K_j are arbitrary. For example, if we want to increase K_1 we could decrease K_3, as long as the product remained constant.

Even when we have completely determined the second-order functions $T_j(s)$, we still have some freedom left in the realization—the sequence of cascading. As an example, two different possibilities for a sixth-order transfer function are shown in Fig. 17.7.

We have just seen that there are different possibilities as to how we choose the pole-zero pairs, how we choose the gain constants, and how we choose the cascade sequence. We can use the various degrees of freedom to provide an "optimum" realization for $T(s)$. What is meant by optimum will depend on the specific example. For instance, the active filter might have to process signals of large amplitude. The input signal $V_{\text{in}}(j\omega)$ and the output signal $V_o(j\omega)$ would be specified, but the signals interior to the active filter would not. The signal level at the output of each second-order function $T_j(s)$ would be influenced by (1) the grouping of pole-zero pairs, (2) the scale factors K_j, and (3) the cascade sequence. Thus there is some optimum realization for $T(s)$ as far as overload considerations are concerned.

If the input signal is very small, we would be concerned about degradation due to noise. This could present another "optimum" realization, different from the one produced by considering overload. Finally, if

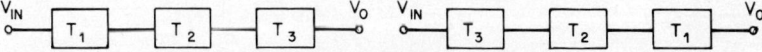

Fig. 17.7 Second-order transfer functions can be cascaded in different orders.

we considered the sensitivity of the realization, we might get a third "optimum" solution.

In any specific optimization problem we shall proceed as follows: First we shall pick pole-zero pairs, next we shall select the constant multipliers K_j, and finally we shall select the cascade sequence. It should be noted that this process does not guarantee an optimum solution. For example, the optimum set of K_j might be a function of the cascade sequence. However, the approach of picking pole-zero pairs, then K_j, and then the cascade sequence has the advantage of separating the problem into discrete parts. If the resulting solution is not good enough, then we can use it as the starting point for an iterative procedure which would vary the pole-zero pairs, the constant multipliers, and the cascade sequence.

To illustrate how the various degrees of freedom can be utilized, we shall assume that in our particular active-filter problem we are most concerned about overload. We shall thus try to make the output of each second-order section as small as possible.

Example 17.2

This example is a continuation of the work started in Example 17.1, which found the transfer function

$$T(s) = \frac{1}{220.4} \prod_{i=1}^{3} \frac{s^2 + \omega_{0i}^2}{s^2 + (\omega_{pi}/Q_i)s + \omega_{pi}^2}$$

where $\omega_{01} = 2\pi \times 26.58$ $\omega_{02} = 2\pi \times 33.28$ $\omega_{03} = 2\pi \times 82.57$
$Q_1 = 0.625$ $\omega_{p1} = 2\pi \times 12.98$
$Q_2 = 1.79$ $\omega_{p2} = 2\pi \times 18.33$
$Q_3 = 7.88$ $\omega_{p3} = 2\pi \times 20.83$

In the previous section we discussed second-order frequency-rejection functions and mentioned that the larger the quality of the pole, the larger the gain bump. Thus the second-order section that has the highest Q ($Q_3 = 7.88$) will tend to overload the most. We can minimize the magnitude of the gain bump by making the zero of this second-order section be as close to the pole as possible. For this reason we shall group the zero ω_{01} with the pole ω_{p3}. Similarly we shall group ω_{02} with ω_{p2}, and ω_{03} with ω_{p1}. Thus the second-order functions we want to realize are

$$T_1(s) = K_1 \frac{s^2 + \omega_{03}^2}{s^2 + (\omega_{p1}/Q_1)s + \omega_{p1}^2}$$

$$T_2(s) = K_2 \frac{s^2 + \omega_{02}^2}{s^2 + (\omega_{p2}/Q_2)s + \omega_{p2}^2}$$

$$T_3(s) = K_3 \frac{s^2 + \omega_{01}^2}{s^2 + (\omega_{p3}/Q^3)s + \omega_{p3}^2}$$

where all parameters except K_1, K_2, K_3 were given at the beginning of this problem.

Because we are concerned about overload in this example, the gain constants K_j were chosen such that each section has the same maximum amplitude. From (17.19) this maximum is located at

$$\omega_m = \omega_p \left[\frac{1 - (\omega_0/\omega_p)^2(1 - 1/2Q^2)}{(1 - 1/2Q^2) - (\omega_0/\omega_p)^2} \right]^{1/2}$$

Applying this to each of the second-order sections yields*

$$\omega_{m1} = 0 \qquad \omega_{m2} = 2\pi \times 15.63 \qquad \omega_{m3} = 2\pi \times 20.48$$

Substituting these values into the second-order functions leads to

$$|T_1(j\omega_{m1})| = 40.466 K_1 \qquad |T_2(j\omega_{m2})| = 4.68 K_2 \qquad |T_3(j\omega_{m3})| = 5.123 K_3$$

In order that these maximums all have the same value and, at the same time, satisfy $K_1 K_2 K_3 = 1/220.4$, we must have

$$K_1 = 0.0405 \qquad K_2 = 0.350 \qquad K_3 = 0.320$$

Thus far, we have determined the pole-zero grouping and the constant multipliers. The cascade sequence can be chosen with the aid of Table 17.1. For each possible cascade sequence, the table identifies the filter section T_m that has the maximum output, the frequency F_m at which this occurs, and the maximum gain G_m produced by the cascade sequence. For example, suppose a 0-dB signal is applied to the cascade sequence $T_2 T_3 T_1$. Table 17.1 indicates that a maximum signal level occurs if the signal frequency is $f = 20.05$. For this frequency the level at the output of section T_3 would be $+5.12$ dB.

Table 17.1 indicates that either cascade sequence $T_1 T_3 T_2$ or $T_3 T_1 T_2$ could be used to optimize the overload performance.

17.6 SOME PRACTICAL ACTIVE CIRCUITS

Active filters are usually synthesized by cascading second-order filter sections. There have been many different circuits described in the

* This formula was not valid for $T_1(s)$ because its maximum is at the boundary edge $s = 0$.

TABLE 17.1 Aid for Determining Cascade Sequence

Cascade sequence	T_m	F_m	G_m, dB
$T_1 T_2 T_3$	T_2	0	5.53
$T_1 T_3 T_2$	T_1	0	4.29
$T_2 T_1 T_3$	T_1	0	5.53
$T_2 T_3 T_1$	T_3	20.05	5.12
$T_3 T_1 T_2$	T_3	20.48	4.29
$T_3 T_2 T_1$	T_2	20.05	5.12

literature; only a few will be discussed here. These circuits have been chosen to represent practical configurations that can be used.

The first circuit we shall study uses three operational amplifiers, while the others need only one. This raises the question, "Why use three operational amplifiers when it is possible to use just one?" The reason is that the additional amplifiers offer us quite a few things: a very versatile network, an insensitive network, and an easily adjusted network.

The three-operational-amplifier circuit will be a generalization of the circuit shown in Fig. 17.8, which will be discussed first because it is simplest.* For ease of analysis the operational amplifiers are assumed to be ideal devices, which implies that they have infinite input impedance and infinite voltage gain. This results in a concept termed *virtual ground*:[9] The negative input terminal of the operational amplifier is at the same potential as the positive input terminal and is thus said to be virtual ground.

The principle of virtual ground can be applied to help analyze the circuit in Fig. 17.8. For example, if we use the fact that the negative input terminal of amplifier A_2 is at ground potential (along with the fact that no current can flow into this terminal because of infinite input impedance), then it is easy to show that

$$\frac{V_2}{V_4} = -\frac{R_8}{R_7} \tag{17.22}$$

Amplifier A_2 is said to be used as an *inverting* amplifier; it produces a voltage gain which is proportional to the ratio of R_8 to R_7, and it inverts the sign of the input voltage.

* Circuits similar to this one have been used on analog computers for a long time. Some references pertinent to this approach are included at the end of the chapter.[10,11]

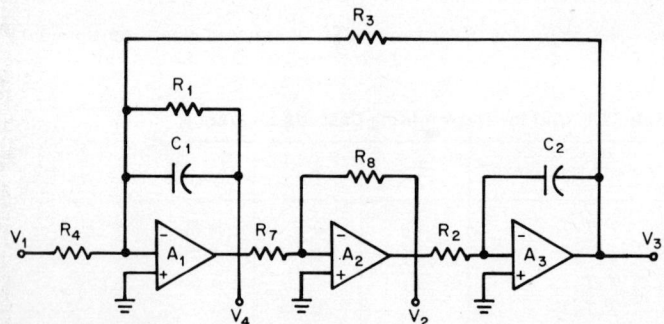

Fig. 17.8 "Analog-computer" type of active filter, noted for its excellent sensitivity performance.

Amplifier A_3 is an *integrator* because its operation can be described by

$$\frac{V_3}{V_2} = -\frac{1}{sR_2C_2} \qquad (17.23)$$

That is, the Laplace operation $1/s$ corresponds to integration in the time domain.

The first operational amplifier acts as a *summing device;* its output V_4 is a function of V_1 and V_3. Summing currents at its input terminal yields

$$\frac{V_1}{R_4} + \frac{V_3}{R_3} + \frac{sC_1R_1 + 1}{R_1}V_4 = 0 \qquad (17.24)$$

The three equations given above can be solved to yield a transfer function for the circuit in Fig. 17.8. The result is

$$\frac{V_3}{V_1} = \frac{-R_8/R_2R_4R_7C_1C_2}{s^2 + (1/R_1C_1)s + R_8/R_2R_3R_7C_1C_2} \qquad (17.25)$$

This transfer function is in the form of the general second-order lowpass function

$$T(s) = \frac{C}{s^2 + (\omega_p/Q_p)s + \omega_p^2} \qquad (17.26)$$

The lowpass transfer function was obtained by using V_3 as the output voltage. If V_2 is instead defined to be the output, then

$$\frac{V_2}{V_1} = \frac{(R_8/R_4R_7C_1)s}{s^2 + (1/R_1C_1)s + R_8/R_2R_3R_7C_1C_2} \qquad (17.27)$$

This is a second-order bandpass transfer function.

The transfer functions in (17.25) and (17.27) both have the same denominator. The reason for this is that the denominator determines the network's natural frequencies, and these are not affected by what we call the output. In fact, the natural frequencies are not even affected by the input resistor R_4; thus, if we add other input resistors as shown in Fig. 17.9, the natural frequencies will still be the same.

The resistors R_5 and R_6 are *feedforward* resistors;[10] they feed the input voltage V_1 forward to amplifiers A_3 and A_2, where the voltage is summed along with the other voltages.* The result of this summation is indicated by the transfer function

$$-\frac{V_2}{V_1} = \frac{R_8}{R_6}\frac{s^2 + (1/R_1C_1)(1 - R_1R_6/R_4R_7)s + R_6/R_3R_5R_7C_1C_2}{s^2 + (1/R_1C_1)s + R_8/R_2R_3R_7C_1C_2} \qquad (17.28)$$

* This circuit is often called a *feedforward biquad.*

334 Approximation Methods for Electronic Filter Design

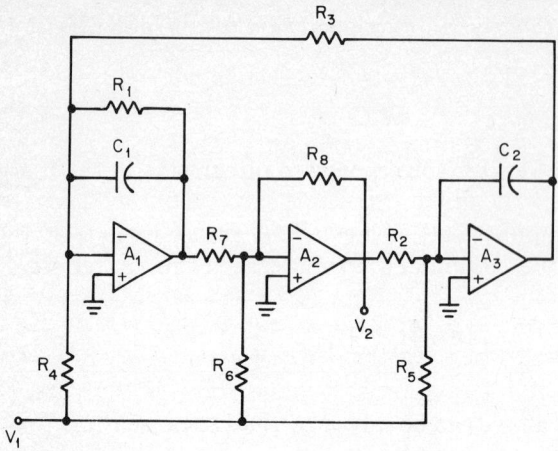

Fig. 17.9 Feedforward circuit that can be used to produce any type of second-order transfer function.

With this one circuit it is possible to adjust the feedforward resistors so as to produce any of the second-order filter sections that we generally encounter in active filter synthesis. If R_4 and R_6 are replaced by open circuits, then the above transfer function represents a lowpass second-order function [this can be seen by letting R_4 and R_6 approach infinity in Eq. (17.28)]. A bandpass function can be produced by replacing R_5 and R_6 with open circuits, whereas a highpass function can be produced by replacing R_5 with an open circuit and requiring that

$$\frac{R_1 R_6}{R_4 R_7} = 1 \tag{17.29}$$

If R_5 is not an open circuit, but the resistors are still adjusted so that (17.29) is valid, then the circuit produces a second-order frequency-rejection function.

Example 17.3

In Example 17.2, one of the second-order transfer functions was*

$$T_3(s) = 0.32 \frac{s^2 + (2\pi \times 2{,}658)^2}{s^2 + (2\pi \times 2{,}083/7.88)s + (2\pi \times 2{,}083)^2} \tag{17.30}$$

A set of element values for the circuit in Fig. 17.9 can be found by equating the coefficients in (17.28) with those in (17.30). Since there are more available

---

* The frequencies have been scaled by 100 in order to make them of more practical interest.

parameters than unknowns, many of the elements can be chosen arbitrarily. These degrees of freedom can be utilized so that reasonable element values are used in the network. A possible set of element values for this example is

$$R_1 = 30104.2 \quad R_2 = 3820.33 \quad R_3 = 1910.16 \quad R_4 = 47037.8$$
$$R_5 = 7331.95 \quad R_6 = 3125 \quad R_7 = 2000 \quad R_8 = 1000$$
$$C_1 = C_2 = 0.02$$

where the resistor values are in ohms, and the capacitor values are in microfarads.

In the example just presented the pole Q was only 7.88, and thus the circuit would be very insensitive to variations due to nonideal operational amplifiers. If the pole Q were instead high (of the order of 100), then the circuit would be much more sensitive. This sensitivity can be minimized by utilizing the degrees of freedom that are available in the selection of element values.

For a low-Q pole we can use a circuit that has only one operational amplifier instead of three as in Fig. 17.9. Thus (for reasons of economics) the frequency-rejection function in Example 17.3 would not usually be realized with a three-operational amplifier circuit. We shall now discuss a more practical low-Q circuit.

The circuit to be discussed is a modification of a bandpass circuit described by T. Deliyannis.[13] This modification was done by J. J. Friend, and the following material closely follows his description of that work.[14]

Figure 17.10 shows the bandpass circuit proposed by Deliyannis. It was desired to modify this circuit to allow a generalized numerator in the transfer function without changing the admittance matrix and, consequently, the low-sensitivity characteristics. Figure 17.11 shows the addition of two inputs (K_1, K_2, and K_a are numbers between 0 and 1). The one labeled $K_2 i_s$ was necessary for generality but has changed the admittance matrix by the addition of G_3. The three-input circuit may be

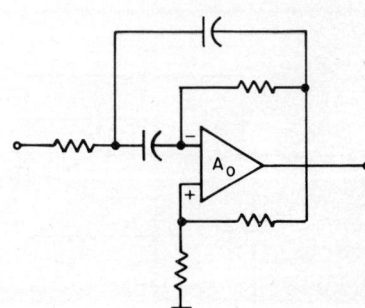

Fig. 17.10 Bandpass circuit of T. Deliyannis.

realized by a single-input circuit as shown in Fig. 17.12. There are four characteristics of this circuit worth noting. Firstly, the input configuration, determined by the magnitudes of K_1, K_2, and K_a, affects the numerator (zeros) only. Secondly, the amplifier gain A_0 affects the denominator (poles) only. Thirdly, the gain at infinite frequency K_a is constrained to be between zero and unity. Lastly, this circuit is not capable of producing a lowpass function because of its inability to provide a -12-

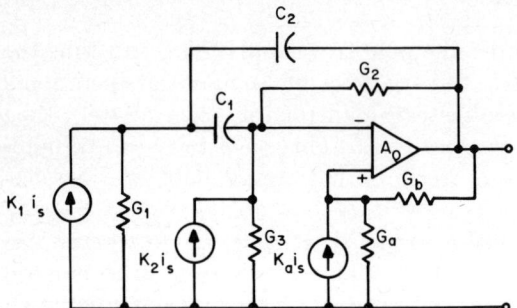

Fig. 17.11 Expanded circuit for the realization of a generalized biquadratic transfer function.

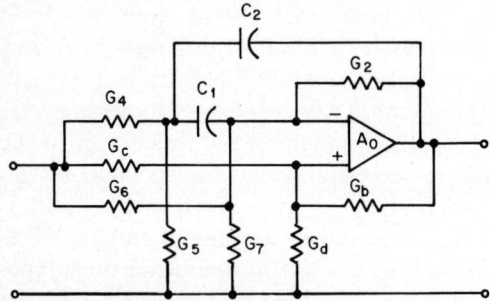

Fig. 17.12 Final generalized circuit configuration for all cases except the lowpass case.

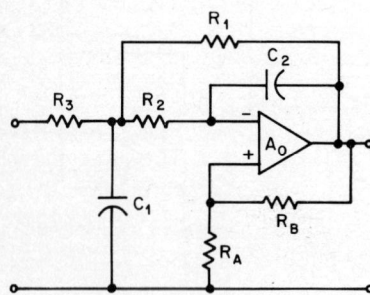

Fig. 17.13 Circuit for the realization of the lowpass transfer function.

dB/octave asymptote at high frequencies. For this particular case, the circuit of Fig. 17.13 is used. However, the circuit shown in Fig. 17.12 can be used to produce bandpass, highpass, and frequency rejection functions.

The elements of the circuit in Fig. 17.12 have the following relations to the variables shown in Fig. 17.11:

$$K_a = \frac{G_c}{G_a} \qquad G_a = G_c + G_d$$

$$K_1 = \frac{G_4}{G_1} \qquad G_1 = G_4 + G_5 \qquad (17.31)$$

$$K_2 = \frac{G_6}{G_3} \qquad G_3 = G_6 + G_7$$

Each K is a ratio of two conductances, and each must have a value between zero and unity. The transfer function of this circuit can be written as

$$T(s) = \frac{E_0}{E_s} = \frac{K_a s^2 + Es + D}{s^2 + As + B}$$

where
$$A = \frac{(C_1 + C_2)(G_a G_2 - G_b G_3) - C_1 G_1 G_b}{C_1 C_2 G_a} \qquad (17.32)$$

$$B = \frac{G_1(G_a G_2 - G_b G_3)}{C_1 C_2 G_a} \qquad (17.33)$$

$$D = \frac{G_1 G_a K_a (G_2 + G_3) - G_1 G_3 K_2 (G_a + G_b)}{C_1 C_2 G_a} \qquad (17.34)$$

$$E = \frac{C_1 G_a K_a (G_1 + G_2 + G_3) + C_2 G_a K_a (G_2 + G_3) - C_1 G_1 K_1 (G_a + G_b) - (C_1 + C_2) G_3 K_2 (G_a + G_b)}{C_1 C_2 G_a} \qquad (17.35)$$

An ideal operational amplifier was assumed; thus its gain is infinite and does not appear in the above equations.

Note that the K's affect only the numerator (or zeros) of the transfer function. K_a is the magnitude of the gain at infinite frequency, and D/B determines the gain at dc. The solution to these equations with the constraints imposed by the K's is not unique, since the number of unknowns is larger than the number of equations. The approach used here is to select arbitrarily some of the element values, and then solve for the remaining ones.

The four coefficient equations (17.32) to (17.35) were solved in terms of the unknowns, so that, using the above input parameters, one could solve for the elements sequentially. These step-by-step design equations are

$$G_1 = \frac{C_2 G_a}{2G_b}\left\{-A + \left[A^2 + 4\left(1 + \frac{C_1}{C_2}\right)B\frac{G_b}{G_a}\right]^{1/2}\right\} \quad (17.36)$$

$$K_1 = \frac{K_a + (1 + C_1/C_2)DC_2{}^2/G_1{}^2 - EC_2/G_1}{1 + G_b/G_a} \quad (17.37)$$

$$G_3 = \frac{C_1 C_2 G_a B(D/B - K_a)}{G_1(G_a + G_b)(K_a - K_2)} \quad (17.38)$$

$$G_2 = \frac{C_1 C_2 B}{G_1} + \frac{G_b G_3}{G_a} \quad (17.39)$$

Assuming that K_1 from Eq. (17.37) was found to be between zero and unity, we solve the next two equations for G_3 and G_2. The rest of the resistors are solved for from the relations in Eq. (17.31).

As previously mentioned, the circuit in Fig. 17.12 can be used to realize second-order bandpass, highpass, or frequency-rejection functions. We now describe the program in Fig. 17.14, which can be used to design networks for the frequency-rejection function

$$T(s) = K\frac{s^2 + (2\pi ZF)^2}{s^2 + (2\pi PF/Q)s + (2\pi PF)^2} \quad (17.40)$$

Similar programs could be written for bandpass or highpass functions.

Step 1.1 asks for the constants K, ZF, PF, and Q which are identified in (17.40). These constants are then translated into the parameters A, B, D, and KA which are identified in (17.32) to (17.35). It should be noted that $E = 0$ because we are studying the frequency-rejection case.

Next, the constant K_2 is set equal to 0 or 1, depending on whether the zero frequency is greater than the pole frequency, or the pole frequency is greater than the zero. This provides a minimum value for G_3 which improves the circuit's sensitivity performance.

There is not a unique set of parameter values for the circuit, so some must be chosen arbitrarily. The first arbitrary choice is for R_c, which determines the impedance level of the circuit. Once R_c has been chosen, R_d can be calculated. Next, R_b, C_1, and C_2 are arbitrarily chosen. The remaining elements are then calculated with the help of Eqs. (17.36) to (17.39). If the element values are not reasonable, then new values should be chosen for R_b, C_1, and C_2. These choices also determine how sensitive the circuit is.

An example of the use of the program is given at the bottom of Fig. 17.14. The frequency-rejection function realized in Fig. 17.14 is the same one that was previously realized in Example 17.3.

Approximation Methods and Active Filter Synthesis

```
1.0; J.J. FRIEND FREQUENCY REJECTIØN NETWØRK
1.1 DEMAND K,ZF,PF,Q
1.11 TYPE #,#,#
1.12 A=2*$PI*PF/Q,B=(2*$PI*PF)^2,D=K*(2*$PI*ZF)^2,KA=K
1.2 K2=0
1.21 K2=1 IF PF>ZF
1.31 DEMAND RC
1.32 GC=1/RC,GA=GC/KA,GD=GA-GC,RD=1/GD
1.33 TYPE RD
1.41 DEMAND RB,C1,C2
1.42 GB=1/RB
1.5 G1=C2*GA/2/GB*(-A+SQRT(A*A+4*(1+C1/C2)*B*GB/GA))
1.6 K1=(KA+(1+C1/C2)*D*(C2/G1)^2)/(1+GB/GA)
1.62 G4=K1*G1,G5=G1-G4,R4=1/G4,R5=1/G5
1.7 G3=C1*C2*GA*B*(D/B-KA)/(G1*(GA+GB)*(KA-K2))
1.8 G6=K2*G3,G7=G3-G6
1.81 TYPE R6 FØR R6=1/G6 IF G6>0
1.82 TYPE R7 FØR R7=1/G7 IF G7>0
1.9 G2=C1*C2*B/G1+GB*G3/GA,R2=1/G2
1.91 TYPE R2,R4,R5
>


       DØ PART 1
            K=.32
            ZF=2658
            PF=2083
            Q=7.88



            RC=3000
            RD=      1411.76471
            RB=3000
            C1=10^-8
            C2=5*10^-8
            R7=    28065.2546
            R2=    11592.686
            R4=     2199.47033
            R5=     1450.44955
>
```

Fig. 17.14 Program that can be used to design frequency rejection networks.

It was mentioned that the circuit in Fig. 17.12 could not be used to realize lowpass functions. For this, one can use the network in Fig. 17.13. The design equations for this circuit are

$$\frac{V_2}{V_1} = \frac{-D}{s^2 + As + B}$$

$$G_2 = \frac{AC_1 \pm [A^2C_1^2 - 4(1 - G_bC_1/G_aC_2)(B + D)C_1C_2]^{1/2}}{2(1 - G_bC_1/G_aC_2)}$$

$$G_1 = \frac{C_1C_2(B + DG_b/G_a + G_b)}{G_2}$$

$$G_3 = \frac{DC_1C_2G_a}{G_2(G_a + G_b)} \tag{17.41}$$

17.7 COUPLED ACTIVE FILTERS

Until recently, most active filters have been synthesized by cascading second-order transfer functions:

$$T_i(s) = \frac{A_i s^2 + B_i s + C_i}{s^2 + (\omega_i/Q_i)s + \omega_i^2} \tag{17.42}$$

to form the overall transfer function

$$T(s) = \prod_{i=1}^{m} T_i(s) \tag{17.43}$$

The transfer functions can be readily cascaded because the popular second-order sections have operational amplifiers at their outputs.

Simply cascading second-order sections can result in sensitive designs. To see why this is true consider the following sixth-order bandpass filter:

$A_1 = 0.36$ $A_2 = 0$ $A_3 = 0.317$
$B_1 = 0$ $B_2 = 1{,}315$ $B_3 = 0$
$C_1 = 47{,}140{,}200$ $C_2 = 0$ $C_3 = 61{,}578{,}900$
$Q_1 = 31.78$ $Q_2 = 13.36$ $Q_3 = 31.78$
$\omega_1 = 2\pi 1{,}938.8$ $\omega_2 = 2\pi 2{,}010$ $\omega_3 = 2\pi 2{,}083.8$

The response of each of the three second-order transfer functions is shown in Fig. 17.15, which also shows the passband response of the cascaded transfer function.

Now consider what happens if element variations cause Q_1 to decrease by 1 percent and Q_2 to increase by 1 percent (Q_3 will be assumed constant). The result is shown in Fig. 17.16. The gain bumps due to Q_1 and Q_2 shifted in amplitude, but the passband response was not badly distorted.

However, next examine the consequence of element variations that cause ω_1 to decrease by 1 percent and ω_2 to increase by 1 percent. This result is also in Fig. 17.16. The ω_1 and ω_2 gain bumps shifted, causing a large distortion in the passband response.

Generalizing, it can be shown that transfer functions realized by cascading second-order sections are a factor of Q more sensitive to frequency shifts than to Q shifts. The amount of Q shift can be controlled by using a good second-order section such as a feedforward biquad. However, the pole frequencies are determined by RC products. Thus the available technology usually determines the frequency shift of a second-order section.

What can be done about this susceptibility to frequency shifts? We can turn to passive filters for the solution. It is well known that lossless ladder networks have passbands that are relatively insensitive to element variations. This is because if an element shifts, not just one natural frequency but all natural frequencies are affected, that is, the pole shifts are coupled together, helping to avoid large passband distortion as in Fig. 17.16.

With this as motivation, Girling and Good[19] simulated a lowpass ladder of the form shown in Fig. 17.17a by active filters. To see how this was done, consider the equations in Fig. 17.17b, which describe the ladder network. These equations also describe the active RC structure of Fig. 17.17c. The terms $1/s$ correspond to ideal integrators, and the terms $1/(s + a)$ correspond to lossy integrators.

The feedback paths identified by -1 couple together the active filter sections. Because the passive filter and coupled active filter are both described by the same set of equations, they are equally insensitive to element variations.

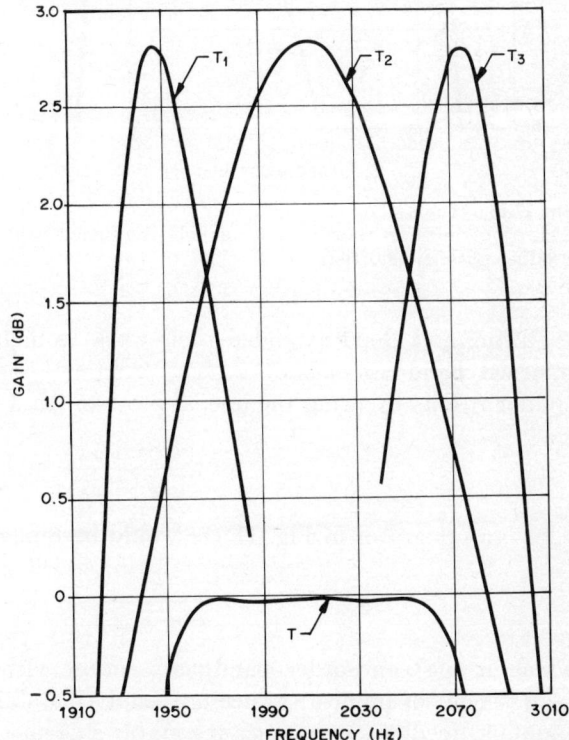

Fig. 17.15 Three second-order transfer functions T_1, T_2, T_3 and the composite bandpass function T.

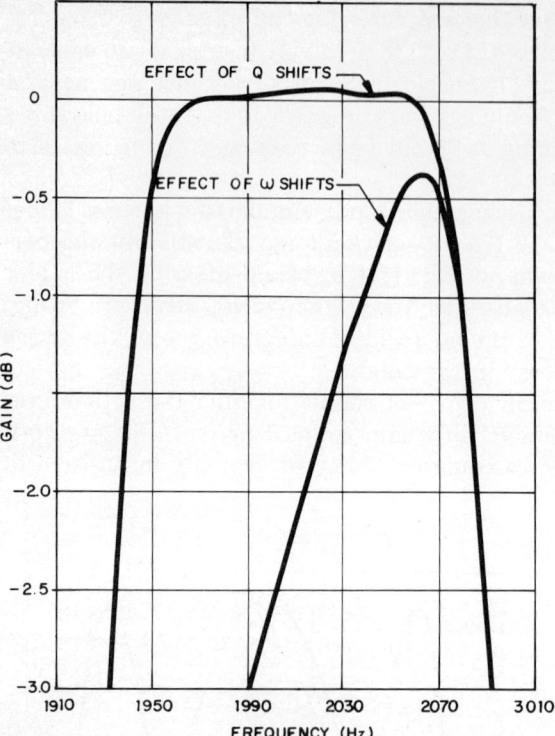

Fig. 17.16 Small shifts in pole quality produce negligible distortion; however, small shifts in pole frequency can produce drastic distortion.

Girling and Good extended their work to include geometrically symmetrical bandpass networks also. Adams[17] independently arrived at similar results by using the lowpass-to-bandpass transformation

$$S \to \frac{s^2 + \omega_A \omega_B}{(\omega_B - \omega_A)s} \qquad (17.44)$$

Thus an integrator in Fig. 17.17c would be replaced by

$$\frac{ks}{s^2 + \omega_A \omega_B} \qquad (17.45)$$

which is a second-order bandpass section with infinite pole Q. The requirement of infinite Q does not imply instability because the overall coupling (feedback) guarantees a stable system.

Using the standard lowpass-to-bandpass transformation yields filter responses that are geometrically symmetrical. Szentirmai[20] was able to remove this restriction by developing an active filter synthesis procedure that is completely analogous to the zero-shifting technique used in the synthesis of passive ladder filters. Szentirmai's structure, which is shown in Fig. 17.18, is a generalization of the circuit of Fig. 17.17c. The functions T_i are second-order transfer functions* that can be realized by any of the standard active filter sections such as feedforward biquads or single op-amp biquads. In either case it is possible to use the op-amp to produce both the second-order transfer function and the summing node.

The coupled active filters of the Girling and Good topology have been termed "leapfrog" active filters. Other types of coupled active filters also exist. Gunnar Hurtig[21] has developed a primary resonator block (PRB) technique of filter synthesis while Laker and Ghausi[22] have

* As in Adams' work, the center sections have infinite pole Q (corresponding to lossless LC resonant circuits).

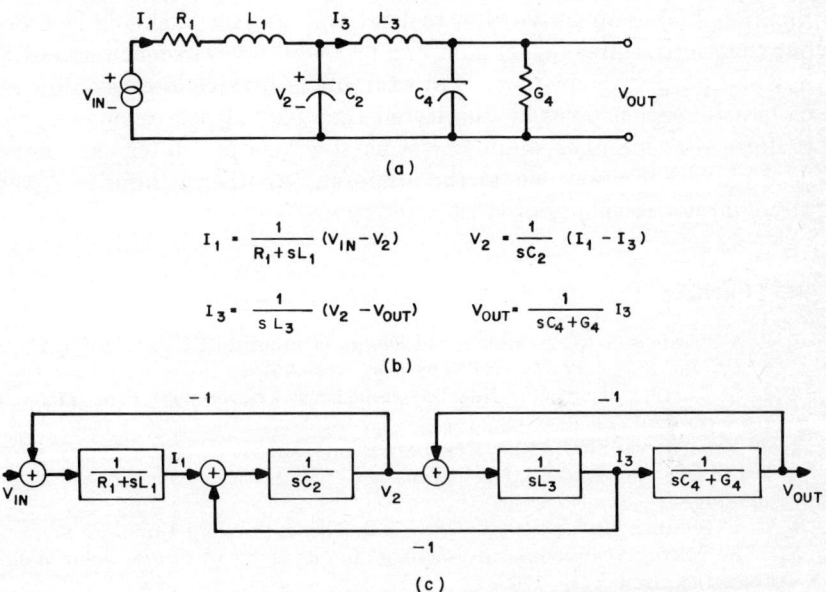

Fig. 17.17 (a) Passive ladder network; (b) equations that describe either a passive or active network; (c) representation of a coupled active filter.

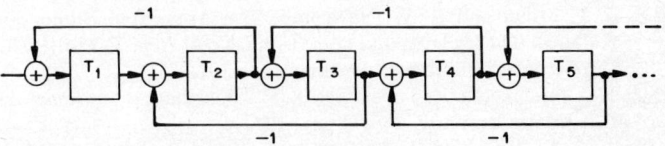

Fig. 17.18 Leapfrog active filter structure.

described a follow-the-leader feedback (FLF) which is very similar to the PRB. Also, Tow and Kuo[18] have described a state space approach which leads to coupled biquadratic sections.

17.8 CONCLUSIONS

Active circuits can be used to realize any of the transfer functions that are derived in the first part of this book. The most common realization procedure is to split the overall transfer function $T(s)$ into a product of second-order terms T_1, T_2, T_3, ... These functions can be synthesized directly by cascading second-order active filter sections; or the transfer functions can be coupled together to give improved sensitivity performance.

Because the solutions to the approximation problem can be given as a product of second-order terms, it is an easy matter to go from a transfer function $T(s)$ to an active filter realization. In fact, it should be pointed out that active filter realizations can be found for transfer functions that have no passive realization. For example, a passive lowpass filter must have attenuation poles at infinity so that partial pole removals can be performed at infinity. Similarly, a passive bandpass filter must have at least one attenuation pole at the origin and another at infinity. Active filters do not have any of these restrictions.

REFERENCES

1. R. A. Friedenson, "Computer Aided Design of Practical RC Active Filter," *1972 IEEE Int. Symp. on Circuit Theory Dig.*, April 1972.
2. J. G. Linvill, "Transistor Negative-Impedance Converters," *Proc. IRE*, June 1953, pp. 725–729.
3. T. Yanagisawa, "RC Active Networks Using Current Inversion Type Negative-Impedance Converters," *IRE Trans. on Circuit Theory*, September 1957, pp. 140–144.
4. D. F. Sheahan, *Inductorless Filters*, Ph.D. Thesis, Stanford Univ., 1968.
5. S. K. Mitra, *Analysis and Synthesis of Linear Active Networks*, John Wiley & Sons, Inc., New York, 1969.
6. R. W. Newcomb, *Active Integrated Circuit Synthesis*, Prentice-Hall, Inc., Englewood Cliffs, N.J., 1968.
7. L. P. Huelsman, *Theory and Design of Active RC Circuits*, McGraw-Hill Book Company, New York, 1968.
8. J. A. Miller and R. W. Newcomb, "An Annotated Bibliography on Gyrators in Network Theory: Circuits and Uses," *Tech. Rep. R-72-01*, Electrical Engineering Department, University of Maryland, 1972.
9. J. G. Graeme, G. E. Tobey, and L. P. Huelsman, *Operational Amplifiers*, McGraw-Hill Book Company, New York, 1971.
10. A. J. L. Muir and A. E. Robinson, "Design of Active RC Filters Using Operational Amplifiers," *Syst. Technol.*, April 1968, pp. 18–30.

11. J. Tow, "Design Formulas for Active RC Filters Using Operational Amplifier Biquad," *Electron. Lett.*, July 24, 1969, pp. 339–341.
12. J. J. Friend, "Biquad Filters—Design and Performance," *Southwest. IEEE Conf.* Dallas, Texas, April 1972.
13. T. Deliyannis, "High-Q Factor Circuit with Reduced Sensitivities," *Electron. Lett.*, December 1968, p. 557.
14. J. J. Friend, "A Single Operational-Amplifier Biquadratic Filter Section," *1970 Int. Symp. on Circuit Theory*.
15. E. Lueder, "A Decomposition of a Transfer Function Minimizing Distortion and In-band Losses," *Bell Syst. Tech. J.*, March 1970, pp. 455–470.
16. S. Halfin, "An Optimization Method for Cascading Filters," *Bell Syst. Tech. J.*, February 1970, pp. 185–190.
17. R. L. Adams, "On Reduced Sensitivity Active Filters," *Fourteenth Midwest Symp. on Circuit Theory*, University of Denver, May 1971, pp. 14.3-1 to 14.3-8.
18. J. Tow and Y. L. Kuo, "Coupled-Biquad Active Filters," *1972 Int. Symp. on Circuit Theory Dig.*, April 1972.
19. F. E. J. Girling and E. F. Good, "The Leapfrog or Active-Ladder Synthesis," *Wireless World*, July 1970, pp. 341–345; "Applications of the Active-Ladder Synthesis," *Wireless World*, September 1970, pp. 445–450; "Bandpass Types," *Wireless World*, pp. 505–510.
20. G. Szentirmai, "Synthesis of Multiple-Feedback Active Filters," *Bell Syst. Tech. J.*, April 1973, pp. 527–555.
21. G. Hurtig, "The Primary Resonator Block Technique of Filter Synthesis," *Int. Filter Symp.*, April 1972, p. 84.
22. K. R. Laker and M. S. Ghausi, "A Low Sensitivity Multiloop Feedback Active RC Filter," *Proc. Int. Symp. on Circuit Theory*, April 1973, pp. 126–129.

CHAPTER EIGHTEEN

Approximation Methods and Digital Filter Synthesis

18.1 INTRODUCTION

The first part of this book focused attention on the approximation problem; much time was spent in developing methods that yield a transfer function satisfying specified attenuation criteria. This chapter demonstrates how that material can be used in the design of digital filters. Thus at the end of this chapter we will have another way (in addition to passive filters and active filters) in which to synthesize a transfer function.

It was pointed out in the previous chapter that active filters face strong competition from passive filters; digital filters, of course, face the same competition. Again they fare best in this competition at the lower frequencies, where inductors are bulky and of poor quality. Also, at the low frequencies the speed requirement on digital filters is not as stringent. As large-scale integration becomes more and more practical, the cost of digital filters will fall, and they will become even stronger competitors to the real-time filters of the analog world.

This chapter does not attempt an all-inclusive treatment of the approximation problem of digital filters; instead it demonstrates how many of the continuous-filter results can be applied to digital filters. The bilinear transformation described in Sec. 18.7 usually serves as an excellent link between the analog approximation problem and the digital one. In fact, it can be considered as a solution of the approximation problem for digital filters.[3,8] However, it should be pointed out that it is a good solution for recursive* filters, but other techniques are used for non-recursive digital filters.[4,9]

The passive filters we examined are devices that take an input signal $x(t)$ and shape it to a desired output waveform $y(t)$. A digital system, as indicated in Fig. 18.1, can also be used to process the band-limited signal $x(t)$ and yield the desired output $y(t)$.

In our model of a digital filtering system, the signal first encounters a sampler. The output of this sampler is written in terms of the unit impulse function $\delta(t)$ as†

$$x^*(t) = T \sum_{k=0}^{\infty} x(kT) \, \delta(t - kT) \qquad (18.1)$$

In Fig. 18.1, the sampled waveform $x^*(t)$ is manipulated by the digital filter to yield a signal labeled as $y^*(t)$, which we assume to be of the form

$$y^*(t) = T \sum_{k=0}^{\infty} y(kT) \, \delta(t - kT) \qquad (18.2)$$

This digital filter can be thought of as taking a sequence of numbers $x(0)$, $x(T)$, $x(2T)$, ... and producing another sequence of numbers $y(0)$, $y(T)$, $y(2T)$, ... It will be our task to design the filter so that the output has the desired form.

A sampled-data system can be considered to be a generalization of the digital system that is shown in Fig. 18.1. In the study of a sampled-data system no reference to time need be made. That is, we can consider the input sequence to be $x(0)$, $x(1)$, $x(2)$, ..., and the output sequence to be $y(0)$, $y(1)$, $y(2)$, ... Thus sampled-data systems can be used to process information which has no time dependence. For example, it can be used to process data that describe the contrast of a picture.

The "reconstruction device" in Fig. 18.1 uses the information contained

---

* This term is defined in Sec. 18.4.

† Some prefer instead to write $x^*(t) = \sum_{k=0}^{\infty} x(kT) \, \delta(t - kT)$. The constant multiplier T is included in this discussion so that the Laplace transform of $x^*(t)$ can be interpreted as a frequency response.

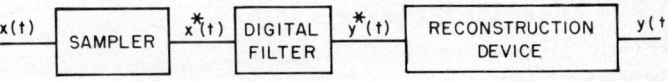

Fig. 18.1 Model of a digital filtering system.

in $y^*(t)$ to reconstruct an analog signal $y(t)$. Such a reconstruction device usually consists of a holding network and a lowpass filter.

In our study of classical filter theory we made extensive use of the Laplace transformation. For example, the frequency variable $s = \sigma + j\omega$ is the Laplace frequency term. Because of the sampling process, in digital filtering systems it will be more convenient to use the Z transformation, which is discussed briefly in the next section. For a more thorough discussion one can consult additional references.[1,2]

18.2 THE Z TRANSFORMATION

The Z transformation can be most easily understood by comparing it with the more familiar Laplace transformation. Consider an arbitrary function of time $f(t)$ that is assumed to be zero for $t < 0$. The Laplace transformation of $f(t)$ is $F(s)$, which is defined as

$$F(s) \triangleq \mathcal{L}[f(t)] \triangleq \int_0^\infty f(t)e^{-st}\,dt \tag{18.3}$$

We now define a sampled version of $f(t)$ and write it as $f^*(t)$, where

$$f^*(t) = T \sum_{k=0}^\infty f(kT)\,\delta(t - kT) \tag{18.4}$$

The Laplace transformation of the sampled waveform is

$$F^*(s) \triangleq \mathcal{L}[f^*(t)] \triangleq T \sum_{k=0}^\infty f(kT)\mathcal{L}[\delta(t - kT)] \tag{18.5}$$

Since $\mathcal{L}[\delta(t)] = 1$ and $\mathcal{L}[f(t - \tau)] = e^{-s\tau}\mathcal{L}[f(t)]$, it follows that the above equation can be rewritten as

$$F^*(s) = T \sum_{k=0}^\infty f(kT)e^{-kTs} \tag{18.6}$$

In our study of digital filters we shall frequently encounter the quantity e^{sT}, which we shall define to be a new variable z. Thus the Laplace transformation of the sampled waveform can be written as

$$F^*(s) = T \sum_{k=0}^\infty f(kT)z^{-k} \tag{18.7}$$

where
$$z = e^{sT} \tag{18.8}$$

By definition, the Z transform of $f(t)$ is

$$\mathbf{F}(z) = \sum_{k=0}^\infty f(kT)z^{-k} \tag{18.9}$$

The symbol $\mathbf{F}(z)$ is used here so that the Z transform of $f(t)$ will not be confused with $F(s)$, which is the Laplace transform of $f(t)$. $\mathbf{F}(z)$ cannot be found simply by replacing the variable s with z in $F(s)$. However, we can find $\mathbf{F}(z)$ by considering the Laplace transform of $f^*(t)$. That is, it follows from (18.7) and (18.9) that

$$F^*(s) = T\mathbf{F}(z) \tag{18.10}$$

where $z = e^{sT}$.

18.3 DESIGN OF DIGITAL FILTERS FROM CONTINUOUS FILTERS

A general representation of a digital filtering system is given in Fig. 18.1. In this chapter we shall assume that we have solved the approximation problem for a continuous filter and have obtained the transfer function

$$H(s) = \frac{X(s)}{Y(s)} = \frac{1}{T(s)} \tag{18.11}$$

We want to obtain a digital filtering system that acts in the same way as the continuous filter.

We are going to focus our attention on the part of the system labeled "digital filter" in Fig. 18.1. That is, we shall assume that we have an ideal sampler that produces the function

$$x^*(t) = T \sum_{n=0}^{\infty} x(nT)\,\delta(t - nT) \tag{18.12}$$

It is our task to find a digital filter that will convert $x^*(t)$ to

$$y^*(t) = T \sum_{n=0}^{\infty} y(nT)\,\delta(t - nT) \tag{18.13}$$

where $y^*(t)$ is such that the digital filtering system closely models the continuous system and thus produces the transfer function $H(s)$.

Our goal may be stated another way: given a sequence of numbers $x(0)$, $x(T)$, $x(2T)$, ..., we want to find a digital filter that will produce the output sequence $y(0)$, $y(T)$, $y(2T)$, ... The digital filter will be described in terms of its Z-transformation transfer function

$$\mathbf{H}(z) = \frac{\mathbf{X}(z)}{\mathbf{Y}(z)} \tag{18.14}$$

where $\mathbf{X}(z) = \sum_{k=0}^{\infty} x(kT)z^{-k} \quad \mathbf{Y}(z) = \sum_{k=0}^{\infty} y(kT)z^{-k}$

The following sections discuss methods that can be used to solve this approximation problem. Most of the material assumes that we have a transfer function $H(s)$ in the standard frequency variable and that we want to find a satisfactory transfer function $\mathbf{H}(z)$. However, sometimes digital filters are not meant to simulate analog filters; for example, the nonrecursive digital filters (briefly mentioned in the next section) are often used for data manipulation.

18.4 NONRECURSIVE DIGITAL FILTERS[4,11]

By definition, nonrecursive digital filters have transfer functions that can be written as a polynomial

$$\mathbf{H}(z) = a_0 + a_1 z^{-1} + a_2 z^{-2} + \cdots \qquad (18.15)$$

This implies that they can be realized by configurations that have no feedback paths (see Sec. 18.8) and thus do not encounter stability problems as indicated by limit cycles.[7,10] Like their counterpart analog polynomial filters, they require very high degrees to produce a sharp attenuation shape.

Because nonrecursive digital filters require very high degrees to produce a sharp attenuation shape, they are not often used for real-time filtering of waveforms. For this reason they will not be discussed in detail in this book. This is not to imply that nonrecursive filters are not useful; for example, they are often programmed on computers to process information. They can be used for integration, interpolation, or differentiation. Like their counterpart analog polynomial filters, they can have excellent phase response.

The next three sections discuss the most common methods that are used to design recursive digital filters. Because recursive digital filters are not restricted to be polynomial filters, they can produce much sharper attenuation characteristics than can nonrecursive filters.

18.5 IMPULSE-INVARIANT METHOD*

If a continuous filter has an mth-order transfer function $T(s)$, it can be expanded in terms of partial fractions as

$$T(s) = \sum_{i=1}^{m} \frac{R_i}{s + p_i} \qquad (18.16)$$

---

* Instead of using the input/output transfer function $H(s)$, this section is written mainly in terms of its reciprocal $T(s)$. The reason for this is that the impulse response of a system is usually considered to be an output/input quantity. Also, it might be worthwhile to mention that impulse-invariant digital filters are also called standard Z-transform filters.

where we are assuming that all poles are first order. The impulse response of this filter is

$$i(t) = \mathcal{L}^{-1}[T(s)] = \sum_{i=1}^{m} R_i e^{-p_i t} \tag{18.17}$$

The impulse-invariant method of designing digital filters finds a transfer function $\mathbf{T}(z)$ (in terms of the Z transformation) that has the same impulse response as the continuous filter. That is, we want

$$i^*(t) = \sum_{k=0}^{\infty} i(t) \, \delta(t - kT) \tag{18.18}$$

where $i(t)$ is as given by (18.17). Note that (18.18) does not contain the usual constant multiplier T as in (18.4). Thus, to find $\mathbf{T}(z)$ we must introduce the constant:

$$\mathbf{T}(z) = T \sum_{k=0}^{\infty} i(t) z^{-k} = T \sum_{k=0}^{\infty} \left(\sum_{i=1}^{m} R_i e^{-p_i kT} \right) z^{-k}$$

Thus
$$\mathbf{T}(z) = T \sum_{i=1}^{m} \frac{R_i}{1 - e^{-p_i T} z^{-1}} \tag{18.19}$$

In summary, assume we are given a prototype continuous filter that has a transfer function $T(s)$. This filter has an impulse response which we can denote as $i(t)$. We can design a digital filter that has the impulse response given by (18.18) if $\mathbf{T}(z)$ is as given by (18.19).

The continuous filter will usually have complex poles. The above material can be applied to this case by realizing that for every complex pole p_i with residue R_i there is also a conjugate pole p_i^* with residue R_i^*. Applying (18.19) thus yields the following substitutions for impulse-invariant filters:

$$\begin{aligned}
\frac{1}{(s+a)^2 + b^2} &\to \frac{e^{-aT}(\sin bT) z^{-1}}{b D(z)} \\
\frac{s}{(s+a)^2 + b^2} &\to \frac{1 - e^{-aT}[\cos bT + (a/b) \sin bT] z^{-1}}{D(z)} \\
\frac{s+a}{(s+a)^2 + b^2} &\to \frac{1 - e^{-aT}(\cos bT) z^{-1}}{D(z)}
\end{aligned} \tag{18.20}$$

where $D(z) = 1 - 2e^{-aT}(\cos bT) z^{-1} + e^{-2aT} z^{-2}$.

Example 18.1

In Chap. 16 we found a passive realization for a third-degree elliptic lowpass filter that has a passband ripple of 0.25 dB and a passband edge of 10 Hz. In this section we shall find a digital-filter realization that approximates the frequency response of this analog filter. As a first step we shall find the transfer function **T**(z) for an impulse-invariant filter.

From Example 16.2 the continuous filter has the transfer function

$$T(s) = \frac{s^2 + (2\pi \times 22.7007)^2}{0.0719(s + \sigma)[s^2 + (\omega Z_1/ZQ_1)s + \omega Z_1^2]}$$

where $\sigma = 54.0546$
$\omega Z_1 = 2\pi \times 11.515$
$ZQ_1 = 1.8021$

Substituting and making a partial-fraction expansion yield

$$T(s) = \frac{13.907(s^2 + 20{,}344.05)}{(s + 54.0546)(s^2 + 40.148s + 5{,}234.64)}$$

$$= \frac{54.0546}{s + 54.0546} - \frac{40.148s}{(s + 20.074)^2 + (69.5102)^2}$$

To find **T**(z) for the impulse-invariant filter we can use (18.19) and (18.20). These yield

$$\mathbf{T}(z) = T\left\{\frac{\sigma}{1 - e^{-\sigma T}z^{-1}} - 40.148\,\frac{1 - e^{-aT}[\cos bT + (a/b)\sin bT]\,z^{-1}}{1 - 2e^{-aT}(\cos bT)z^{-1} + e^{-2aT}z^{-2}}\right\} \quad (18.21)$$

where $\sigma = 54.0546$
$a = 20.074$
$b = 69.5102$

If the sampling time is chosen as $T = \frac{1}{80}$, evaluation of the above yields

$$\mathbf{T}(z) = \frac{0.6757}{1 - 0.5088z^{-1}} + \frac{-0.5019 + 0.3382z^{-1}}{1 - 1.00479z^{-1} + 0.60541z^{-2}}$$

We shall now discuss the frequency response of a digital filter that has been designed by the impulse-invariant method. The digital filter has an impulse response that can be represented by the sequence of numbers $i(0), i(T), i(2T), \ldots$. Or, alternatively, we can write

$$i^*(t) = T\sum_{k=0}^{\infty} i(kT)\,\delta(t - kT) \quad (18.22)$$

If we recall that the Laplace transform of the impulse response is the transfer function $T(s)$, then the Laplace transformation of the above equation can be written as*

$$T^*(s) = \sum_{k=-\infty}^{\infty} T\left(s + j\frac{2\pi k}{T}\right) \quad (18.23)$$

---

*See page 166 of Ref. 2. In Eq. (18.23), a term $Ti(0^+)/2$ was ignored, as it is equivalent to an unimportant dc quantity.

where T is the sampling period, and $T^*(s)$ is the Laplace transformation of $i^*(t)$.

The frequency response $T^*(s)$ can be written in terms of $\mathbf{T}(z)$ by recalling from (18.10) that $F^*(s) = T\mathbf{F}(z)$, where z is evaluated at e^{sT}. Since $\mathbf{T}(z)$ is a ratio of polynomials, the constant T cancels and the loss can be expressed as

$$\mathbf{A}(\omega) = -20 \log |T^*(j\omega)| = -20 \log |\mathbf{T}(z)| = 20 \log |\mathbf{H}(z)| \quad (18.24)$$

where $z = e^{j\omega T}$.

We shall generally write $\mathbf{T}(z)$ in one of two ways: as a summation of second-order functions or as a product of second-order functions. These two expansions may be written as

$$\mathbf{T}(z) = \sum_{i=1}^{n} \frac{A_i + B_i z^{-1} + C_i z^{-2}}{D_i + E_i z^{-1} + F_i z^{-2}} \quad (18.25)$$

$$\mathbf{T}(z) = \prod_{i=1}^{n} \frac{A_i + B_i z^{-1} + C_i z^{-2}}{D_i + E_i z^{-1} + F_i z^{-2}} \quad (18.26)$$

where the coefficients in each expansion are, of course, different. The summation form for $\mathbf{T}(z)$ is especially convenient for digital filters designed by the impulse-invariant method. The product form will also be

```
1; LØSS ØF DIGITAL FILTER            (SUM TYPE)
1.01; EXPLANATIØN ØF NØTATIØN
1.02; T(Z)=SUM (A[I]+B[I]/Z+C[I]/ZZ)/(D[I]+E[I]/Z+F[I]/ZZ)
1.03; T(Z)=(X1+J Y1)/(X2+J Y2)=X3+J Y3
1.1  DEMAND T,N
1.2  DEMAND A[I],B[I],C[I],D[I],E[I],F[I] FØR I=1:1:N
1.3  TP=2*$PI
1.5  DEMAND F
1.6  TYPE #
1.7  TØ PART 6

6: DETERMINATIØN ØF LØSS
6.1  WT=TP*F*T,TR=0,TI=0,I=0
6.2  C=CØS(WT),S=SIN(WT)
6.3  I=I+1
6.41 X1=A[I]+C*B[I]+(C*C-S*S)*C[I]
6.42 X2=D[I]+C*E[I]+(C*C-S*S)*F[I]
6.43 Y1=-S*B[I]-2*C*S*C[I]
6.44 Y2=-S*E[I]-2*C*S*F[I]
6.45 D=X2*X2+Y2*Y2
6.46 X3=(X1*X2+Y1*Y2)/D
6.47 Y3=(X2*Y1-X1*Y2)/D
6.5  TR=TR+X3,TI=TI+Y3
6.6  TØ PART 6.3 IF I<N
6.7  A=-10*LØG(TR*TR+TI*TI)
6.8  TYPE F,A IN FØRM 1

>
```

Fig. 18.2 Program for the loss of a digital filter that has its transfer function expressed as a sum of second-order terms.

useful to us in later sections. Because we shall encounter these two forms of $T(z)$ very often, it is worthwhile to have some general programs to calculate the loss $A(\omega)$ for either case. The program in Fig. 18.2 performs the calculations implied by (18.24) and (18.25), and the program in Fig. 18.3 performs the calculations implied by (18.24) and (18.26).

Example 18.2

In Example 18.1 we studied a prototype continuous filter that had

$$A(\omega = 20\pi) = 0.25 \text{ dB} = \text{passband ripple}$$
$$A(\omega = 40\pi) = 28.05 \text{ dB} = \text{minimum stopband loss}$$

For a sampling time of $T = \frac{1}{80}$ we found the impulse-invariant filter

$$T(z) = \frac{0.6757}{1 - 0.5088z^{-1}} + \frac{-0.5019 + 0.3382z^{-1}}{1 - 1.00479z^{-1} + 0.60541z^{-2}}$$

Using the program in Fig. 18.2 yields

$$A(\omega = 20\pi) = 0.65 \text{ dB} \qquad A(\omega = 40\pi) = 21.14 \text{ dB}$$

The impulse-invariant filter in Example 18.2 does not model the prototype filter very well. For example, the impulse-invariant filter has a passband-edge loss of 0.65 dB, while the prototype passband ripple is 0.25 dB. These filters act differently because of aliasing. As indicated in (18.23), the frequency response of the impulse-invariant filter is related

```
1; LØSS ØF DIGITAL FILTER     (PRØDUCT TYPE)
1.01; EXPLANATIØN ØF NØTATIØN
1.02; T(Z)=PRØD (A[I]+B[I]/Z+C[I]/ZZ)/(D[I]+E[I]/Z+F[I]/ZZ)
1.03; T(Z)=PRØD T[I]
1.1 DEMAND I,N
1.2 DEMAND A[I],B[I],C[I],D[I],E[I],F[I] FØR I=1:1:N
1.3 TP=2*$PI,A1=-20*LØG(T)
1.5 DEMAND F
1.6 TYPE #
1.7 TØ PART 6

6; DETERMINATIØN ØF LØSS
6.1 WT=TP*F*T,A=A1,I=0
6.2 C=CØS(WT),S=SIN(WT)
6.3 I=I+1
6.41 X1=A[I]+C*B[I]+(C*C-S*S)*C[I]
6.42 X2=D[I]+C*E[I]+(C*C-S*S)*F[I]
6.43 Y1=-S*B[I]-2*C*S*C[I]
6.44 Y2=-S*E[I]-2*C*S*F[I]
6.45 D=X2*X2+Y2*Y2
6.46 X3=(X1*X2+Y1*Y2)/D
6.47 Y3=(X2*Y1-X1*Y2)/D
6.5 A=A-10*LØG(X3*X3+Y3*Y3)
6.6 TØ PART 6.3 IF I<N
6.8 TYPE F,A IN FØRM 1

>
```

Fig. 18.3 Program for the loss of a digital filter that has its transfer function expressed as a product of second-order terms.

not only to the prototype $T(s)$, but also to translates of $T(s)$. Because the sampling time T was not small enough, the translates overlapped and caused aliasing. The aliasing can be reduced by choosing a higher sampling frequency. For example, if $f_s = 800$ instead of 80, then (18.21) yields

$$\mathbf{T}(z) = \frac{0.067568}{1 - 0.9347z^{-1}} + \frac{-0.05185 + 0.04998z^{-1}}{1 - 1.943z^{-1} + 0.951z^{-2}}$$

Using the program in Fig. 18.2 yields

$$\mathbf{A}(\omega = 20\pi) = 0.27 \text{ dB} \qquad \mathbf{A}(\omega = 40\pi) = 29.31 \text{ dB}$$

These losses are much closer to the loss of the prototype filter, which has

$$A(\omega = 20\pi) = 0.25 \text{ dB} \qquad A(\omega = 40\pi) = 28.05 \text{ dB}$$

We have just seen that digital filters designed by the impulse-invariant method can have aliasing problems. This can usually be reduced by making the sampling frequency high enough. That is, at $0.5\omega_s$ the magnitude of $T(s)$ should be negligibly small. If $T(s)$ has a transmission zero at $s = \infty$, then $T(0.5\omega_s)$ can be made arbitrarily small by making ω_s large enough. However, some transfer functions (for example, even-order elliptic lowpass filters) do not have a transmission zero at infinity. If aliasing is a problem, it can be reduced by cascading a "guard filter" with the original prototype filter $T(s)$. The guard filter would be a lowpass filter that gives negligible attenuation in the passband of $T(s)$ but gives added attenuation at $0.5\omega_s$. This added attenuation would reduce aliasing; however, an even more attractive solution is simply to redesign the original prototype filter so that it gives sufficient attenuation at $0.5\omega_s$.

18.6 MATCHED Z-TRANSFORM METHOD

The impulse-invariant method of designing digital filters started with the analog transfer function

$$T(s) = \sum_{i=1}^{m} \frac{R_i}{s + p_i} \qquad (18.16)$$

and expressed the digital transfer function as

$$\mathbf{T}(z) = \sum_{i=1}^{m} \frac{R_i}{1 - e^{-p_i T} z^{-1}} \qquad (18.19)$$

Thus the analog pole located at $s = -p_i$ was transformed to a digital pole located at $z = e^{-p_i T}$.

The matched Z-transform method of designing digital filters is quite similar to the impulse-invariant method. In the matched Z-transform method, $\mathbf{T}(z)$ is formed by replacing every s in $T(s)$ by $s = (\ln z)/T$. An analog pole located at $s = -p_i$ will again be transformed to a digital pole located at $z = e^{-p_i T}$. Similarly, an analog zero located at $s = -u_i$ will be transformed to a digital zero located at $z = e^{-u_i T}$. In the matched Z-transform method it is thus very easy to express $\mathbf{T}(z)$ in terms of its poles and zeros. The constant multiplier of $\mathbf{T}(z)$ can be chosen to adjust the overall gain of the filter.

In the matched Z-transform method, the following substitutions are helpful [these can be found by replacing the analog frequency s with $(\ln z)/T$]:

$$s + a \to 1 - e^{-aT} z^{-1}$$
$$(s + a)^2 + b^2 \to 1 - 2e^{-aT}(\cos bT) z^{-1} + e^{-2aT} z^{-2} \quad (18.27)$$

Since this is the same as $D(z)$ in (18.20), it follows that digital filters designed by the impulse-invariant method have the same poles as those designed by the matched Z-transform method. The methods differ in that the numerator of $\mathbf{T}(z)$ in the matched Z-transform method is chosen to match the numerator of $T(s)$, whereas in the impulse-invariant method the numerator of $\mathbf{T}(z)$ is chosen so that the two impulse responses match.

Example 18.3

In Example 18.1 we found an impulse-invariant approximation for the continuous filter described by

$$T(s) = \frac{s^2 + (2\pi \times 22.7007)^2}{0.0719(s + \sigma)[s^2 + (\omega z_1/ZQ_1)s + \omega z_1^2]}$$

where $\sigma = 54.0546$
$\omega z_1 = 2\pi \times 11.515$
$ZQ_1 = 1.8021$

Using the transformation in (18.27), the matched-Z transfer function is

$$\mathbf{T}(z) = k \frac{1 - 2(\cos cT) z^{-1} + z^{-2}}{(1 - e^{-\sigma T} z^{-1})[1 - 2e^{-aT}(\cos bT) z^{-1} + e^{-2aT} z^{-2}]}$$

where $\sigma = 54.0546 \quad a = 20.074$
$b = 69.5102 \quad c = 142.633$

Thus, for $T = 1/80$,

$$\mathbf{T}(z) = k \frac{1 + 0.42105 z^{-1} + z^{-2}}{(1 - 0.5088 z^{-1})(1 - 1.00479 z^{-1} + 0.60541 z^{-2})}$$

Choosing $k = 9.75$ adjusts the filter loss to be zero at the origin. Using the program in Fig. 18.3 yields

$$\mathbf{A}(\omega = 20\pi) = 0.10 \text{ dB} \quad \mathbf{A}(\omega = 40\pi) = 27.46 \text{ dB}$$

which is approximately the same as the loss of the prototype filter.

18.7 BILINEAR DIGITAL FILTERS

This section discusses a method of designing digital filters that allows us to avoid the aliasing problem. The method also allows us to use a pole-placer program to locate optimally the attenuation poles of a digital filter. In essence, this section can be considered as a solution of the approximation problem for digital filters.[3]

Digital filters designed by the impulse-invariant method have aliasing problems because sampling causes the analog characteristic to be repeated in the frequency domain, as indicated by

$$T^*(s) = \sum_{k=-\infty}^{\infty} T\left(s + j\frac{2\pi k}{T}\right) \qquad (18.28)$$

To find out how we can avoid the aliasing problem, assume we want to design a digital filter that satisfies the loss requirements shown in Fig. 18.4a. If this were an analog filter, we could find a set of attenuation poles by using a pole-placer program. For example, if we wanted an equiripple passband, we would use the pole placer in Fig. 8.15. For a digital filter we shall *prewarp* the requirements by the following trans-

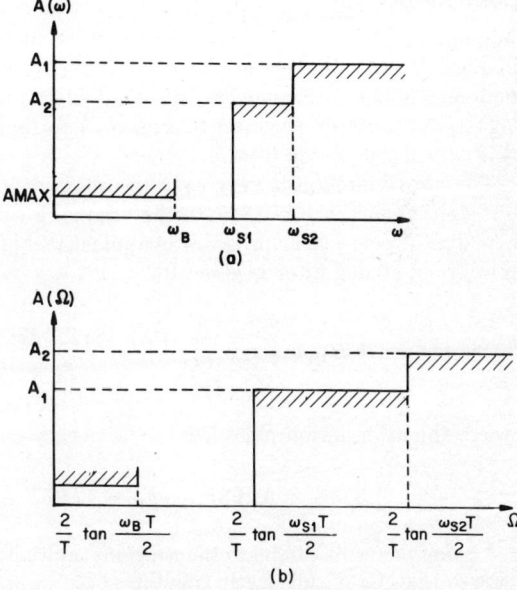

Fig. 18.4 (a) Typical lowpass digital filter requirements; (b) their prewarped equivalent.

formation, which is illustrated in Fig. 18.4b:

$$\Omega = \frac{2}{T} \tan \frac{\omega T}{2} \qquad (18.29)$$

The important thing to note about this transformation* is that for any finite frequency Ω the corresponding frequency ω is such that

$$0 \le \omega < \frac{\omega_s}{2}$$

where ω_s is the sampling frequency. Thus, if we consider $A(\Omega)$ to represent the loss of an analog filter, then the transformed loss of the digital filter $\mathbf{A}(\omega)$ is restricted to the above frequency range, and there will be no aliasing.

In summary, if we are given digital filter requirements as in Fig. 18.4a we prewarp them to analog filter requirements as in Fig. 18.4b. We can find the input/output transfer function $H(S)$ by solving the approximation problem by normal methods. It remains to be seen how we obtain the digital transfer function $\mathbf{H}(z)$; but before discussing that, an example might be useful.

Example 18.4

Assume we want to find an equiripple digital filter that satisfies the requirements in Fig. 18.5a.† By use of the prewarping transformation of (18.29), these requirements can be changed to those given in Fig. 18.5b. We can then use the lowpass pole-placer program of Fig. 8.15 to find a set of attenuation poles for this prewarped analog filter.

The above method is very easy to program; in fact, we only need to add the few steps shown in Fig. 18.5c to the original lowpass pole-placer program. This new digital pole-placer program is applied in Fig. 18.6, which indicates that the prewarped analog filter is given by

$$H(S) = C_H \prod_{i=1}^{3} \frac{S^2 + (2\pi ZF_i/ZQ_i)S + (2\pi ZF_i)^2}{S^2 + (2\pi F_i)^2}$$

where the attenuation poles are

$$F_1 = 34.681 \qquad F_2 = 42.977 \qquad F_3 = 76.605$$

---

* Some prefer to eliminate the constant multiplier $2/T$ in (18.29). It is included here so that, for T sufficiently small, $\Omega \approx \omega$.

† These requirements were also examined in Fig. 8.17. Note that these requirements differ from those in Example 18.1.

The natural modes and constant multiplier C_H can be found by using the lowpass zero-finder program of Fig. 10.3. The results are

$$ZQ_1 = 6.8612 \qquad ZF_1 = 24.188$$
$$ZQ_2 = 1.6637 \qquad ZF_2 = 20.777$$
$$ZQ_3 = 0.6191 \qquad ZF_3 = 14.264$$
$$C_H = 256.65$$

We shall now investigate how we can find the digital transfer function $\mathbf{H}(z)$ from the prewarped analog transfer function $H(S)$. Recall that the digital frequency response is given by

$$\mathbf{A}(\omega) = 20 \log |\mathbf{H}(z)| \tag{18.30}$$

where $z = e^{j\omega T}$.

This loss function $\mathbf{A}(\omega)$ should satisfy the loss requirements as they are specified in the normal frequency domain (e.g., see Fig. 18.4a). We

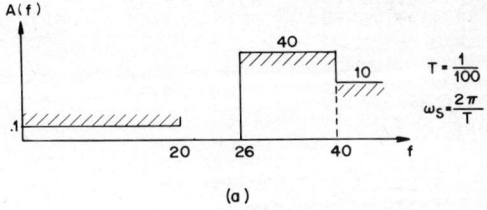

(a)

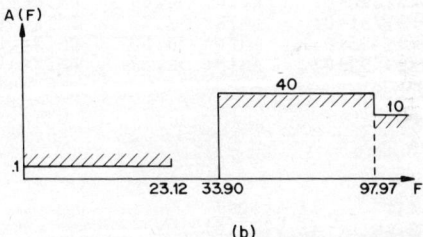

(b)

```
1; THE FØLLØWING STEPS SHØULD BE ADDED TØ LØWPASS PØLE PLACER
1.26 DEMAND T
1.27 TP=2*SPI,Ø=TP*FB*T/2,FB=2/T/TP*SIN(Ø)/CØS(Ø)
1.28 Ø=TP*FS[I]*T/2,FS[I]=2/T/TP*SIN(Ø)/CØS(Ø) FØR I=1:1:S
1.29 TYPE "ALL REMAINING CALCULATIØNS ARE FØR PREWARPED FILTER",#,#
1.291 TYPE FS[I] FØR I=1:1:S
1.292 TYPE FB,#,#
```

(c)

Fig. 18.5 (a) Specific example of digital filter requirements; (b) their prewarped equivalent. The Steps in (c) help form a digital pole-placer program.

360 Approximation Methods for Electronic Filter Design

```
DØ PART 1
        S=2
     FS[1]=26
      A[1]=40
     FS[2]=40
      A[2]=10
      AMAX=.1
        FB=20
       NIN=0
         N=3


        T=1/100
ALL REMAINING CALCULATIØNS ARE FØR PREWARPED FILTER


     FS[1]=     33.89656
     FS[2]=     97.9657096
       FB=     23.1265669


     F[1]=34.5
     F[2]=40
     F[3]=80

FMIN[ 1]=3.390+01    AMIN= 62.90    DMIN= 22.90
FMIN[ 2]=3.653+01    AMIN= 62.26    DMIN= 22.26
FMIN[ 3]=5.256+01    AMIN= 54.39    DMIN= 14.39
FMIN[ 4]=9.797+01    AMIN= 58.82    DMIN= 18.82


>DØ PART 49


     F[1]=      34.6808624
     F[2]=      42.8597861
     F[3]=      76.4967

FMIN[ 1]=3.390+01    AMIN= 58.69    DMIN= 18.69
FMIN[ 2]=3.731+01    AMIN= 58.75    DMIN= 18.75
FMIN[ 3]=5.423+01    AMIN= 58.50    DMIN= 18.50
FMIN[ 4]=9.797+01    AMIN= 58.48    DMIN= 18.48


>DØ PART 49


     F[1]=      34.681299
     F[2]=      42.9773163
     F[3]=      76.6046101

FMIN[ 1]=3.390+01    AMIN= 58.61    DMIN= 18.61
FMIN[ 2]=3.734+01    AMIN= 58.61    DMIN= 18.61
FMIN[ 3]=5.434+01    AMIN= 58.61    DMIN= 18.61
FMIN[ 4]=9.797+01    AMIN= 58.61    DMIN= 18.61


>
```

Fig. 18.6 Example of the use of the digital pole-placer program.

can find the proper **H**(z) for the above equation by using

$$\mathbf{H}(z) = H(S) \tag{18.31}$$

where
$$S = \frac{2}{T}\frac{z-1}{z+1} \tag{18.32}$$

That is, to form the digital transfer function **H**(z), we replace every S in $H(S)$ with $(2/T)(z-1)/(z+1)$.

To see why this procedure produces a digital transfer function **H**(z) that has the proper frequency response, we must return to the prewarping transformation.

$$\Omega = \frac{2}{T}\tan\frac{\omega T}{2} \tag{18.33}$$

which is equivalent to

$$S = \frac{2}{T}\tanh\frac{sT}{2} \tag{18.34}$$

However, we also have $z = e^{sT}$. This relation and (18.34) imply that

$$S = \frac{2}{T}\frac{z-1}{z+1}$$

which is the transformation we used in (18.32). The fact that S is a bilinear transformation of z explains why filters derived by this method are referred to as *bilinear* digital filters.

Example 18.5

This example continues the work started in Example 18.4, which found a prewarped transfer function $H(S)$ for the requirements in Fig. 18.5. Applying

$$\mathbf{H}(z) = H(S) \quad \text{and} \quad S = \frac{2}{T}\frac{z-1}{z+1}$$

to the transfer function $H(S)$ yields

$$\mathbf{H}(z) = C_H \prod_{i=1}^{3} \frac{D_i + E_i z^{-1} + F_i z^{-2}}{A_i + B_i z^{-1} + C_i z^{-2}}$$

where $C_H = 256.65$, $A_i = 1$, $C_i = 1$, and

$B_1 = 0.17108$ $D_1 = 0.77189$ $E_1 = -0.38642$ $F_1 = 0.67061$
$B_2 = 0.58304$ $D_2 = 0.64415$ $E_2 = -0.40663$ $F_2 = 0.36619$
$B_3 = 1.4111$ $D_3 = 0.28338$ $E_3 = -0.23534$ $F_3 = 0.07023$

The loss of this digital filter can be found by using the program in Fig. 18.3. For example, at $F = 20$, the program yields a loss of 0.10 dB, which is the loss at the passband edge. It should be noted that the frequencies used in this program are the normal frequencies and not the prewarped ones.

The previous material describes how the bilinear transformation can be used to produce digital filters to meet specified attenuation requirements. The same approach can be applied to yield digital filters that meet specified delay requirements.[8] To see how this can be done, it will be easiest to proceed by analogy to the approach used for attenuation requirements.

The loss of the digital filter is given by

$$\mathbf{A}(\omega) = 20|\mathbf{H}(z)| \qquad (18.35)$$

where $\mathbf{H}(z)$ is related to the prewarped analog filter by

$$\mathbf{H}(z) = H(j\Omega) \qquad (18.36)$$

Thus the attenuation of the digital filter can be rewritten as

$$\mathbf{A}(\omega) = 20 \log |H(j\Omega)| \qquad (18.37)$$

But the bilinear transformation implies that

$$\Omega = \frac{2}{T} \tan \frac{\omega T}{2} \qquad (18.38)$$

which finally yields

$$\mathbf{A}(\omega) = 20 \log \left| H \left(j \frac{2}{T} \tan \frac{\omega T}{2} \right) \right| \qquad (18.39)$$
$$= A \left(\frac{2}{T} \tan \frac{\omega T}{2} \right)$$

This equation implies that by prewarping the frequency axis in the analog domain, we can find a digital filter that has the correct attenuation response. It should be noted that only the frequency axis had to be prewarped—there was no change required for the attenuation axis.

We shall now determine how the delay of a digital filter is related to the delay of its prewarped analog counterpart. The delay of an analog filter was defined in Chap. 14 as

$$D(\omega) = \frac{d}{d\omega} \arg H(j\omega) \qquad (18.40)$$

For a digital filter the corresponding definition is

$$\mathbf{D}(\omega) = \frac{d}{d\omega} \arg \mathbf{H}(z) \qquad (18.41)$$

where $\mathbf{H}(z)$ is related to the prewarped analog filter by

$$\mathbf{H}(z) = H(j\Omega) \tag{18.42}$$

Thus the delay of the digital filter can be rewritten as

$$\mathbf{D}(\omega) = \frac{d}{d\omega} \arg H(j\Omega) \tag{18.43}$$

But the bilinear transformation implies that

$$\Omega = \frac{2}{T} \tan \frac{\omega T}{2} \tag{18.44}$$

which finally yields

$$\mathbf{D}(\omega) = \left(\sec^2 \frac{\omega T}{2}\right) D\left(\frac{2}{T} \tan \frac{\omega T}{2}\right) \tag{18.45}$$

Assume that we used prewarping of the frequency axis and the bilinear transformation to obtain a transfer function for a digital filter. Equation (18.39) implies that the attenuation of the digital filter is the same as that of the prewarped filter; however, (18.45) indicates that the delay of the digital filter is different. If one is designing a delay equalizer in which this difference would be objectionable, then one can also predistort the delay axis by the factor $\sec^2(\omega T/2)$.

18.8 REALIZATIONS FOR DIGITAL FILTERS

For any specific transfer function $\mathbf{T}(z)$ there are many possible digital filter realizations. In this section we discuss the two most common: the parallel type and the series type. The references at the end of this chapter can be consulted for additional realizations.

Equations (18.25) and (18.26) express $\mathbf{T}(z)$ in two different forms: as a sum of second-order factors and as a product of second-order factors. The sum of second-order factors can be rewritten as

$$\mathbf{T}(z) = \sum_{i=1}^{n} \frac{A_i + B_i z^{-1}}{1 + E_i z^{-1} + F_i z^{-2}} \tag{18.46}$$

If $\mathbf{T}(z)$ is originally given as a rational function in z, it can be put into the form of (18.46) by performing a partial-fraction expansion.

We shall now obtain a digital filter for a general term of the sum of second-order factors

$$\mathbf{T}(z) = \frac{\mathbf{Y}(z)}{\mathbf{X}(z)} = \frac{A + Bz^{-1}}{1 + Ez^{-1} + Fz^{-2}} \tag{18.47}$$

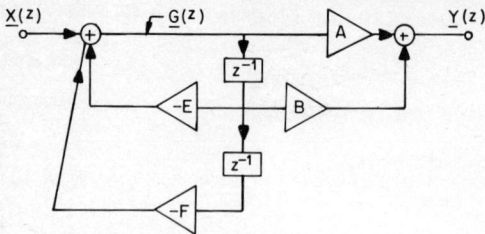

Fig. 18.7 General section for use in the parallel realization of digital filters.

From (18.47) it follows that we can write

$$\mathbf{Y}(z) = (A + Bz^{-1}) \frac{\mathbf{X}(z)}{1 + Ez^{-1} + Fz^{-2}} \qquad (18.48)$$
$$= (A + Bz^{-1})\mathbf{G}(z)$$

Considering the function $\mathbf{G}(z)$ to be an internally generated function leads to the realization shown in Fig. 18.7.

In the time domain, the realization of Fig. 18.7 can be considered to generate a solution for the following second-order difference equation.*

$$y(kT) + Ey(kT - T) + Fy(kT - 2T) = Ax(kT) + Bx(kT - T) \qquad (18.49)$$

Comparing (18.47) and (18.49), we see that the term z^{-1} is equivalent to a delay of time T. Thus, the rectangles in Fig. 18.7 represent devices that delay the input by time T. For a sequence of numbers, the device would simply be a shift register.

Example 18.6

In Example 18.1 we found the following transfer function for an impulse-invariant digital filter:

$$\mathbf{T}(z) = \frac{0.6757}{1 - 0.5088z^{-1}} + \frac{-0.5019 + 0.3382z^{-1}}{1 - 1.00479z^{-1} + 0.60541z^{-2}}$$

A realization for the transfer function is given in Fig. 18.8. The numbers shown in that drawing have been rounded off for ease of notation. In practice, such roundoff errors would yield an unacceptable filter.

The example just considered is a specific case of the parallel form for digital filters. That is, if $\mathbf{T}(z)$ is written as the sum of second-order functions, then it can be realized by connecting second-order networks

---

* This assumes that the initial conditions are zero.

in parallel. Similarly, if $\mathbf{T}(z)$ is written as a product of second-order terms,

$$\mathbf{T}(z) = A_0 \prod_{i=1}^{n} \frac{1 + B_i z^{-1} + C_i z^{-2}}{1 + E_i z^{-1} + F_i z^{-2}} \qquad (18.50)$$

then it can be realized as a cascade of second-order networks, as indicated in Fig. 18.9. The cascade realization is especially convenient for bilinear digital filters that are written as a product of second-order terms, as was the case in Example 18.3. As was true in that example, if the analog filter has an attenuation pole on the imaginary axis, then the corresponding coefficient C_i is unity.* This means that the C_i multiplier in Fig. 18.9 can be replaced with a short circuit.

The realizations in Figs. 18.7 and 18.9 have feedback as well as feedforward. The feedback terms are due to the denominator coefficients

* As, if the s-plane roots are on the imaginary axis, the Z-plane roots are on the unit circle.

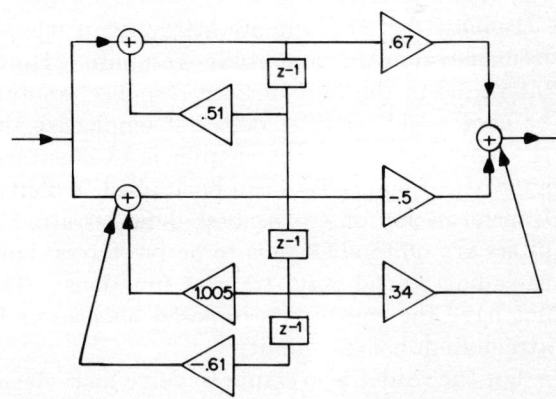

Fig. 18.8 Example of a parallel-type digital filter.

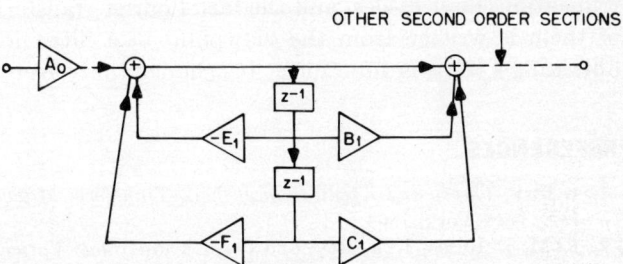

Fig. 18.9 General section for use in the cascade realization of digital filters.

E_i and F_i. A digital filter that has such terms is said to be *recursive*. Nonrecursive digital filters have transfer functions that can be written as polynomials in z^{-1}. Like their counterpart analog polynomial filters, they require a much higher degree to produce a sharp attenuation shape. On the other hand, they do not encounter stability problems as indicated by limit cycles.[7,10]

18.9 CONCLUSION

This chapter has shown how an analog solution to the approximation problem can be used to find a transfer function for a digital filter. One of the most satisfactory methods[3] utilizes a bilinear transformation. Since this transformation distorts the frequency scale, the original requirements are usually prewarped. The transfer function for this prewarped characteristic can be obtained by using the methods developed in the first part of this book. In particular, the pole-placer programs and zero-finder programs can be applied to yield a suitable transfer function.

Bilinear digital filters are attractive if one is interested mainly in the frequency-domain amplitude response. However, if one is instead interested in the time-domain response, another approach may have to be used. This chapter does not emphasize the other methods because the main purpose of the chapter is to illustrate how our solution to the approximation problem can be applied to digital filters—and the bilinear transformation offers the best demonstration. In fact, bilinear digital filters are quite analogous to active filters: both are usually realized by cascading second-order transfer functions. These transfer functions do not have the constraints imposed on passive filters, such as having an attenuation pole at infinity.

For the reader who wants to delve more deeply into the area of digital filters, there are many texts[4,5] and articles[6,9] available.* These cover topics not even considered here, such as numerical inaccuracies due to truncation, limit cycles, and the fast Fourier transform. However, none of them is written from the viewpoint of a filter designer interested in obtaining a transfer function. It is hoped that this text will fill that void.

REFERENCES

1. I. Jury, *Theory and Application of the Z-Transform Method*, John Wiley & Sons, Inc., New York, 1964.
2. P. M. DeRusso, R. J. Roy, and C. M. Close, *State Variables for Engineers*, John Wiley & Sons, Inc., New York, 1965.

* These references contain many additional references.

3. M. Sablatash, "Approximation Theory for Digital Filters," *IEEE Trans. on Circuit Theory*, November 1971, pp. 741–743.
4. F. F. Kuo and J. F. Kaiser, *System Analysis by Digital Computer*, John Wiley & Sons, Inc., New York, 1966.
5. B. Gold and C. M. Rader, *Digital Processing of Signals*, McGraw-Hill Book Company, New York, 1969.
6. Special Issue on Digital Filtering, *IEEE Trans. on Audio and Electroacoust.*, September 1968.
7. C-Y Kao, "An Analysis of Limit Cycles Due to Sign-Magnitude Truncation in Multiplication in Recursive Digital Filters," *Proc. Fifth Asilomar Conf. on Circuits and Syst.*, Pacific Grove, Calif., November 1971.
8. A. J. Gibbs, "The Design of Digital Filters," *Aust. Telecommun. Res.*, May 1970, pp. 29–34.
9. A. J. Gibbs, "An Introduction to Digital Filters," *Aust. Telecommun. Res.*, November 1970, pp. 3–14.
10. S. R. Parker, S. F. Hess, "Limit-Cycle Oscillations in Digital Filters," *IEEE Trans. on Circuit Theory*, November 1971, pp. 687–697.
11. L. R. Rabiner, "Design Techniques for Finite Impulse Response Digital Filters," *1973 Int. Symp. on Circuit Theory Dig.*

APPENDIX A

Telcomp

Introduction

This appendix describes the Telcomp language, which is designed for writing interactive computer programs. The appendix does not attempt to explain all the features of Telcomp; however, it does go into enough detail so that the reader will be able to understand the programs contained in this book. Parts of this appendix were obtained from the Telcomp Manual.*

Steps and Parts

Each stored command in a Telcomp program is called a *step* and is identified by a step number. Within a program, steps are grouped together to form parts. The part that a step belongs to is indicated by the integer portion of its step number.

Input/Output Information

The easiest way to input data to the computer is simply to type it in. For example,

$$X = 4$$

---

* *Telcomp Computer Services Manual for Users*, Bolt Beranek and Newman Inc., Cambridge, Mass., August 1967.

sets the variable X equal to four. Because the above statement was not preceded by a step number, it would be executed when the carriage return is pressed.

The DEMAND command causes the computer to ask what value a variable should have. A typical step in a program might be

$$3.1 \ \text{DEMAND X}$$

When the computer encounters this step, it will print

$$X =$$

at which time the user should enter the value.

The TYPE command causes the computer to type information, as illustrated in the next section.

A Simple Program: The DO Command

A simple program is shown at the top of Fig. A.1. To start the program one must use the DO command. Typing

$$\text{DO PART 1}$$

causes the computer to execute, in numerical order, each step in Part 1. On the other hand, a command such as

$$\text{DO STEP 1.1}$$

would cause the computer to execute only that step.

Use of Form

In the previous example, the numbers were typed out in a format determined by Telcomp. The *form* feature lets the user select the output format, as illustrated in Fig. A.2. The command

$$\text{DO PART 1}$$

caused the computer to demand values for X and Y. A and B were then computed to be 17 and 247, respectively. These values were typed in a format

```
                            1.1 DEMAND X,Y
                            1.2 A=2*X+3
                            1.3 B=X+80*Y
                            1.4 TYPE A,B IN FORM 1

    1.1 DEMAND X           FORM 1
    1.2 Y=2*X+3              ##.###    #.##^^^
    1.3 TYPE Y
   >DO PART 1             >DO PART 1
         X= 4                     X=7
         Y=    11                 Y=3
   >                         17.000    2.47+02
                           >
```

Fig. A.1 Illustration of the DO command.

Fig. A.2 Output format can be determined by the FORM feature.

```
1.1 DEMAND F
1.2 DO PART 6

6.1 A=2*F
6.2 TYPE F,A IN FORM 2


FORM 2
    ##.#    ##.#

>DO PART 1
           F=1
     1.0    2.0
>
>DO PART 6 FOR F=1,3,5,6,8
     1.0    2.0
     3.0    6.0
     5.0   10.0
     6.0   12.0
     8.0   16.0
>
>DO PART 6 FOR F=1:2:11
     1.0    2.0
     3.0    6.0
     5.0   10.0
     7.0   14.0
     9.0   18.0
    11.0   22.0
>
```

Fig. A.3 Illustration of the FOR command.

identified as form 1. Form 1 indicates that the first number to be printed should have a maximum of two digits before the decimal point, and should have three digits after the decimal point. Form 1 also indicates that the value of B should be printed on the same line as A. It should have one digit before the decimal point, and two after. The "up arrows" in the form indicate that the number should be in exponential form.

FOR Command

A FOR clause causes the command in question to be executed repeatedly, for a specific set of values. This is illustrated in Fig. A.3. The statement

$$\text{DO PART 1}$$

caused the computer to demand F; then step 1.2 told the computer to DO PART 6. In Part 6, A was calculated, and the values of F and A were typed.

The statement

$$\text{DO PART 6 FOR F} = 1,3,5,6,8$$

caused Part 6 to be executed repeatedly as shown. The statement

$$\text{DO PART 6 FOR F} = 1:2:11$$

caused Part 6 to be executed first for $F = 1$; F was then increased in increments of 2 until F was equal to 11.

TO Command

When Telcomp executes stored steps, it does so under control of a DO command, either a DO stored in the program or the direct DO that started the program. Normally, when Telcomp obeys a DO, it executes steps in numerical order in the part referred to by DO.

The TO command changes the order in which Telcomp executes steps by causing it to transfer from one point to another in the program. As a consequence, TO has the effect of modifying DO. This is demonstrated in Fig. A.4.

IF Command

In executing a program, Telcomp can make decisions as to what action to take, depending on the current values of variables or expressions within the program. Decisions are specified with the IF clause. Like FOR, an IF clause modifies a Telcomp command. Some simple uses of IF are illustrated in Fig. A.5.

Boolean Operators

The command IF can be expanded by using another kind of operator, the Boolean operator, to modify or link together comparison subclauses. Two Boolean operators available in Telcomp are AND and OR. As shown below, these operators mean just what their names imply:

$$\text{DO PART 10 IF A = B AND C > D}$$

Part 10 is done if A equals B and C is greater than D, that is, if both comparison subclauses are true.

$$\text{DO PART 10 IF A = B OR C > D}$$

Part 10 is done if either or both of the comparison subclauses are true.

```
1.1 TYPE 1.1
1.2 DO PART 2
1.3 TYPE 1.3

2.1 TYPE 2.1
2.2 TO STEP 2.4
2.3 TYPE 2.3
2.4 TYPE 2.4
2.5 TO STEP 3.2
2.6 TYPE 2.6

3.1 TYPE 3.1
3.2 TYPE 3.2
3.3 TO PART 4
3.4 TYPE 3.4

4.1 TYPE 4.1
4.2 TYPE 4.2
>DO PART 1
     1.1=     1.1
     2.1=     2.1
     2.4=     2.4
     3.2=     3.2
     4.1=     4.1
     4.2=     4.2
     1.3=     1.3
>
```

Fig. A.4 Illustration of the TO command.

```
1.05 TYPE A
1.1 TYPE 1 IF A=B
1.2 TYPE 2 IF A<B
1.3 TYPE 3 IF A>B

2.1 B=2
2.2 DO PART 1 FOR A=1:1:3
>DO PART 2
     A=     1
     2=     2
     A=     2
     1=     1
     A=     3
     3=     3
>
```

Fig. A.5 Boolean operators can be used to make logical decisions.

Multiple Clauses

IF and FOR clauses can be combined in modifying a Telcomp command. Any clauses that modify the command are read from right to left. For example:

3.7 DO PART 10 FOR A = 1:1:N IF C > D IF M < 1

Telcomp would first check M < 1, and then, if that were true, C > D. If either were false, Telcomp would go on to the next step. Otherwise, Part 10 would be executed for the indicated values of A.

The order in which multiple clauses are executed is different from the order in which the arguments of a command are executed. Telcomp reads the arguments of a command from left to right. For example:

1.5 DO PART 5, PART 7

Telcomp would do Part 5 first, and then Part 7.

Subscripts

A subscripted variable is denoted by square brackets. A variable can have more than one subscript; for example, X[2,3] is an allowable subscripted variable. The value of a subscript can be any integer from 0 to 249. The value of this integer can be computed from an expression; for example, X[C + D] is an allowable subscripted variable.

Summary of Symbols

Some of the more obvious symbols (such as =) will not be explained here; only those with which the reader may be unfamiliar are mentioned.

*	indicates multiplication
^	indicates exponentiation
' '	absolute value
#	causes Telcomp to skip a line
;	in a step, Telcomp ignores all statements after a semicolon
$PI	pi = 3.14159 . . .

Some Functions Available in Telcomp

SQRT(A)	positive square root
LOG(A)	base-10 logarithm
LN(A)	natural logarithm
EXP(A)	e raised to the power A
'A'	absolute value of A
ATN(A,B)	arctan (A/B), radians
ATN(A)	arctan (A), radians
IP(A)	integer part of A
SIN(A)	sine of A, A in radians
COS(A)	cosine of A, A in radians

APPENDIX B

Filter Design by the "Cookbook" Approach

A handbook of filter designs can only describe a finite number of networks and thus cannot always provide the optimum solution to a particular design problem. Computer programs (such as pole-placer programs) can be written that provide the engineer with more flexibility than is offered by handbooks, but it is not unusual for a filter designer to present requirements that cannot be handled by the existing program. Recognizing that no program will be foolproof, the approach taken in this book was to provide enough theoretical background to enable an engineer to modify the sample programs given in the text. This is in contrast to a cookbook approach, in which one uses tabulated results or mechanically follows a prescribed procedure.

While no cookbook approach will solve all possible approximation problems, this does not imply that such an approach does not have merits. Many engineers do not have the luxury of sufficient time to delve deeply into approximation theory—but they still need filters. Also, it is often preferable to use an existing program for quick results, instead of returning to fundamental equations to get a slightly better solution.

The purpose of this appendix is to help those filter designers who want the option of using a cookbook approach. The equations given in the text can be used for a cookbook approach to most filter approximation problems. The rest of this appendix indicates which equations should be used for each type of filter. A program that can be used for the filter and a sample use are cited in most cases.

The cookbook is divided into sections. The first section deals with classical lowpass filters; these filters can be transformed into other types of filters (e.g., bandpass) by using the material in Chap. 7. The other sections describe equiripple, maximally flat, and parametric filters.

I. Classical lowpass filters
   A. Butterworth

   The degree and loss of the filter can be found by using (2.15) and (2.13). These equations can be easily solved without resorting to a computer program, as illustrated in Example 2.1.

   Since this is a polynomial filter, the attenuation poles are all located at infinity. The natural frequencies are on a circle, as indicated (for a normalized lowpass) by (2.17). Example 2.2 contains an application of this formula.

   B. Tschebycheff

   The degree and loss of the filter can be found by using (3.24), (3.20) or (3.22), and (3.14). A program based on these equations is given in Fig. 3.7, and a sample use is given in Fig. 3.8.

   Since this is a polynomial filter, the attenuation poles are all located at infinity. The zeros of $H(s)$ are on an ellipse, as indicated (for a normalized lowpass) by (3.34). Instead of describing the ellipse, it is easier to find the zeros by using the zero-finder program of Fig. 10.3. Sample uses of this program for Tschebycheff lowpass filters are given in Fig. 10.4.

   C. Inverse Tschebycheff

   The degree and loss of the filter can be found by using (4.10), (3.20) or (3.22), and (4.7). A program based on these equations is given in Fig. 4.4, and a sample use in Example 4.1.

   From (4.7) it follows that the attenuation poles are the solutions to $T_n(\omega_H/\omega) = 0$. Instead of finding the roots to this equation, one can use the maximally flat pole-placer program of Fig. 11.9, as illustrated in Fig. 11.10. The zeros of the transfer function can be found by using the zero-finder program of Fig. 11.14.

   D. Elliptic

   The degree and loss of the filter can be found by using (5.51), (5.49), (5.66), and (5.5). The attenuation poles can be determined by using (5.67) and (5.68). A program based on these equations is given in Fig. 5.12, and a sample use in Example 5.3.

   The attenuation poles can also be determined by using the equiripple pole-placer program of Fig. 8.15, as demonstrated in Fig. 8.16. The zeros of the transfer function can be found by using the zero-finder program of Fig. 10.3, as demonstrated in Fig. 10.5.

II. Equiripple filters
   A. Lowpass

   The lowpass pole-placer program of Fig. 8.15 can be used to find the degree, attenuation poles, and loss of a lowpass filter that has an equi-

ripple passband. It is most useful for filters that have arbitrary stopbands (e.g., see Fig. 8.17), but can also be used for elliptic filters that have equiminimum stopbands (e.g., see Fig. 8.16).

The lowpass zero-finder program of Fig. 10.3 can be used to find the zeros of the transfer function and its constant multiplier, as demonstrated in Figs. 10.4 to 10.6.

B. Bandpass

The bandpass pole-placer program is shown in Fig. 8.11, and sample uses in Figs. 8.12 to 8.14. The bandpass zero-finder program is shown in Fig. 10.1, and a sample use in Fig. 10.3.

C. Highpass

Highpass filters can be designed by transforming lowpass filters by using (6.3), as demonstrated in Example 6.2. Another way to design highpass filters is to assume the requirement is for a bandpass filter by letting the upper passband edge approach infinity, as is demonstrated in Fig. 8.18.

III. Maximally flat filters

Filters with maximally flat passbands can be designed by modifying the programs for equiripple filters. Modifications for the pole-placer programs are shown in Figs. 11.7, 11.9, and 11.11. Modifications for the zero-finder programs are shown in Figs. 11.13 to 11.15.

IV. Parametric filters

Bandpass filters with parametric passbands can be designed by modifying the programs for equiripple filters. The pole-placer program can be modified as shown in Fig. 12.5, and the zero finder as shown in Fig. 12.9.

Answers to Selected Problems

Chapter 2

2.1 Frequency-scaling the numerator in Eq. (2.23) yields $\alpha^2 s^2 + \omega_z/Q_z \; \alpha s + \omega_z^2$. Thus the new natural frequency is ω_z/α, but the quality remains Q_z.

2.2 Equation (2.21) can be manipulated to yield $Q_p(\omega/\omega_p - \omega_p/\omega) = \pm 1$. Calling ω_1 and ω_2 the solutions to this equation, it follows that $\omega_{1,2} = 0.5\omega_p(\sqrt{1/Q_p^2 + 4} \mp 1/Q_p)$, from which one can obtain the answer.

2.3 The roots are given by Eq. (2.25). The magnitude of the roots can be shown to be ω_p. Therefore, the roots are located on a circle.

2.4 $H(s) = s^5 + 3.236s^4 + 5.236s^3 + 5.236s^2 + 3.236s + 1$

2.5 $Q_1 = 0.5; \; Q_2 = 0.618; \; Q_3 = 1.618$

Chapter 3

3.1(a) -1 **(b)** 26
3.2(a) 30.6 dB **(b)** 23.4 dB
3.3(a) $32x(x^2 - 0.5)$ **(b)** $8(x^2 - 0.5)^2$ and $8x^2(x^2 - 1)$
3.5 $2(4x^4 - 4x^2 + 0.5)$
3.6 12
3.8 $T_n^2(\omega) = \cos^2(n \cos^{-1} \omega) = \frac{1}{2}[\cos(2n \cos^{-1} \omega) + 1] = \frac{1}{2}[T_{2n}(\omega) + 1]$

Chapter 5

5.1 207.1
5.2(a) $u(\pi + \theta, k) = 2k + u(\theta, k)$. Letting $\theta = \phi - \pi/2$ and using $u(\phi - \pi/2, k) = -u(\pi/2 - \phi, k)$ yields the answer. **(b)** $u(\pi/2 + \phi, k) = K + \delta = 2K - u(\pi/2 - \phi, k)$, from which the answer follows.

5.3 From Prob. 5.2 if $u(\pi/2 + \phi, k) = K + \delta$, then $u(\pi/2 - \phi, k) = K - \delta$. This implies sn $(K + \delta, k) = \sin(\pi/2 + \phi) = \sin(\pi/2 - \phi) = \text{sn}(K - \delta, k)$.
5.4(a) $u(\pi + \phi, k) = 2K + u(\phi, k)$, which implies sn $[2K + u(\phi), k] = \sin(\pi + \phi) = -\sin\phi = -\text{sn}(u,k)$ **(b)** $\text{sn}(u + 4K, k) = -\text{sn}(u + 2K, k) = \text{sn}(u,k)$
5.6 $\text{sn}[i(u + K'), k]$
$= i[\text{sn}(u + K', k')/\text{cn}(u + K', k')] = i[\text{sn}(K' - u, k')/-\text{cn}(K' - u, k')]$
$= -\text{sn}[i(K' - u), k]$
5.9 $u(190°, 0.5) = 2K + u(10°, 0.5) = 3.5464$; $\text{sn}(u,k) = \sin(190°) = -0.1737$; $\text{cn}(u,k) = \cos(190°) = -0.9848$
5.10(a) This is a direct application of Eq. (5.32).

Chapter 6

6.3 1.1387 and 5.2690
6.4(a) $f'_L = 1.2$ or $f'_A = 0.9375$ **(b)** $f'_A = 1.1859$ and $f'_L = 0.9487$
6.5(a) $-0.265 \pm j1.063$ and $-0.442 \pm j1.771$ **(b)** 2.065
6.6 N poles at the origin and N poles at infinity.
6.7(a) $s^2 = \Omega_B{}^2(s^2 + \Omega_0{}^2)/(\Omega_B{}^2 - \Omega_0{}^2)$ where $\Omega_0 = 0.195\Omega_B$ and $\Omega_B = 4\pi$.
(b) $A(2.6) = 29.4$ dB, which does not meet the requirement.

Chapter 7

7.1 $Z = 1.15 + 2.39j$
7.2 $F(Z) = (48/81)(Z^2 + 1.25)(Z^2 + 0.5)$; $Q(Z) = (55/81)(1 - 0.64Z^2)^{1/2}(Z^2 - 1)^{1/2}(Z^2 - 2/11)$
7.4(a) 0.0002   **(b)** $a = 2.030$ and $b = 1.030$ **(c)** 0   **(d)** 0.0002

Chapter 8

8.3(a) IA = 69.2 dB; C = 22.35 dB; loss = 46.8 dB   **(b)** 46.8¹ dB   **(c)** 54.0 dB
8.4 Exact answer is 10.14 dB; approximate answer is 9.69 dB.
8.5(a) $Z_1 = 1.326$; $Z_2 = 1.525$; $Z_3 = 0.719$; $Z_4 = 0.931$ **(b)** 1.447   **(c)** 4.371 and 3.714   **(d)** $Z = 0.778$; $f = 1.373$

Chapter 9

9.1(a) 1   **(b)** If $ZN \neq 0$, then the denominator in Eq. (9.13) must be zero at $s = 0$. Using this fact and $C_1{}^2 = 1$ yields the answer.
9.2(a) $K(s) K(-s) = \epsilon^2[\text{Ev}(Z + 1)^{\text{NIN}}]^2/(-1)^{\text{NIN}}(Z^2 - 1)^{\text{NIN}}$ **(c)** $4\omega^3 - 3\omega$
9.4(a) 205.12   **(b)** 0.01834   **(c)** 40.49 dB   **(d)** $-1,385.8$ **(e)** 40.49 dB
9.5 0.084 dB
9.6(a) $-3350j$ and $0.7375 - 0.6754j$ **(b)** 48.15 dB and 0.055 dB

Chapter 10

10.1 $(1 + \epsilon^2)Z^4 - 2(1 - \epsilon^2)Z_1{}^2Z^2 + (1 + \epsilon^2)Z_1{}^4$
10.2(a) $C_H(s^2 + 0.7832s + 4.0339)$ **(b)** 2.565
10.3 Equations (10.10) and (10.12) imply $2(a_i - b_i/2)^{1/2} = 2\pi ZF_i/ZQ_i = (a_i/ZQ_i)^{1/2}$ from which the answer follows.

Chapter 11

11.2 The first $m - 1$ derivatives at $\omega = \omega_0$ are equal to zero; that is, the filter is maximally flat at $\omega = \omega_0$.

Answers to Selected Problems

11.3 If $f_i = 1/f_j$, it follows that $Z_i{}^2 Z_j{}^2 = f_B{}^4$. This can be used in Corollary 11.1 to yield $Z_0 = f_B$. Substituting this and $f_A = 1/f_B$ into $f_o = [(f_B{}^2 + Z_0{}^2 f_A{}^2)(1 + Z_0{}^2)]^{1/2}$ yields the answer.

Chapter 12

12.2(a) 0.095 dB **(b)** 20.91 dB **(c)** 28.61 dB
12.3(a) 0.00817 **(b)** 0.09918 dB **(c)** 29.79 dB
12.4(a) 0.091 dB **(b)** 0.091 dB

Chapter 14

14.2(a) $H(s) = (15 + 15Ts + 6T^2s^2 + T^3s^3)/15$ **(b)** $D(0) = T$
14.3(a) $V_{\text{out}} = \cos[\omega_1(t - T) - \phi] + \cos[\omega_2(t - T) - \phi]$
14.4(a) h_1/h_0 **(b)** $h_0 = 1$ and $h_1 = 1$; therefore $D(0) = 1$
14.5(c) $(945 + 945s + 420s^2 + 105s^3 + 15s^4 + s^5)/945$ **(d)** 0.979
14.6 $H(s) = 1 + s + s^2/2 + s^3/3! + s^4/4! + s^5/5!$ The even and odd parts of $H(s)$ are $m = 1 + s^2/2 + s^4/24$ and $n = s + s^3/6 + s^5/120$. If $H(s)$ is Hurwitz, then the zeros of m and n must be on the $j\omega$ axis. However, n has complex zeros; therefore $H(s)$ is not Hurwitz.
14.7(a) 6 **(b)** 1.55 rad/s **(c)** 4; 0.3 rad/s.
14.8 $(105 + 1.05s + 0.0045s^2 + 10^{-5}s^3 + 10^{-8}s^4)/1.05$
14.9 $1 + s$
14.11(a) 0.63 dB **(b)** $[105 + 105(s + 12/s) + 45(s + 12/s)^2 + 10(s + 12/s)^3 + (s + 12/s)^4]/105$ **(c)** 2.33; 2; 1.75
14.12(a) $h(s) = H(s_+)H(s_-)$ where $H(s) = (105 + 105s + 45s^2 + 10s^3 + s^4)/105$, $s_+ = 2[s + (7/2)j]$, and $s_- = 2[s - (7/2)j]$. **(b)** 2.12; 2.104; 2.09
14.15 A series combination of a resistance R^2/R_1 and a reactance $-R^2/X_1$
14.16 $f(s) = 1 + s + s^2/2 + s^3/3! + \cdots$

Chapter 15

15.1(a) 1 **(b)** $x(0) = \lim_{s \to \infty} sX(s)$; therefore $x(0) = 0$
15.2 For Eq. (15.13) $x(0) = \lim_{s \to \infty} sX(s) = a_{m-1}$. However, if the degree of the numerator is equal to or greater than the degree of the denominator, then the initial value would be infinite.
15.4 At the time $t = 0$, Eq. (15.15) yields $x^{(0)} = 0$, $x^{(1)} = 0$, and $x^{(2)} = 1$, and Eq. (15.16) yields $x^{(3)} = -\sqrt{2}$ and $x^{(4)} = 1$. Thus $x(t) = \sum_{j=0}^{4} x^{(j)} \frac{T^j}{j!} = 0.004768$.

15.5(a) 0.09516; 0.18127; 0.25918 **(b)** 0.09517; 0.18129; 0.25924

Index

Index

Index

Abele, T. A., 260
Abramowitz, M., 60, 62
Achieser, N. I., 1
Active filter synthesis, 319–345
Adams, R. L., 342, 343n.
Admittance (Y), 2
Aliasing, 354
Allpass networks, 268–275
Amplitude, elliptic integral (ϕ), 60
Amplitude equalizer, 265–268
Analytic continuation, 13
Aprille, T. J., 230n.
Arbitrary passband, 234–237
Arc, pole-placer program, 134
Arithmetically symmetrical, 262
Aronhine, P., 190, 260
Attenuation (A), 3
  of Butterworth filters, 10
  of elliptic filters, 55
  of inverse Tschebycheff filters, 46
  of Tschebycheff filters, 27
  (*See also* Attenuation programs; Characteristic function)
Attenuation function (α), 248
Attenuation poles, 122–155
  (*See also* Pole-placer program)
Attenuation programs:
  Bessel, 252
  elliptic, 79
  equiripple: bandpass, 168
    lowpass, 169
  inverse Tschebycheff, 48
  maximally flat: bandpass, 193
    lowpass, 194, 195
  parametric, 216, 218
  Tschebycheff, 35
Attenuation step, pole-placer program, 133
Available parameters, 21, 230

Bairstow's iterative procedure, 175
Balabanian, N., 292
Bandpass filter, 4
  attenuation program, 168
  delay, 261–264

Bandpass filter (Cont.):
  frequency transformation, 88–97, 261–264
  pole-placer program, 138, 196, 220
  zero-finder program, 174, 203, 225
  (See also Equiripple filters; Maximally flat passbands; Parametric filters)
Bandpass second-order transfer function, 326
Bandreject filter, 4
  (See also Bandstop filter)
Bandstop filter, frequency transformation, 97–98
Barnes, J. L., 285
Beletskiy, A. F., 264
Bell, W. W., 30
Bennett, W. R., 186
Bessel approximation, 249–256, 289
Bilinear digital filters, 357–363
Bingham, J. A. C., 120, 222
Blinchikoff, H., 263
Bode, H. W., 268
Bridged T, 266
Bucker, W. M., 260
Budak, A., 190, 260, 264
Butterworth filter, 7–19
  attenuation formula, 10
  characteristic function, 10
  degree, calculation of, 12
  delay, 242–244, 254–255
  examples, 12, 13
  natural modes, 12–14
  quality of roots, 14–18
  step response, 285
  transient response, 289
Butterworth polynomial (B_n), 9
Butterworth-Thomson filter, 259–260, 289

Calahan, D. A., 230n., 266
Cascade sequence of active filters, 329
Caslin, J. C., 56, 62
Cauer, W., 21, 51, 58n., 98n.
Chang, C. Y., 200
Characteristic function (K), 7–8, 156–171
  of Butterworth filters, 10
  of elliptic filters, 55
  of equiripple bandpass filters, 161
  of equiripple lowpass filters, 164
  of inverse Tschebycheff filters, 45

Characteristic function (K) (Cont.):
  of maximally flat bandpass filters, 188
  of maximally flat lowpass filters, 190, 191
  of parametric filters, 214, 215
  of Tschebycheff filters, 27
Chebyshev (see Tschebycheff)
Christian, E., 102
Close, C. M., 348
Complementary complete elliptic integral (K'), 61, 67
Complementary modulus (k'), 61
Complete elliptic integral (K), 60, 67
Complex frequency, 2
Conning, S. W., 311
Constant resistance network, 265
Correction factor (C), 127
Corrington, M. S., 286
Coupled active filters, 340–344
Coupling network, 300–303
Critical frequency (ω_0), 269
Crout reduction, 142
Cunningham, V. R., 271

Darlington, S., 51n.
Davey, J. R., 186
Decomposition into second-order functions, 328–331
Delay, 238–281
  of bandpass filters, 261–264
  bandwidth product, 252–253
  of Butterworth filters, 242–244, 254–255
  definition of, 239
  of digital filters, 362–363
  of elliptic filters, 203–207
  equalizer, 268
  of inverse Tschebycheff lowpass filters, 49–50
  and lowpass-to-bandpass transformation, 261–264
  of maximally flat filters, 203–207
  program, 242
  time, 283
  of Tschebycheff lowpass filters, 49–50
Deliyannis, T., 264, 335
DeRusso, P. M., 348
Digital filter synthesis, 346–367
Dishal, M., 256, 259
Dutta Roy, S. C., 272

Index 385

Eisenmann, E., 102
Elliptic filter, 51–85
  attenuation formula, 55
  attenuation program, 79, 89, 92, 94
  characteristic function, 55
  degree, calculation of, 72
  delay, 203–207
  examples, 72, 75, 79, 82, 88, 91, 93, 101, 143, 145, 150, 182, 206
  natural modes, 82, 181–182
  transformation to produce loss peak at infinity, 99–102
Elliptic functions, 62–66
  cosine (cn), 62
  difference (dn), 62
  sine (sn), 62
Elliptic integral (u), 59–62, 66–67
  (See also Complete elliptic integral)
Equalizer:
  amplitude, 265–268
  delay, 268
Equiripple filter, 21, 202–207
  bandpass filter: attenuation program, 168
    characteristic function, 161
    natural modes, 172–184
    passband loss, 166–170
    pole-placer program, 138
    stopband loss, 165–166
    zero-finder program, 174
  lowpass filter: attenuation program, 169
    characteristic function, 164
    natural modes, 172–184
    passband loss, 167–168
    pole-placer program, 148
    stopband loss, 165–166
    zero-finder program, 178
  (See also Elliptic filter; Tschebycheff filter)
Error criteria, 5, 230–231
Even part (Ev), 161

Feedforward biquad, 333n.
Fettis, H. E., 56, 62
Final-value theorem, 289
Fleischer, P. E., 230n.
Fletcher, R., 233
Follow-the-leader feedback, 344
Frequency rejection, 327

Frequency transformation, 86–107
  for delay, 261–264
  lowpass-to-bandpass, 88–97, 261–264
  lowpass-to-bandstop, 97–98
  lowpass-to-highpass, 87–88
  lowpass-to-multiple-bandpass, 98–99
  nonreactance, 99–104
Friedenson, R. A., 320
Friedman, M., 233
Friend, J. J., 335

Gardner, M. F., 285
Gaussian magnitude approximation, 256–259, 289
Geffe, P. R., 262, 263
Generalized inverse Tschebycheff filter, 200
Geometrically symmetrical bandpass, 90
Geometrically symmetrical transformation, 262
Gibbs, A. J., 347, 366
Girling, F. E. J., 341–343
Gold, B., 4, 366
Golden-section optimization technique, 233
Good, E. F., 341–343
Gpop (general-purpose optimization program), Bell Telephone Laboratories, 234
Gradient, 233
Graeme, J. G., 332
Greenhill, A. G., 64
Group delay, 239n.
Guillemin, E. A., 2, 4, 38, 76, 161n., 239n., 248n., 282
Gyi, M., 235
Gyrator, 322–324

Half-power frequency, 15
Halfin, S., 328
Hellerstein, S., 271
Henderson, K. W., 285, 289
Hepner, C. F., 275
Herrero, J. L., 211n.
Hess, S. F., 350, 366
Hessian, 233
Highpass filter, 4
  frequency transformation, 87–88
  pole-placer program, 152–154
  second-order transfer function, 326

Hilbert transformation, 248–249
Hildebrand, F. B., 142
Hodgman, C. D., 57, 60
Huelsman, L. P., 324, 332
Humpherys, D. S., 260, 292
Hurtig, G., 343
Hurwitz, 3, 161n.

Image attenuation (IA), 127
Image parameters, 127
Impedance (Z), 2
Impulse-invariant digital filters, 350–355
Impulse response, 283
Initial parameters, 232
Input/output transfer function (H), 4
Insertion loss (IL), 293
Integer part, 75
Inverse Tschebycheff filter, 43–50, 198–202
 attenuation formula, 46
 attenuation program, 48
 characteristic function, 45
 degree, calculation of, 46
 delay, 49–50
 examples, 47, 198
 natural modes, 49, 202–205
 quality of roots, 49–50

Jacobian elliptic functions, 62
Jury, I., 348

Kaiser, J. F., 266, 271, 347, 350, 366
Kao, C-Y, 350, 366
Karnaugh, M., 264
Kautz, W. H., 285, 289
Kelly, L. G., 175
Kuo, F. F., 211n., 264, 266, 271, 311, 347, 350, 366
Kuo, Y. L., 344

Ladder network, 302
Laplace transformation, 2, 284–289, 348
Laurent, T., 126
Leapfrog active filters, 343
Lee, H. B., 233, 304
Legendre filters, 289
Linvill, J. G., 275, 320

Liou, M. L., 286
Loss (*see* Attenuation)
Loss equalizer, 266
Loss function, 124–126, 164–165
Lowpass filter, 4
 attenuation program, 169
 frequency transformation, 86–107
 pole-placer program, 148, 198, 200, 201
 zero-finder program, 178, 204, 205
 (*See also* Equiripple filters; Maximally flat passbands; Parametric filters)
Lowpass second-order transfer function, 325
Lueder, E., 328

MacNee, A. B., 260
Magnus, W., 59
Magnuson, W. G., 211n., 311
Match (optimization program), 233
Matched Z-transform digital filters, 355
Maximally flat at origin, 9, 189, 246–247
 (*See also* Maximally flat passbands, lowpass filter)
Maximally flat beyond origin, 191, 246–247
 (*See also* Maximally flat passbands, lowpass filter)
Maximally flat delay, 252
Maximally flat filter, 9, 45, 185–209
 bandpass filter: attenuation program, 193
  characteristic function, 188
  natural modes, 202–205
  pole-placer program, 196
  zero-finder program, 203
 lowpass filter: attenuation program, 194, 195
  characteristic function, 190, 191
  delay, 203–207
  natural modes, 201–205
  pole-placer program, 198, 200, 201
  zero-finder program, 204, 205
 (*See also* Butterworth filter; Inverse Tschebycheff filter)
Maximum available power (P_m), 294
Maximum passband loss ($A_{\max}$), 9
Mead, R., 233
Mean-square error, 5, 231
Miller, J. A., 324

Minimum difference (D_{min}), pole-placer program, 140
Minimum frequency (FM), pole-placer program, 140
Minimum-phase transfer function, 248
Minimum stopband loss (A_{min}), 9
Min-max approximation, 21
  (*See also* Equiripple filters)
Mitra, S. K., 4, 323, 324
Möbius transformation, 100n.
Modular angle of elliptic integral (θ), 60
Modulus of elliptic integral (k), 60
Moschytz, G. S., 272
Muir, A. J. L., 332n., 333
Murakami, T., 259, 260, 289

Natural frequency, 16
  (*See also* Natural modes)
Natural modes [zeros of $H(s)$], 172–184
  of Butterworth filters, 12–14
  of elliptic filters, 82, 181–182
  of equiripple filters, 172–184
  of inverse Tschebycheff filters, 49, 202–205
  of maximally flat filters, 202–205
  of parametric filters, 224–226
  of Tschebycheff filters, 36–40, 180–181
Negative-impedance converter (NIC), 320–322
Neirnck, J. J., 260
Nelder, J. A., 233
Newcomb, R. W., 323, 324
Newton's iterative procedure, 137
Nielsen, G., 274
Nonreactance transformation, 99–104
Nonrecursive digital filters, 350
Normalized lowpass filter, 86

Oberhettinger, F., 59
Optimization, 229–237
Orchard, H. J., 109, 186, 311

Padé approximation, 274–275
Papoulis, A., 85, 231
Parametric filter, 210–228
  bandpass filter: attenuation program, 216, 218
    characteristic function, 214–215
    natural modes, 224–226
    pole-placer program, 220

Parametric filter, bandpass filter (*Cont.*):
    zero-finder program, 225
  lowpass filter, 223–224
Parametric multiplier, 214
Parametric pole:
  at infinity, 211–214
  at the origin, 214–215
Parker, S. R., 350, 366
Partial fraction expansion, 285
Partial pole removal, 305
Passband, 4
Passive network synthesis, 292–318
Peless, Y., 259, 260, 289
Percent overshoot, 284
Periodic rectangle, 65, 69–71
Phase delay, 239n.
Phase function (β), 248
Pierre, D., 233
Piloty, H., 260
Pole-placer program, 122–155
  for digital filters, 359
  for equiripple filters: bandpass, 138
    highpass, 152–155
    lowpass, 148
  for maximally flat filters: bandpass, 196
    lowpass, 198, 200, 201
  for parametric bandpass filters, 220
Polynomial filter, 40
Powell, M. J. D., 233
Predistortion, 106
Prewarping for digital filters, 357
Primary resonator block (PRB), 343
Pth-error criteria, 231

Quadrantal symmetry, 240
Quality of roots (Q), 14, 96–97
  Butterworth, 14–18
  inverse Tschebycheff, 49–50
  Tschebycheff, 39–40, 49–50

Rabiner, L. R., 350
Rader, C. M., 4, 366
Reactance transformation, 99
Recursive digital filters, 366
Reflected power (P_r), 298
Reflection coefficient, 298
Reflection function (T_1), 298–300
Relative change in ripple (R_C), 212
Rise time (t_r), 283

Robinson, A. E., 332n., 333
Ross, R. I., 286
Roy, R. J., 348

Saal, R., 309
Sablatash, M., 347, 366
Sampled waveform, digital filters, 347
Sampler, digital filters, 347
Savage, L. S., 233
Second-order transfer functions, 324–328
Sheahan, D. F., 323
Signal delay, 239n.
Simplex optimization technique, 233
Skwirzynski, J. K., 271n.
Smith, B. R., 133, 146
Steepest-descent optimization technique, 233
Stegun, I., 60, 62
Step (see Attenuation step)
Step response, 283
Stiffness (b), delay equalizer, 270
Stopband, 4
Storch, L., 250–252
Storer, J. E., 274, 275
Superposition integral, 283
Suprox (successive-approximation program), Bell Telephone Laboratories, 234
Szentirmai, G., 72n., 109, 235, 262, 266n.

Talbot, A., 1
Taylor approximation, 6, 9n.
Taylor series, 6, 286
Telcomp, 368–372
Temes, G. C., 109, 133, 146, 186, 230n., 235, 311
Template method, 126–129
Thomson, W. E., 252, 286
Time-domain response, 282–291
Tobey, G. E., 332
Tow, J., 332n., 344
Transformation (see Frequency transformation)
Transformed functions, $F(Z)$, $Q(Z)$, and $E(Z)$, 112–119
Transformed variable (Z), 108–121, 311–314
Transient response, 282–291
Transition ratio, 81

Transition region, 4
Transitional Butterworth-Thomson (TBT) filter, 259–260, 289
Transmission function (H), 294–298
Tschebycheff approximation:
 of constant delay, 260
 definition of, 20
Tschebycheff filter, 20–42
 attenuation formula, 27
 attenuation program, 35
 characteristic function, 27
 degree, calculation of, 33–34
 delay, 49–50
 examples, 34, 103, 181
 inverse (see Inverse Tschebycheff filter)
 natural modes, 36–40, 180–181
 quality of roots, 49–50
 transient response, 289
Tschebycheff polynomial (T_n), 27–30, 170
Tschebycheff rational function (R_n), 52–59, 68–71, 73–78

Ulbrich, E., 260, 309
Unit impulse function (δ), 283
Unit step function (u), 283
Universal loss curve, template method, 127–128

Valley, G. E., 257
Virtual ground, operational amplifier, 332

Wallman, H., 257
Wantanabe, H., 222
Weinberg, L., 14, 98n., 292
Willoner, G., 211n.
Woods, F. S., 59

Z-transform filters, digital filters, 350n.
Z transformation, 348
Zero-finder program:
 for equiripple filters: bandpass, 174
  lowpass, 178
 for maximally flat filters: bandpass, 203
  lowpass, 204, 205
 for parametric bandpass filters, 225
Zero-shifting technique, 302–307
Zobel, O. J., 265
Zolotarev filter, 51
Zverev, A. I., 259, 260, 289, 292